Consumer-Resource Dynamics

MONOGRAPHS IN POPULATION BIOLOGY
EDITED BY SIMON A. LEVIN AND HENRY S. HORN

Complete series list follows Index.

Consumer-Resource Dynamics

WILLIAM W. MURDOCH,

CHERYL J. BRIGGS, AND

ROGER M. NISBET

PRINCETON UNIVERSITY PRESS

Princeton and Oxford

Learning Resources
Centre
12615765

Copyright © 2003 by Princeton University Press
Published by Princeton University Press, 41 William Street, Princeton,
New Jersey 08540
In the United Kingdom: Princeton University Press, 3 Market Place,
Woodstock, Oxfordshire OX20 1SY
All Rights Reserved

Library of Congress Cataloging-in-Publication Data
Murdoch, William W,
Consumer-resource dynamics / William W. Murdoch, Cheryl J. Briggs, and
Roger M. Nisbet.
p. cm. – (Monographs in population biology ; 36)
Includes bibliographical references (p.).
ISBN 0-691-00658-X (alk. paper) – ISBN 0-691-00657-1 (pbk. : alk. paper)
1. Population biology–Mathematical models. I. Briggs, Cheryl J., 1963- II. Nisbet, R. M.
III. Title. IV. Series.
QH352 .M966 2003
577.8′8–dc21   2002029273

British Library Cataloging-in-Publication Data is available
This book has been composed in Times Roman
Printed on acid-free paper. ∞
www.pupress.princeton.edu
Printed in the United States of America
10 9 8 7 6 5 4 3 2 1
10 9 8 7 6 5 4 3 2 1 (pbk.)

The cover illustration was drawn by Paul Chadwick, the writer/artist of
the *Concrete* series of comics and graphic novels that
often touch on environmental issues.

To Joan, John, and Mary

Contents

Preface

This book covers a wide range of theory describing the dynamics of interacting populations of consumer and resource species. Our central focus is how the properties of real organisms affect population dynamics. The core of the book (chapters 5–8), therefore, is a synthesis and elaboration of mainly our own stage-structured models that use insect parasites (parasitoids) and their hosts as illustrative organisms. It develops a general theory for such interactions that takes into account how parasitoids respond to differences among individual hosts.

The rest of the book resulted from our reaching out from this core to other models of consumer-resource dynamics. Chapters 3 and 4, on simple models, provide necessary background for understanding and evaluating the more complex models of chapters 5–8. (Chapter 8, on competing parasitoids, contains its own review of simple competition models.) What began in these chapters as the explication of simple models and their known results, however, came to include an interesting exercise in uncovering novel insights and connections. We hope even those familiar with simple theory will find something of interest in these early chapters. Chapters 3 and 4, and especially chapter 11, result also from our desire to connect our models to the larger body of population-dynamics theory. We are convinced that consumer-resource theory is not only a coherent whole but connects seamlessly to theory for single-species dynamics, and we try to make that case in chapter 11.

Chapters 1 and 2 provide conceptual and empirical contexts. Chapter 9 examines implications of model results for biological control of pests in agriculture, a subject of long and continuing interest for both empirical and theoretical ecologists and entomologists. A book on consumer-resource dynamics cannot ignore spatial processes, though chapter 10 makes no pretence to being a representative survey. Instead it focuses on questions closely related to those analyzed in the rest of the book. Finally, chapter 12 draws together some insights gained in the preceding chapters and looks briefly at a few promising research areas.

The models are developed and analyzed for their potential insights into a range of dynamic behavior. We are interested in how various processes affect equilibrium densities, stability, and non-equilibrium behavior. In this last regard, we explore throughout the various types of cycles that arise in the different models. The results on cycles play a key role when we ask, in chapter 11, whether there is justification in nature for using the few-species models that characterize most of ecological theory.

This book is primarily a research monograph that presents both existing and new theory and results, so it focuses mainly on our own work. Although we reach out to the literature for background and larger synthesis, sometimes extensively as in chapters 3, 4, 8, 10, and 11, the book is not intended as, nor is it, a survey of consumer-resource theory. There is, for example, a large literature on spatial processes, and as noted we deal only with a few aspects that we think particularly relevant to our central focus. The book contains no theory for generalist predators. We deal with true parasite-host and disease-host dynamics only in passing. We do not cover statistical and other issues that arise in confronting models with data.

We have written primarily for ecologists, including entomologists, who are interested in the dynamics of real populations in the field. We have tried to make the theory as clear and understandable as possible, and hope readers can see how their system might be represented in some of the models. We have also provided in the various appendixes enough explanation of techniques for the neophyte to get a start on developing and analyzing such models, for we believe the techniques have potentially broad application. We hope also, however, that theorists will find something of interest even in areas with which they are familiar.

The gap between theorists and field ecologists is a serious impediment to progress in ecology. On the one hand, we think models of some sort are essential for investigating real ecological systems and for developing general understanding. Yet field ecologists often reject models, and their potential insights, on the grounds they are too simple, i.e., too unrealistic. So in this book we go from simple to quite realistic models, in the hope of showing both that models can approach reality, and that simple models have much to offer.

On the other hand, and surprisingly, field ecologists sometimes accept models uncritically, when critical analysis could pinpoint their weaknesses. Such uncritical acceptance is also unfortunate, because theory needs creative feedback from field ecologists and experimentalists. For this reason, we strive to make transparent the translation of process and mechanism into assumption and then results.

Our efforts, of course, reflect our assumption that mathematical theory can and does contribute to the science of ecology. Indeed, we believe that without theory, and probably mathematical theory, population ecology is not science. For this reason, we work toward a framework that encompasses both generality and testability, which we consider the main features of useful theory. Because we are concerned with testing theory, we are frankly advertising to field ecologists a modeling approach, and a set of results, that we believe have direct relevance to field populations. We hope this stimulates observations and experiments to test models in real systems, and also development of new models.

We are grateful to colleagues and students who read and commented on various chapters: Matt Daugherty, Andre de Roos, Rebecca Doubledee, Martha Hoopes, Eran Karmon, Annette Kilmer, Brian Kinlan, John Latto, Hunter Lenihan, Ed McCauley, Brent Miller, Lara Rachowicz, Shane Richards, and Will White. Allan Stewart-Oaten was particularly vigilant in catching mathematical inconsistencies. We especially thank several colleagues who read and commented on the entire manuscript: Priyanga Amarasekare, Elizabeth Borer, Jamie Lloyd-Smith, and Susan Swarbrick. Susan Swarbrick also helped with the figures. Peter Chesson made helpful comments on the section in chapter 2 on regulation in stochastic populations. Andre de Roos kindly supplied electronic files of results from spatially explicit models, used in chapter 10, and commented on this section. Nick Mills provided useful discussions on parasitoid natural histories. Van-Yee Leung helped prepare the figures. The research in this book was supported by grants from NSF, USDA and EPA. WM is grateful to the National Center for Ecological Analysis and Synthesis where he worked on an early draft.

Consumer-Resource Dynamics

CHAPTER ONE

Introduction

A central challenge, perhaps the central challenge, for population ecology is to explain the persistence of species. In the variable, uncertain, and sometimes catastrophic natural world, species extinction is rare—why? The broad answer is that most natural populations are regulated most of the time at some spatial scale. The study of population dynamics is therefore a search for regulatory mechanisms and their effects.

Regulated populations can exhibit a range of dynamics. These include stability; instability, for example in the form of population cycles; local extinction but regional regulation; and local instability but regional stability (all discussed in more detail in chapter 2). Population ecology therefore also needs to explain when and how dynamics take one rather than another of these forms.

WHY CONSUMER-RESOURCE INTERACTIONS?

The consumer-resource interaction is arguably the fundamental unit of ecological communities. Virtually every species is part of a consumer-resource interaction, as a consumer of living resources, as a resource for another species, or as both. Consumer-resource interactions are, in addition, fundamentally prone to being unstable (chapters 3 and 4). If we are to understand population regulation and its various manifestations, we therefore need to focus on consumer-resource interactions.

Our major focus is either on generic (that is, simple) consumer-resource models, typically developed with predators and prey in mind, or on more detailed models developed initially to portray parasitoid-host interactions. We treat disease-host models only briefly and hardly mention herbivore-plant interactions. In chapter 11, however, we show that models for all of these types of consumer-resource systems have a common origin, and we suggest that many of the insights that emerge from the models examined are likely to apply broadly to different classes of consumers and living resources.

Except in chapter 8, we focus mainly on one consumer and one resource population. The book is thus frankly reductionist. In real communities, of course, consumers frequently attack several species, and resource populations are often attacked by several consumer species. We believe that understanding simpler systems is a useful prelude to understanding more complex ones for reasons explored next; in addition, chapter 11 gives some, perhaps surprising, reasons for believing that few-species models are often appropriate for populations living in many-species food webs.

ON THEORY AND MODELS

This book focuses on mathematical models as a means of understanding population dynamics. Here we give reasons why we think this approach is essential to understanding such real-world dynamics.

The ecological world bristles with particulars. Each species is unique. Most species are represented by many populations, each living in an environment that differs from those of other populations. There are surely at least 100 million such populations. We can study and understand the population dynamics of pitifully few—a miniscule fraction of the total—and if all we can get is an account of these few populations, it is not worth beginning the task. The goal and promise of theory are to reach into this thicket of particularity, grasp what is general, and express it in ways that let us test the predictions in real systems.

Theory expresses our current understanding of the real world. In ecology, theory may never be all-encompassing. Furthermore, theory will never predict, even for one particular population, most details of future dynamics—the exact age distribution on 4 July next year, for example. Theory might nevertheless explain how various types of dynamical behavior (stability, long-period cycles, etc.) can arise and how a range of different natural histories and life histories can give rise to the same kind of dynamics. It might predict new types of dynamics or define field observations and experiments that can tell us which among several potential mechanisms is actually operating in a real system. Theory in this book does just these things, and thus extracts generality from particularity and also achieves testability.

A model is a set of assumptions about how nature works, together with an algorithm for calculating the consequences. The assumptions abstract from the real situation those features one considers "more important" and ignore those that are less important. The model then deduces the consequences of these assumptions. Since none of us can think simultaneously about all the details of any ecological system, we all use models all the time. The model may be verbal, or a vague picture in our head, but it is a model.

We all have models in our heads all the time. The major value of mathematical models, in a subject that is quintessentially quantitative, is explicitness. A mathematical model forces us to say exactly what our ideas/assumptions are. If the calculations are correct, the model establishes the consequences of the assumptions, consequences we often could not derive purely by intuition. A mathematical model may be wrong, but it makes the assumptions explicit, so they can be examined and disputed, and the conclusions follow logically from the assumptions, so with luck they can be tested.

We do not suggest that the models in this book are correct. We know that, like all models, they are at some level wrong. We hope they are not wrong in fundamentally important ways. But if they are, exploring the errors will point the way to better insights.

The key question is not whether a model is true, but whether it is useful. Models are tools to understand the real world and can serve this purpose in various ways. They can sharpen our intuition about ecological mechanisms. Even the simplest model, which may match no living system, can be useful. For example, it can still tell us that delays in density dependence can lead to instability, population cycles, fewer recruits next year if we have more adults this year, and so on.

Simple Models and (or versus) Complex Models

Every model is a judgment about what is more, and what is less, important; i.e., about what to include and what to omit. Theorists typically love simple models. This is partly because they are more mathematically tractable, and because their lack of ecological detail suggests they may be broadly applicable. (A counter argument, of course, is that lacking so much ecology they cannot be relevant.) But simpler models have another key property that is often overlooked: they are easier to understand, so their assumptions and mechanisms are more transparent.

Simple models do of course lack ecological features that may be important to answering the question posed. And there is a school of modeling that argues that as much as possible of the real system should be represented in the model. But whereas more realism might always seem better, increasing complexity poses increasing difficulties. Two seem to us especially important.

First, complex models are harder to understand. It is often not at all clear how each component of a complex model affects the outcome. Since most models in this book are developed as theory, they are the simplest consistent with the need to include the crucial natural history. We also argue that complex models need to be placed in the context of simpler and more easily understood models, which gives rise to the notion of theory as a hierarchy of models (chapter 12).

Second, and perhaps paradoxically, beyond a certain point models become more difficult to test as they become "more realistic" and hence more complex. One reason is simply that the field biologist cannot supply the data needed to estimate the functions and parameters. But in addition, as more complexity is added, the greater is the chance that the model will "explain" the observations for entirely spurious reasons. When fitting a model to real data, the best predictive models have a small to intermediate number of parameters (Burnham and Anderson 1998). We do little model fitting in this book.

THEMES

Here we adumbrate themes that run through the book. It will be useful to keep them in mind, for they can explain why sometimes we may spend time on matters that do not immediately seem of prime importance.

- *Effects of ecological realism.* We are concerned in much of the book about exploring the dynamical effects of adding realism to models. In particular, we show how differences between individuals, associated with growth and development, offer new explanations for existing phenomena, including new routes to stability, and predict new phenomena, including novel kinds of population cycles.

• *Realism, scope, and generality.* Adding realism to models usually implies tying them more explicitly to particular systems or kinds of systems, with a potential loss in the range of systems to which they are relevant—a loss of generality. However, whereas we focus for stretches on particulars, we are consistently concerned with the range of real systems to which models may apply, and with exploring the extent to which they can be generalized.

• *Coherent theory.* An aspect of generality is the extent to which different models are connected in a larger, coherent body of theory. We explore this issue explicitly at several points through the book and especially in chapter 11. We also continually check for properties of the simple models we start with that survive to our ultimate, more complex models. Such persistent properties are the threads that connect the range of models we explore. These connections build to the final notion of coherent theory that takes the form of a hierarchy of models (chapter 12).

• *The role of cycles.* We focus on three main dynamical aspects of the models in this book: equilibrium densities, stability properties, and the form of population cycles the models produce under certain circumstances. We focus on cycles because they provide a strong signal determined by the structure of the interacting populations and the mechanisms through which they interact. If the details of particular types of cycles seem at first sight arcane, we counsel patience. It turns out there are rather few classes of cycles, and we can put them to good use as probes of real systems.

CHAPTER TWO

Population Dynamics:
Observations and Basic Concepts

This chapter has two parts. The first discusses the various types of population dynamics observed in nature. The second looks at basic concepts that relate these observations to mathematical theory. A brief appendix introduces local stability analysis, illustrated by the logistic model.

TYPES OF POPULATION DYNAMICS:
PHENOMENA TO BE EXPLAINED

In this section we present a range of observed dynamics seen in populations. By "dynamics" we mainly mean temporal changes in density or abundance. Average density of a population, and the relative densities of different populations, are of course also of interest. For example, we care greatly about average density in pest control and in populations we harvest, and we would like to be able to explain why some species are rare, others common, and why the general pattern of relative abundance of different species in an area is often maintained over ecologically long periods.

Interestingly, particular types of dynamics do not appear to be restricted to any particular group of organisms or environments. Cyclic populations, for example, occur in terrestrial, freshwater, and marine habitats. They are found in birds, mammals, insects, fish, marine crabs, etc. (Kendall et al. 1998). Only in plants do cycles appear to be rare, perhaps because they often compete with other plant species for non-living resources.

Biological Control of California Red Scale by Aphytis melinus

We first introduce a particular system, California red scale and its "natural enemies," since we use it to illustrate various phenomena we wish to explain. This system also motivated many of the models discussed later in the book, so we briefly summarize its natural history here.

California red scale, a plant-sucking insect (Homoptera), was accidentally introduced to southern California, probably from China via Australia, more than 100 years ago. It is found all over the world wherever lemons, oranges, and other citrus cultivars are grown. Uncontrolled, the scale population in a tree can number in the millions and can kill the tree. Red scale almost destroyed the citrus industry in California in the early and mid–twentieth century and quickly became resistant to a broad range of pesticides.

Over a half-century, about 50 natural enemies were introduced from Asia to control scale in southern California. Most failed to establish. The parasitoid (insect parasite) *Aphytis chrysomphali* somehow arrived in the early twentieth century, persisted in some areas, but did not control scale. In 1947 another parasitoid, *A. lingnanensis*, was introduced and, although this was not studied in any detail, probably competitively displaced *chrysomphali* (DeBach et al. 1950; DeBach and Sisojevic 1960). *A. lingnanensis* achieved some modest success near the coast. Around the same time another parasitoid, *Encarsia perniciosi*, was introduced (DeBach 1965). DeBach and Sundby (1963) concluded that it supplemented the effect of *lingnanensis*, but there is no record of field data to support this notion of a supplementary effect. Finally, *A. melinus* was introduced mainly in 1959 and was an immediate and spectacular success, as it has been in many parts of the world. It also famously competitively displaced *lingnanensis* everywhere except near the coast. We refer to other aspects of this system below and in later chapters.

Population Regulation and Persistence

The fossil record suggests that before humans began large-scale exterminations, species, on average, persisted for at least 2 million years, and probably somewhat longer (Raup 1991; Alroy 2000; J. Alroy, pers. comm.). Thus, if up to the advent of humans there were, say, 10 million species on earth, the expected number going extinct in any year was only 5. Since each species is made up of a few to many populations, a fundamental challenge is to explain the persistence over long ecological periods of populations or collections of populations. This is the problem of population regulation.

We thus equate population regulation with population persistence over a significant ecological time frame. Intuitively, we suggest that the key idea is that the abundance of a regulated population is *bounded*. Since all populations live in a finite environment, each population must be bounded at some maximum abundance. Avoiding extinction means that abundance is bounded at some lower level that exceeds zero.

Extinction

Some populations do go extinct. They are therefore not regulated at the spatial scale at which extinction occurs. If all of the populations in the species go extinct, then of course so does the species. But it seems that local population extinction is a common feature of some species that nevertheless persist because collections of their populations are regulated at a larger spatial scale.

Butterflies provide the best-known examples of regional persistence via extinction and recolonization of local subpopulations (Hanski and Thomas 1994; Thomas and Hanski 1997). For example, frequent local extinctions and colonizations occurred over just a few years in a checkerspot butterfly confined to a collection of about 1500 small meadows in a set of islands off the Finnish coast (Hanski et al. 1995).

Such metapopulation dynamics (in which movement among subpopulations is crucial to local and regional dynamics) are the subject of a great range of studies, and we will not deal with them at much length. We stress here just two points. First, aggregate dynamics across semi-isolated subpopulations may often be important in explaining both local and regional population dynamics. Second, populations can and do go extinct, and the fact that most of the time most of them seem to persist over a long ecological time frame is therefore, as we stress, a central phenomenon to be explained.

Stable versus Cyclic Populations

STABLE POPULATIONS

Some populations are remarkably constant over long periods, and it seems natural to define these as stable populations. Stability in dynamic models has a strict mathematical meaning: for example, an equilibrium is stable when a perturbation from equilibrium dies away with time, i.e., the population returns to equilibrium, as discussed below. Such "convergence" has in fact been demonstrated experimentally in several real populations (Eisenberg 1966; Stimson and Black 1975; Adams and Tschinkel 2001). Thus it seems likely that a real population that is relatively invariant can reasonably be represented by a deterministic model that has a stable equilibrium.

A population can vary quite strongly over time and still be considered stable. For example, a population that is frequently perturbed from equilibrium by a randomly varying environment, but whose equilibrium is stable, should show marked random variation around its average density. Another might be tracking an equilibrium that is changing gradually in response to a change in the environment such as long-term oceanic conditions.

One common class of stable populations that is particularly interesting is those suppressed to low density by their consumers. California red scale and its enormously successful biological control agent, *Aphytis melinus*, provide a dramatic example.

Red scale has been under biological control by *A. melinus* in southern California for almost half a century. Over this period, in any particular grove, the pest probably is always present in very low numbers, probably at less than $\frac{1}{200}$ of the density it can achieve in the absence of control. This stable pattern, and the contrasting potential for massive change in scale density, were demonstrated in the early 1960s. DeBach et al. (1971) sprayed a tree with DDT to kill off the controlling natural enemies. While the population in the grove in general remained remarkably constant, scale density in the sprayed tree increased about 70-fold over 3 years and increased by more than 3-fold per generation during the 15 months of most rapid increase; i.e., the per generation multiplication rate, R, was 3.14 (Fig. 2.1a). Almost 30 years later, scale populations in the same area were found to be

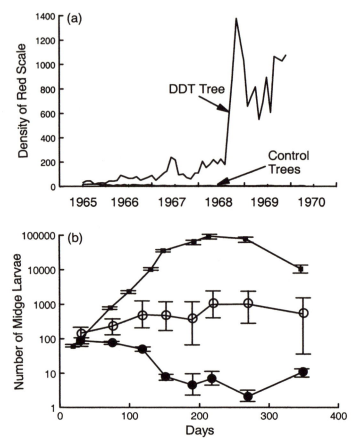

FIGURE 2.1. (a) Abundance of red scale under biological control in a citrus grove in southern California and in a tree sprayed with DDT. Redrawn,
with permission of Kluwer Academic Publishers, from DeBach et al. 1971.
There are two scale generations per year, so the population increased 17.5-
fold in 2.5 generations, giving a factor of increase per generation (R) of
3.14. (b) Abundance of larvae of the midge *Rhopalomyia californica* on
individual *Baccharis pilularis* bushes. Data from an experiment with the
following three treatments: solid squares are "parasitoid removed" (caged
bushes from which all midges and parasitoids were removed and midge populations were re-established by addition of a few midge individuals); open
circles are "caged control" (caged bushes on which natural populations of
midges and parasitoids were unmanipulated); and solid circles are "uncaged
controls" (uncaged bushes on which natural populations of midges and parasitoids were unmanipulated). Time series represents 6–8 generations of the
midge and parasitoids (Briggs 1993b).

among the most constant of recorded animal populations (Murdoch et al. 1995).

This phenomenon—resource populations suppressed to low densities by consumers in a stable interaction—may be rare in biological control (Murdoch et al. 1985) but may be common in natural systems. Gall midges that attack *Baccharis*, a bush species of coastal chaparral in California, provide an excellent example. The midge lays its many eggs at the tip of *Baccharis* stems, and the larvae burrow into the tissue, which responds by forming a gall. The midge is attacked by a range of parasitoid species. Collectively they cause well over 90% parasitism, and experiments excluding the parasitoids show they suppress the gall population to far less than 1% of its potential abundance. Furthermore, in the presence of parasitoids the midge population in a bush varies little in density over many generations (Fig. 2.1b) (Briggs 1993b; Briggs and Latto 2000).

The interaction between the zooplankter *Daphnia* and its edible algal prey provides another natural example we discuss later (chapters 3 and 11). Murdoch and McCauley (1985) and Murdoch et al. (1998b) showed that *Daphnia* can reduce the density of edible algae to around 1% of the level set by algal resources and that, following the spring peak in density, *Daphnia* and algal populations typically fluctuate little (perhaps around a seasonal trend) or show very small-amplitude cycles.

A dilemma exemplified by red scale, gall midges, and zooplankton is that their population densities are relatively constant, whereas simple predator-prey models predict large-amplitude cycles when the predator has the capacity to reduce the prey to very low densities (the "suppression-stability trade-off," chapters 3 and 4). Such cycles do in fact occur in other systems in nature (Fig. 2.2a). It is therefore all the more intriguing that red scale and other severely suppressed prey populations are stable. Explaining this widespread mismatch between simple theory and real predator-prey interactions is a central problem in population dynamics.

CYCLIC POPULATIONS

Cyclic populations can present a startling contrast to the type of dynamics seen in galls and the red scale–*Aphytis* system. They are

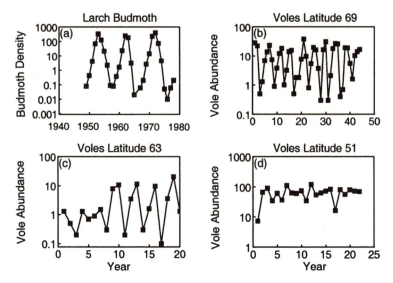

FIGURE 2.2. (a) Observed population cycle in larch budmoth in Swiss forests. Over the three periods of approximately exponential increase, the budmoth population increased by an annual factor (R) of 7.1. (b, c, and d) Vole population time series from three latitudes in Europe. Populations at higher latitudes are more likely to cycle, and the amplitude and period tend to be larger (modified from Turchin and Hanski 1997). Data from P. Turchin.

surprisingly common and, in contrast to a widely held belief, are not especially limited to high latitudes. Kendall et al. (1998) analyzed almost 700 time series of population numbers, each recorded over a minimum of 25 years, and concluded that almost 30% were cyclic. The populations included mammals, birds, fish, and insects and other invertebrates.

Cycles interest population ecologists because, unless driven by some external cyclic force, they provide a strong signal of a density-dependent process. They are thus more likely to yield information about mechanism than are random variations about equilibrium. They are especially interesting in our context since long-period cycles are an inherent property of tightly coupled consumer-resource interactions (chapter 11) and are not characteristic of competitive or mutualistic interactions.

Cycles come in different types. Some have rather small amplitude, whereas a few vary more than 10,000-fold over a single cycle (Fig. 2.2a).

The period also varies: peak density may be reached every generation or only every several to many generations (Fig. 2.2a, b).

Of special interest is that the same species may cycle in some regions and not in others. For example, in Swiss forests the larch budmoth exhibits very large-amplitude cycles (Fig. 2.2a) but does not cycle farther north in Germany. (This population, incidentally, shows the largest amplitude cycles we know of. Increases of about 20,000-fold have been observed over 6 years. During the increase phase of the cycles in Fig. 2.2a, the population grew roughly exponentially at an annual multiplication rate, R, of 7.1). Again in Europe, voles are more cyclic at higher latitudes and the period and amplitude of the cycles are larger than at lower latitudes (Turchin and Hanski 1997) (Fig. 2.2b–d). Clearly, we would like to know what causes such changes in dynamic behavior.

Chaotic Populations

We will say little about chaos. Chaotic populations are those that fluctuate, within bounds, from purely deterministic causes yet do not follow a fixed trajectory such as a cycle (May 1974; May and Oster 1976). The fluctuations appear at first sight to be random. Hassell et al. (1976) argued that chaos was likely to be rare in real populations, and this appears to be the case (Ellner 1991; Ellner and Turchin 1995) (chapter 4). For some time it was thought that measles outbreaks in human populations might exemplify chaotic dynamics, but it now seems more likely that these are cyclic populations whose dynamic pattern changes in response to long-term changes in the human environment (Ellner et al. 1998).

Competition: Competitive Exclusion and Coexistence of Competitors

The red scale system nicely illustrates the three major competition phenomena that require explanation: exclusion; its alternative, i.e., coexistence; and the relationship between exclusion/coexistence and pest density. As noted above, the displacement of *Aphytis lingnanensis* by *A. melinus* is one of the most dramatic examples of competitive exclusion (see chapter 9 for details).

Although no data were ever published on red scale densities before and after *melinus* appeared, it is known that *melinus* improved the level of control. So displacement of the weaker competitor was associated with, and presumably achieved by, substantial reduction in the common resource—red scale. The system thus illustrates the main outcome predicted by competition theory, and we discuss this example further in chapter 8.

Paradoxically, the red scale system also provides a counterexample to competitive exclusion. Based on the studies by DeBach and his colleagues mentioned earlier, the history of control in groves near Santa Barbara, California, is probably close to that diagrammed in Fig. 2.3. In any particular citrus grove, we find *A. melinus*, usually one other parasitoid species, and one or two general predators. Yet competition theory predicts that only the best competitor will survive when consumers compete for a limited resource (chapter 8). This prediction surely should have special force in successful biological control, where the resource is severely depressed. Indeed, it appears from the historical record that *Encarsia* likely went extinct in at least some groves after *A. melinus* was introduced. Currently, however, these two species coexist quite extensively.

Obvious questions are: what made *melinus* so much better than its congeners? How can *Encarsia* persist in the presence of *Aphytis*, since *Aphytis* greatly reduced scale below the level that previously maintained *Encarsia*? How is control affected by the coexistence of enemies other than *Aphytis*? Do they supplement, or interfere with, *melinus*? Would *Aphytis* be even more effective on its own? What processes keep the red scale populations so remarkably invariant (Fig. 2.1a)? We look at some of these questions in later chapters, but mostly they have not yet been answered either for this system or for coexisting consumers in general.

The gall midge provides a natural system in which perplexing coexistence occurs. Fig. 2.1b establishes that the parasitoids reduce the abundance of galls far below the limit set by the gall resources, yet a single clump of bushes typically has five or more coexisting species attacking the galls (Briggs 1993b).

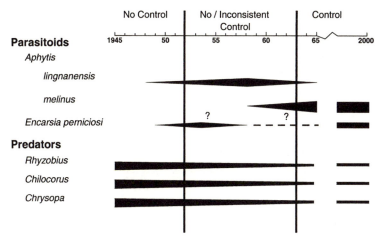

FIGURE 2.3. Diagrammatic representation of the history of biological control agents of California red scale in groves in Ventura County, southern California. Thickness of each line represents a guess at the relative abundance of the species in the area.

SOME ESSENTIAL CONCEPTS

In this section we lay out several mathematical ideas that relate the real dynamics presented earlier to the models we will analyze. We expand on some of them as a basis for understanding material in later chapters. Several recent books provide explanations of simple models and their analysis, including those by Hastings (1997), Gurney and Nisbet (1998), Case (2000), Mueller and Joshi (2000), and Kot (2001).

Equilibrium, Stability, Instability, and Regulation in Deterministic Models

Models in population ecology consist of a description of the current state of the system and a set of rules defining how it will change in the immediate future. Getting a "solution" to the model equations tells us what the system's state will be at any time in the future, given its current state and the rules for change. A key assumption is that we need to know only the current state, not how the system got there.

In the simplest exponential growth model the single *state variable* is current population density, N. The rule for change is that the instantaneous per-head rate of change $(dN/dt/N)$ is constant and does not depend on N, giving

$$\frac{dN}{dt} = rN. \tag{2.1}$$

r greater than zero gives growth without limit, r less than zero gives a constant fractional reduction in density per unit time, and r equal to zero gives a constant population.

The model assumes that population processes—births and deaths—are continuous, all individuals are the same, there are no time lags, and the population is closed to immigration and emigration. N is a continuous variable and can take non-integer values, such as half an organism. This makes sense if we think of N as a density rather than a number of organisms, but it means that extinction is not formally possible in this model since N can become infinitesimally small. The parameter r embodies the effects of the environment as constrained by the organism's life history.

Equilibrium is the state (in this case density) at which no further change in state occurs unless the system is perturbed by an external force. It exists when dN/dt is equal to zero. This equality holds in equation 2.1 only when $r = 0$ (or, trivially, when $N = 0$). However, $r = 0$ is not a reasonable assumption: it implies the population would stay at any density at which it starts. We give other reasons why it is biologically unreasonable, in the section on stochastic populations, below. Equation 2.1 therefore has no non-zero equilibrium. This is because the per-head rate of change does not depend on density. Existence of an equilibrium requires that the per-head rate of change of the population does depend on its density.

NON-LINEARITY

The exponential growth model is the only linear model of a closed system in this book. All others are non-linear. In a non-linear model at least one per-head rate depends on the density of at least one population. It is non-linearity that creates an equilibrium. All reasonable ecological models of closed systems are non-linear.

The source of non-linearity is often clear, for example in the logistic model analyzed in the appendix to this chapter: the per-head rate of change in density depends on density. But occasionally it is less obvious what makes a model non-linear. For example, in the Lotka-Volterra predator-prey model (chapter 3), the prey equation is

$$\frac{dH}{dt} = rH - aHP.$$

This equation is non-linear as it contains the product term aHP. The non-linearity arises because the prey's per-head death rate from predation, namely aP, depends on predator density, P.

All the interesting dynamics in ecology arise from non-linearity. Unfortunately, so do the difficulties in analyzing models.

SOLUTIONS TO DIFFERENTIAL EQUATIONS

Equation 2.1 tells us how density will change from its current state but not what it will be at any time in the future. For this we need to solve the equation. Since this differential equation is linear, it can be solved analytically, meaning we can write an expression for N at all times in the future, given an initial value for N. The analytic solution is

$$N_t = N_0 e^{rt}. \tag{2.2}$$

We can use this formula to calculate the number in the population at any time in the future, t, if we know the number at time $t = 0$ (say, now) and r.

Unlike equation 2.1, non-linear differential equations usually cannot be solved analytically. Numerical solutions, however, can be found for all models using a number of algorithms. Such algorithms (e.g., the Runge-Kutta or Euler methods) are available within software (e.g., Matlab, Mathematica, Madonna, Solver) that allows one to define the model, specify the initial densities, assign parameter values, and calculate the population trajectory over any desired time period.

A numerical solution, however, is for only one particular combination of parameter values and initial conditions, so it is difficult to know if the population dynamics obtained represent the model's behavior in general. Model behavior may change with even small changes in parameter values. One can explore a range of combinations of parameter values and initial conditions, but even so, one cannot be sure there is not some biologically reasonable region of parameter space where the dynamics are quite different. This is especially true for complex models. We can think of each parameter as defining an axis or dimension, going from small to large values of the parameter. A plane then defines the "parameter space" of a model with two parameters, and one can potentially calculate the solution over a grid of possible parameter values to get a thorough picture of how the model behaves over the entire feasible parameter space. But if the model has many parameters, we are not likely to be able to explore thoroughly the many-dimensional parameter space.

This is the primary reason we carry out local stability analysis. It delineates the boundaries between conditions (e.g., different combinations of parameter values) leading to qualitatively different dynamical behavior.

STABILITY AND LOCAL STABILITY ANALYSIS

As noted above, once a model population has reached an equilibrium, or steady state, its density will never again change unless the population is perturbed from the equilibrium. By "perturbed" we mean one of two things. Some event adds or subtracts individuals from the population without changing any of the processes that determine the rate of change of the population. In a model, we would move the population away from equilibrium but not change any of the functions or parameters. Alternatively, we might change one or more parameter(s) to a new value(s), so the current density would be some distance from the new equilibrium.

An equilibrium is stable if the population eventually moves toward it, and stays close to it, after such a perturbation; that is, the perturbation gets arbitrarily small if we wait long enough. The equilibrium is otherwise unstable if the density moves away from it through time.

We would like to know, for each model, whether the population returns to equilibrium after we have disturbed it some arbitrarily large distance from equilibrium. However, we typically cannot get an explicit answer to this question, because models are non-linear, and as noted, we usually cannot solve non-linear equations. Local stability analysis turns a non-linear problem into a linear one. It is a procedure that asks whether, if we perturb the population a small distance (hence "local") from its equilibrium value, the density tends to return to, or move away from, the equilibrium. Since the distance is small, non-linear or "higher order" terms can be ignored. Local stability does not tell us directly about non-linear behavior far from equilibrium, but it turns out in many cases that we can show, by judicious simulations, that it is a reliable guide.

Equation 2.1 can be solved to give equation 2.2. This simple fact is used whenever we do local stability analysis. The behavior of models depends on a quantity—called the eigenvalue—that is an analog of r. We then use the fact that, in an exponential growth model, r greater than zero implies unrestricted growth, r less than zero continuous decline, and r equal to zero no change. The appendix to this chapter illustrates local stability analysis using the logistic model, the simplest non-linear model of population growth.

REGULATION IN DETERMINISTIC MODELS

Above we defined a regulated population as being bounded and hence persisting. Here we stress that regulation can be consistent with an unstable as well as a stable equilibrium.

The equilibrium in the logistic model is stable: perturbations from equilibrium die away with time and the population returns to equilibrium (appendix). The equilibrium is thus an "attractor." This case clearly conforms to regulation. When an equilibrium is unstable, the perturbations increase with time. In some models the perturbations grow without limit, and eventually the population becomes extinct. The Nicholson-Bailey model is an example (Fig. 4.6a). These cases are not consistent with regulation.

In chapter 3 and later chapters, however, we see that in many models, when the equilibrium is unstable, a small perturbation at first increases in size through a series of oscillations, but eventually the size (amplitude) of these oscillations reaches a maximum. This is because there frequently is a *stable limit cycle* that "captures" the perturbations (e.g., Fig. 3.5c). Once on the limit cycle, the population endlessly traces out a cycle with a fixed period and amplitude. Like a stable equilibrium, such a stable cycle is an attractor, and the population is regulated. The population fluctuations are strictly bounded and the population is regulated. Similarly, with bounded chaotic fluctuations (introduced in chapter 4), the population is regulated.

Species Persistence and Regulation in Stochastic Populations

This section reinforces an earlier message: a closed population cannot be regulated if the per-head rate of change is independent of density; there must be an equilibrium or, if the equilibrium is unstable, there must be an attractor (e.g., a stable limit cycle). This section also defines stochastic analogs for stability and population regulation.

Stability and boundedness are straightforward concepts in a deterministic framework. The real world is, however, stochastic, in the sense that no matter how deeply we understand a system, there will be outcomes that are uncertain. For example, it is in practice not possible to predict the weather in a given area for all times in the future. We may nevertheless be able to predict its statistical pattern. Although we largely ignore random processes in this book, it is essential to consider them in thinking about population regulation.

A stochastic framework makes clear why an exponentially growing population cannot persist or be regulated. Consider again equation 2.1. Because this model has no attractor, population regulation, i.e., long-term persistence, is impossible. In a deterministic framework, we might think this is incorrect: the population would stay constant if r were equal to zero. But this requires that r never changes. In real (stochastic) populations, r—the difference between the per-head birth and death rates, $b - d$—must change all the time, and some of that variation must be random, i.e., unpredictable given current information. So r cannot be zero all the time. Still, r might be zero on average; what happens then? To model this variation in the real world we would need to consider the value of r at each point in time, r_t, and now it is the mean, $\bar{r}_t$, that is zero.

A brief digression is needed. In a deterministic model we can talk of a single regulated population. A stochastic model, by contrast, defines what happens if we were to run many replicate populations, all obeying the same "process" but differing because each replicate would experience a different sequence of random conditions. We can then ask what the mean trajectory would look like, and what the probability distribution of replicates ("realizations") would be at any time in the future. In nature any given population is likely to be the only realization of a particular underlying process, but a long enough run may allow equivalent analysis (Chesson 1996). Over very long time periods, of course, the

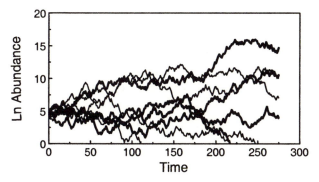

FIGURE 2.4. Random walks generated by a model of geometric growth: $\ln N_{t+1} = \ln N_t + r_t$, where r_t is a random variable, with mean 0, uniformly distributed between -0.5 and 0.5. Initial population size is 100. Abundance is plotted on a logarithmic scale, allowing infinitesimally small fractions of an organism. We therefore define a population as extinct if it declines to 1 individual ($\ln N = 0$) or less. Three of the eight populations went extinct within 250 time steps.

environment may change substantially, thus changing the underlying process.

For calculation purposes, we change equation 2.1 to a discrete-time form and work with the natural log of abundance. Suppose r_t is a random variable with a mean of zero and some statistical distribution. Simulations of populations obeying this rule look like those in Fig. 2.4; they (or more precisely their logarithm) are undergoing *random walk*. They show two properties. First, if we set bounds on population density—an upper and lower limit—and if we allow enough time to pass, the populations will always pass through those bounds: the variance of the population densities increases through time without limit (Murdoch 1994, Fig. 1d). Second, if we run them for long enough, the populations will become extinct, though the time to extinction will vary among populations. Thus, per-capita changes in density that do not depend on density cannot lead to population regulation, even when the expected change from one time to the next is zero.

Indeed, random-walk populations epitomize the absence of regulation. They tell us that *a population cannot be regulated if the per-head rate of change is independent of the state of the system* (Murdoch 1994; Turchin 1995).

What about collections of random-walk populations in heterogeneous space? Perhaps these can be truly populations without equilibria yet still persist as an ensemble? The answer is no. The number of organisms in a collection of random-walk populations in heterogeneous space, and linked by random movement, also randomly walks (Chesson 1981, 1996; Klinkhamer et al. 1983).

Regulated stochastic populations differ from random-walk populations in that the variance of the abundance of *extant* replicate populations does not keep increasing with time. The tricky aspect of a stochastic framework is that even in a process we might agree is regulated, over any long period we expect some populations might go extinct.

As in deterministic models, it is useful to distinguish between models in which there is or is not the equivalent of a stable equilibrium. In the stochastic equivalent of a stable equilibrium, whereas some replicates may have gone extinct, the densities of the remaining replicate populations approach a stationary probability distribution (May 1973b; Chesson 1982). This is true even in the common situation where all replicates eventually go extinct (Nisbet and Gurney 1982; Renshaw 1991) and the distribution is quasi-stationary. Suppose there are many replicate populations, all obeying the same stochastic model but differing in the actual trajectories they trace out because of random differences in the process. Then over the long term there will be a constant fraction of extant populations whose densities will lie between, for example, 10 and 20% of the mean density. (The long term may be very long indeed if there is serial autocorrelation in the population growth rate; Arino and Pimm 1995; Chatfield 1996.) Turchin (1995) argues that this is in fact the stochastic definition of a regulated population. We suggest, instead, that it is the stochastic equivalent of a stable equilibrium (see Nisbet and Gurney 1982) and therefore too narrow for a definition of regulation.

The need for a broader definition arises because, above, we argue that in a deterministic framework, regulation includes non-stationary (cycling and chaotic) populations whose fluctuations are nevertheless bounded. One stochastic analog for this appears to be Chesson's concept of "stochastic boundedness." A rigorous definition can be found in Chesson 1978 (p. 343). In this case, the surviving replicate populations need not approach a stationary probability distribution, just as a cyclic deterministic population will not settle on the unstable equilibrium density. Instead, the variance of population densities, given population persistence, is eventually bounded. Stochastic boundedness can be regarded as one possible definition for persistence (Chesson *in* Turchin 1995), which conforms to our intuitive concept of regulation, defined earlier in this chapter.

Notice that over any long period, we expect some "replicates" of a regulated population to go extinct in a stochastic framework. We expect this also in the real world, where population extinctions do occur. While the population is going extinct, it is of course not regulated.

OPERATIONALIZING THE CONCEPT OF REGULATION

Although stochastic boundedness provides a clear definition of regulation in theory, it has turned out to be difficult to convert this idea into an operational definition that can be used in the field. At any given time, we can define population variance, and we say, as we did above, that a population is regulated if its variance, or better the variance ln of population density (conditional on non-extinction), is bounded. Put less technically, a population is regulated if the amplitude or intensity of its fluctuations does not grow without limit. For further details see Nisbet and Gurney 1982.

Even this definition of regulation, however, is impractical for use on real populations. It turns out that population variance appears bounded in a substantial fraction of long-run simulations of unregulated, random-walk populations, a result we would mistakenly take as evidence of regulation. (This happens because the variance of population variance itself increases rapidly with time: whereas some runs accumulate variance extremely rapidly, the variance in others appears bounded over ecologically substantial periods of time; Murdoch 1994.)

One solution is to create a deterministic approximation to a stochastic model of the real population being studied. We then say the population is regulated if solutions of this deterministic model are bounded (Ellner et al. 1998; Kendall et al. 1999). This allows us to use the unambiguous deterministic definitions of boundedness: mathematically bounded deterministic trajectories include the approach to a stable equilibrium, cycles, and chaos. Alternatively, one could create a full stochastic model and see if it was stochastically bounded (P. L. Chesson, pers. comm.).

Strong and Weak Coupling, and Model Simplification

Throughout this book we look at any one time at interactions among only a few species. In doing so, we follow a hallowed ecological tradition that implicitly asserts that we can understand the dynamics of some particular consumer-resource interaction with little regard to the other species in the community. This is essentially a claim that real ecological communities can be decomposed into components, and that a component is dynamically sufficiently independent that we can, as a first approximation, treat the rest of the community simply as part of the environment of the component of interest. As we explain next, this is equivalent to a claim that such components are weakly coupled to the rest of the community (Murdoch and Walde 1989; Murdoch 1994). The idea of weak coupling will also be important at various points for defining conditions under which consumer-resource dynamics collapse to single-species dynamics, even in many-species food webs (see especially chapter 11).

We first deal with strong coupling. Two species are strongly coupled when there is a complete, strong feedback loop between them. More specifically, a change in the density of species 1 strongly affects the rate of change of species 2, and a change in the density of species 2 affects the rate of change of species 1. For example, in a lake dominated by the zooplankter *Daphnia*, the rate of change in the density of edible algae, on a time scale of days to weeks, is powerfully affected by a change in *Daphnia* density. On a similar time scale, the rate of change of *Daphnia* density is strongly affected by a change in algal density (Fig. 2.5). The strong feedback loop is complete. The important result is that to understand the dynamics of either species, we need to write explicit coupled equations for the dynamics of both. A predator-prey model such as the Lotka-Volterra model is an example. It may be useful to think in the following terms: coupling is strong when a signal sent from one population (a change in density) passes through the second population and then feeds back as a clear signal to the first population.

Fig. 2.5 also illustrates weak coupling. Suppose there is an annually breeding fish population that attacks the *Daphnia* population but also feeds on other species. The fish population may cause substantial mortality on the *Daphnia* population (hence the heavy arrow from the fish to the *Daphnia*-algae box in Fig. 2.5). However, the fish population

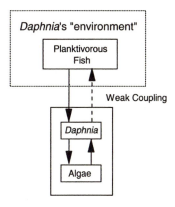

FIGURE 2.5. Strong and weak coupling. *Daphnia* and algal populations are strongly coupled. *Daphnia* population is weakly coupled with its consumer, planktivorous fish. Modified from Murdoch 1993.

cannot respond over the short run to the typical short-term change in the *Daphnia* population, which is on the order of days or weeks. Thus, whereas short-term changes in fish density affect the death rate imposed on *Daphnia*, short-term changes in *Daphnia* feed back only weakly on fish dynamics (so the return arrow from the *Daphnia*-algae box to the fish population is dashed in Fig. 2.5). The feedback loop is incomplete. In other terms, whereas the fish may send a strong signal to the *Daphnia* population, this signal is thoroughly distorted by the time it returns and is expressed in the fish population. As a consequence, a *Daphnia*-algal model that describes plankton dynamics over summer, say, can incorporate the effects of the fish population simply as one component of a "background" *Daphnia* death rate. The rate might even vary seasonally, but *there is no need to write a coupled equation for fish dynamics*. This is weak coupling, and in this case it allows us to define *Daphnia*-algal dynamics with a 2-species model. The fish population in this model is not ignored. It is part of the *Daphnia*-algal environment, just as temperature is.

We put the idea of weak coupling to good use in chapters 3, 4, 6, 10, and 11.

Local Stability Analysis of the Logistic Model

In the logistic model, the per-head rate of growth depends on population density; it declines linearly as density increases. If K is the carrying capacity, the density the environment can support over the long term, the per-head rate of change is

$$\frac{1}{N}\frac{dN}{dt} = r\left(\frac{K - N}{K}\right)$$

and thus declines by an amount $-r/K$ for each individual added to the population. Thus the population grows as

$$\frac{dN}{dt} = rN\left(\frac{K - N}{K}\right).$$

Local stability analysis proceeds in three steps.

1. *Find the equilibrium* by determining the density at which density does not change. In this case, $dN/dt = 0$ when $N = K$. Here and later we denote the equilibrium by N^*.

2. *"Linearize" about the equilibrium* by examining the dynamics of a small perturbation from equilibrium. Linearizing a model is the process of approximating a non-linear differential equation we cannot solve by a linear differential equation we can solve. We change the density by a small amount, n, so the perturbed density is $N^* + n$. We then analyze how n changes, instead of studying the dynamics of the actual population density. Note that a negative value of n corresponds to a perturbation that decreases the density. The rate of change of the

perturbed population density is in fact equal to the rate of change of n because

$$\frac{dN}{dt} = \frac{d(N^* + n)}{dt} = \frac{dN^*}{dt} + \frac{dn}{dt},$$

but

$$\frac{dN^*}{dt} = 0$$

by definition, and hence

$$\frac{dN}{dt} = \frac{dn}{dt}.$$

We want an approximate expression for dn/dt at all times. Taylor's Theorem tells us that if we have a function $f(N)$, and we know the value of f at one value of N, say at N^*, then we can approximate its value at any other value of N to any degree of accuracy we desire by "expanding the function about N^*." The rule is

$$f(N^* + n) = f(N^*) + \frac{df}{dN}\bigg|_* n + \text{"higher order terms."}$$

Here, $\frac{df}{dN}\big|_*$ means the derivative of f with respect to N evaluated at the equilibrium, N^*, and is just a number that can be evaluated once we know the model parameters and have calculated the equilibrium. The phrase "higher order terms" means terms proportional to n^2, n^3, and so on. For the logistic model,

$$f(N) = rN\left(\frac{K - N}{K}\right),$$

and

$$\frac{df}{dN} = r - \frac{2rN}{K} = -r \quad \text{when } N = N^* = K.$$

So the approximate form for dN/dt a small distance away from equilibrium is

$$\frac{dn}{dt} = -rn.$$

3. *Solve the equation.* This can be difficult in complicated models, but in this case it is easy. We have the familiar negative exponential growth, so the solution is

$$n_t = n_0 e^{-rt}, \tag{A2.1}$$

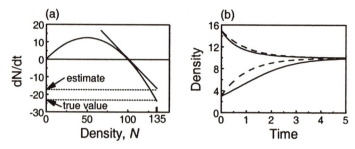

FIGURE A2.1. (a) Approximation to dN/dt in the logistic model, using the first term of Taylor's expansion. The value (-17.5) for $N = 135$, $K = 100$, $r = 0.5$ is predicted by the tangent to dN/dt drawn through the equilibrium (at $dN/dt = 0$) and projected to $N = 135$. The true value is -23.625. (b) Comparison of return to equilibrium in the logistic model (solid curves) and in the linear approximation (dashed curves). Parameter values are $K = 100$, $r = 0.5$.

where n_0 is the initial perturbation. We know that n will grow smaller with time, that is N will approach K, if the exponent is negative. Since the exponent is $-r$, and r is always positive, the logistic model has a stable equilibrium for all positive values of r and K, i.e., the equilibrium is locally stable everywhere in parameter space.

A graphical description of step 2 is in Fig. A2.1a. From the equilibrium, we extrapolate the value of dN/dt by extending the tangent to dN/dt (which has slope $-r$ at the equilibrium) a distance n along the x-axis. To determine if the equilibrium is stable, we only need to know if the approximate value of dN/dt is negative when N is greater than K. We need the approximate value to be positive when N is less than K.

The projected value is only approximate but has the correct sign (negative). In the logistic model, the graph of dN/dt versus N behaves well; for example, no matter how much N exceeds K, dN/dt never becomes positive. Thus, local stability analysis correctly predicts the model's qualitative behavior arbitrarily far from equilibrium. The model is thus globally as well as locally stable. This is not the case for all models. In the graph in Fig. A2.1a, we project a substantial distance from the equilibrium, simply to illustrate the process. In the analysis, the distance must be small so we can drop the higher order terms.

The approximation in local stability analysis will reliably indicate the qualitative behavior of the model if the original equations are not too horribly non-linear. When these conditions are not met, a locally stable equilibrium may be unstable to large perturbations. This occurs when the model has more than one possible equilibrium condition for a given set of parameter values (chapter 6 has an example).

The logistic is one of the few non-linear differential equations that can be solved analytically. Fig. A2.1b shows that our approximation of the behavior of the logistic is qualitatively correct though, as we would expect, wrong in detail.

CHAPTER THREE

Simple Models in Continuous Time

In this chapter we explore, in simple models, processes that tend to stabilize and those that tend to destabilize and cause cycles. Although more realistic and hence more complex models are the focus of much of the book, the simple models are central to our goal of a coherent set of theory. Many properties of simple models survive in complex models; simple models thus serve as a preliminary guide to the more complex. Indeed, complex models are often best understood in light of our necessarily more complete and deeper understanding of simple models. Finally, although the models in this chapter are largely familiar, we extract some fresh and potentially useful insights.

This chapter also has the frankly didactic aim of introducing basic concepts in modeling predator-prey interactions and the tools to analyze such models. This is useful background for the more complex models of chapters 5–8.

We begin with the Lotka-Volterra model and use it to lay out local stability analysis of simple consumer-resource models. We then survey a series of destabilizing and stabilizing mechanisms that affect dynamics in a very explicit way: they cause some parameter, defining a per-head rate, to depend on population density (*direct density dependence*). We then show how models combining a stabilizing and destabilizing mechanism can have either stable or unstable equilibria, can produce limit cycles, and hence can potentially explain the range of dynamics in real populations discussed in chapter 2. We also deal with the "paradox of enrichment," which predicts that dynamics should be more unstable in more productive environments. Finally, we examine a series of spatially structured and stage-structured models. We show a fundamental similarity between these two classes of models: their structure can lead to stability in the same way, through *indirect density dependence*.

The model protagonists in this chapter are generic consumer and resource species. We call them predators and prey, however, because the models are variations on the Lotka-Volterra model. The reader may wish to think of them as true predators and prey (e.g., fish and zooplankton or ladybugs and aphids), as parasitoids and host insects, or even as herbivores and plants (e.g., zooplankton and edible algae).

THE LOTKA-VOLTERRA MODEL

This well-known model portrays a prey population, of density H: there are on average H prey per unit area or volume. All individuals have identical vital (per-head) rates and are distributed randomly in space. The prey population can grow exponentially in the absence of the predator, at rate rH; r is the per-head rate of increase including the fraction dying from causes other than predation. Thus the prey itself has unlimited resources and its per-head rate of change is unaffected, at least directly, by its own density.

The predator population, of density P, is likewise assumed to consist of identical individuals, all fully mature when born. Predators search for and kill prey at a rate a (defined as the area or volume cleared of prey per predator per unit time). Thus each predator eats prey at a rate aH, and the total prey death rate is aHP. Each killed prey gives rise, instantaneously, to e new predators (e is predator conversion efficiency), so the number of new predators produced per unit time is $eaHP$. Finally, the predator per-head death rate has a constant value, d, which is independent of prey density, so the total predator death rate is dP. This assumption is not as stupid as it might appear. For example, a parasitoid might be able to survive on sugar, which is always available, but need to eat prey to obtain protein for egg development. Assuming such "exponential" survivorship, however, leads to the unrealistic feature that there is no upper limit to the predator's lifetime.

The state variables are H and P, and the parameters are r, a, e, and d. The model is obtained by equating the rate of change of each population density to the difference between the birth and death rates, thus

$$\frac{dH}{dt} = rH - aHP$$

$$\frac{dP}{dt} = eaHP - dP.$$

(3.1)

There are many implicit assumptions in the model. Some of these are as follows.

1. The populations are large enough that it makes sense to regard the state variables as continuous and to think of births and deaths as rates rather than as discrete, individually unpredictable, birth and death events (in modelers' jargon, we ignore "demographic stochasticity"). Moreover, because we work with densities, there is no lower bound (other than zero) for ecologically acceptable values of the state variables: H is a density, so it can be a billionth of a prey per square meter.

2. The two populations are completely mixed (sometimes called the "mass action" assumption by analogy with elementary chemical kinetics). Therefore the number of prey successfully encountered by an individual predator depends on the prey density averaged over the entire population, not in some small local area.

3. The populations are "closed" and receive no input from outside.

4. The dynamics are fully determined by the unique and constant values assigned to each parameter and by the initial conditions—there is thus no allowance for seasonality or for "random" changes in parameter values that might be induced by unpredictable processes such as weather.

The Model's Equilibrium

As discussed in chapter 2, the equilibrium is the state at which H and P do not change with time, that is when $dH/dt = dP/dt = 0$. We denote the equilibrium densities as H^* and P^*. The "point" equilibrium (H^*, P^*) is thus a combination of prey and predator densities from which, if it is ever achieved, the populations will not stray unless some external force perturbs them.

From equation 3.1, when prey density is positive and does not change, $rH - aHP = 0$, so (provided H is non-zero) $P^* = r/a$. Similarly, predator density does not change when $H^* = d/(ea)$. So, for example, if P is greater than r/a, then dH/dt is less than zero and, irrespective of its density, the prey population declines because there are "too many" predators.

Notice that prey equilibrium, H^*, depends on parameters that describe attributes of individual predators, not of individual prey, whereas predator equilibrium depends on prey per-head growth rate and predator per-head attack. Although these relationships are at first sight counterintuitive, they do make ecological sense. For example, if there are two similar systems, equal in all respects except for the predator death rates, prey equilibrium will be higher in the system with the higher predator death rate; otherwise predator intake rate would be too low to enable the predator's birth rate to balance its death rate.

These relationships between parameters and equilibria are a basic feature of Lotka-Volterra-type models in which the per-head growth rate of the predator population does not depend directly on predator density, and we will see that they tend to persist in more complex models. The resulting predictions are striking. For example, we do not expect that the system with a higher prey equilibrium will consequently have a higher predator equilibrium: predator density at equilibrium is determined by the *rate* at which new prey are produced, not by prey density. Prey density does determine the rate at which each predator takes in food, but prey per-head rate of increase, r, determines (with predator attack rate) how many predators are needed to keep the prey at equilibrium.

A related consequence is that enrichment of the prey's environment (an increase in r) leads to an increase in the predator's equilibrium density but leaves the prey equilibrium density unaffected. This result extends to analogous models with three or four trophic levels (Oksanen et al. 1981).

The properties of the Lotka-Volterra equilibrium are summarized in Table 3.1, which also records other properties of the model and of the variants discussed below.

TABLE 3.1. Properties of Lotka-Volterra-type equilibria

---

1. Prey equilibrates at density (d/ea) that just allows predator birth rate $(eaHP)$ to equal its death rate (dP).
   $\rightarrow$ So prey population equilibrium is set by predator death rate, attack rate, and conversion efficiency.

2. Predator population equilibrates at density (r/a) that induces just enough prey deaths (aHP) to equal prey population's growth rate (rH).
   $\rightarrow$ So predator equilibrium is set by prey productivity (r), (and by predator attack rate), and not by prey density.

3. Although equilibrium prey density does not determine equilibrium predator density, it does determine rate of intake by each predator and hence predator fecundity.
   $\rightarrow$ So more prey does not imply more predators, but more prey implies more predator births.

4. If we "enrich" the system by improving the prey environment (higher r), predator's equilibrium density, not prey's, increases. So if we enrich lakes that contain only phytoplankton and zooplankton, we expect only the latter to increase in abundance.

The Model's Dynamics

Several points emerge from the simulations plotted in Fig. 3.1.

1. The populations cycle again and again through the same set of densities: the size ("amplitude") and period of the cycles get neither smaller nor larger with time (Fig. 3.1a).
2. The amplitude of the cycles is determined by the initial densities: it is larger when the initial densities are farther from equilibrium (compare Figs. 3.1a and b).
3. The predator population lags behind the prey population by approximately one-quarter of the period of the cycle.

Fig. 3.1c plots the cycles in phase space (P at time t is plotted against H at time t). The populations move anticlockwise around and around the closed curve determined by the parameters and initial conditions.

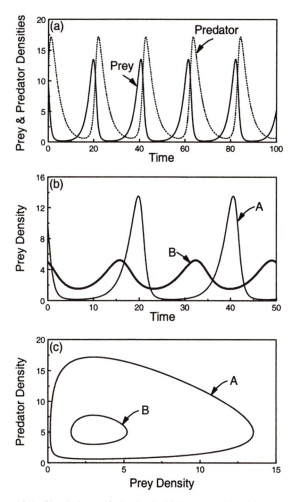

FIGURE 3.1. Simulations of the Lotka-Volterra model with $r = 0.5$, $a = 0.1, e = 1, d = 0.3$. (a) Initial densities were H(prey) $= 10$, P(predator) $= 12$. (b) Prey population from (a) is shown (designated A in figure) together with prey population (B) from a simulation in which parameter values were unchanged but initial densities were lower: $H = 5$, $P = 6$. (c) Prey and predator populations from the two simulations in (b), plotted in phase space; i.e., $P(t)$ is plotted against $H(t)$.

LOCAL STABILITY ANALYSIS

Local stability analysis of most of the two-dimensional models in this chapter is a simple procedure that provides insights into the dynamical effect of various ecological processes. We use it first to show that the features observed in the particular simulations in the previous section are characteristic of the Lotka-Volterra model, regardless of the particular parameter values and starting conditions.

As with the logistic model (chapter 2), local stability analysis proceeds in three steps: (1) find the non-zero equilibrium densities; (2) linearize the equations about the equilibrium; i.e., make initial density perturbations, h and p, small enough that h^2, p^2, and hp are negligibly small and can be ignored; (3) solve the resulting linear equations and test for stability. The process is explained in the appendix to this chapter, and its application to the Lotka-Volterra model is illustrated in Box 3.1.

The appendix shows that stability of the equilibrium in two-equation predator-prey interactions lacking time lags is determined mathematically by the four elements of the *Jacobian matrix*. This is simply an array of four numbers, whose values are the partial derivatives that define how each species' rate of change responds to a small change in its own density or that of the other species. For prey-predator models of the type under consideration here, a matrix of these partial derivatives always has the following pattern of signs:

$$\begin{bmatrix} A_{11} & A_{12} \\ A_{21} & A_{22} \end{bmatrix} = \begin{bmatrix} ? & - \\ + & ? \end{bmatrix}, \qquad (3.2)$$

reflecting the fact that an increase in predator density invariably decreases the prey growth rate, and an increase in prey density enhances the predator growth rate. We simply need to replace the question marks, each indicating the effect of an increase in prey or predator density on the species' own growth rate, with the appropriate sign. In the Lotka-Volterra model the question marks are both equal to zero, and the equilibrium is neutrally stable. If either question mark changes from zero to negative, or if both do, the equilibrium will be stable. If either changes to more than zero when the other remains equal to zero, or if both are more than zero, it will be unstable. The answer needs to be

BOX 3.1.

LOCAL STABILITY ANALYSIS FOR THE LOTKA-VOLTERRA
PREDATOR-PREY MODEL

The model is

$$\frac{dH}{dt} = rH - aHP = F(H, P)$$
$$\frac{dP}{dt} = eaHP - dP = G(H, P).$$

(B3.1)

The mechanics of local stability analysis are described in the appendix.
There are three steps.

1. Find the non-zero equilibrium.

This is achieved by looking for non-zero values of H^* and P^* that
satisfy the equations $F(H^*, P^*) = 0$ and $G(H^*, P^*) = 0$. Thus:

$$0 = F(H^*, P^*) = rH^* - aP^*H^* = H^*(r - aP^*),$$

implying

$$P^* = \frac{r}{a}$$

(B3.2a)

$$0 = G(H^*, P^*) = eaP^*H^* - dP^* = P^*(eaH^* - d),$$

implying

$$H^* = \frac{d}{ea}.$$

(B3.2b)

2. Linearize about the equilibrium (H^*, P^*).

In the appendix, it is shown that small deviations, h and p, from
equilibrium obey the linear differential equations

$$\frac{dh}{dt} = A_{11}h + A_{12}p$$
$$\frac{dp}{dt} = A_{21}h + A_{22}p$$

(B3.3)

where the four quantities A_{11}, A_{12}, A_{21}, and A_{22} are evaluated from the
formulas

$$A_{11} = \left.\frac{\partial F}{\partial H}\right|_*, \quad A_{12} = \left.\frac{\partial F}{\partial P}\right|_*, \quad A_{21} = \left.\frac{\partial G}{\partial H}\right|_*, \quad A_{22} = \left.\frac{\partial G}{\partial P}\right|_*$$

(B3.4)

(Box 3.1. continued)

in which the asterisk indicates that the partial derivatives are evaluated with H and P set equal to their equilibrium values. For the Lotka-Volterra model,

$$A_{11} = \frac{\partial}{\partial H}\left[H(r - aP)\right]\Big|_* = (r - aP)\Big|_* = 0$$

$$A_{12} = \frac{\partial}{\partial P}\left[H(r - aP)\right]\Big|_* = -aH^*\Big|_* = -\frac{d}{e}$$

$$A_{21} = \frac{\partial}{\partial H}\left[P(eaH - d)\right]\Big|_* = eaP\Big|_* = er \qquad (B3.5)$$

$$A_{22} = \frac{\partial}{\partial P}\left[P(eaH - d)\right]\Big|_* = (eaH - d)\Big|_* = 0.$$

3. Test for stability.

The next step in stability analysis is to evaluate the two coefficients (B_1 and B_2) in the characteristic equation that must be strictly positive for local stability. From the appendix and equation B3.5,

$$B_1 = -(A_{11} + A_{22}) = 0$$
$$(B3.6)$$
$$B_2 = A_{11}A_{22} - A_{12}A_{21} = rd.$$

Since B_1 is not strictly positive, the equilibrium is not locally stable. However, having B_1 always exactly equal to zero implies that the system lies right at the knife edge between stability and instability; it is said to be *neutrally stable*.

One can now determine the cycle period. In the appendix (immediately after equation A3.17), we note that on the stability boundary B_1 is equal to zero and the period of the oscillations on the boundary is equal to $2\pi/\sqrt{B_2}$. We know from equation B3.6 that B_2 is equal to rd. Consequently, the period of small amplitude Lotka-Volterra oscillations is $2\pi/\sqrt{rd}$.

calculated explicitly if one is less than zero and one is more than zero, a situation we meet below.

In general, the text omits the mechanics of stability analysis. However, stability analysis is relatively easy for many of the models examined in this chapter. We will therefore do the above step in several cases, because it helps one understand how the ecological process under discussion has its effect, and therefore gives insight into how such processes affect more complicated models.

Neutral Stability of the Lotka-Volterra Model

The Lotka-Volterra equilibrium is neutrally stable, as illustrated by the simulations in Fig. 3.1. Neutral stability in the Lotka-Volterra model is an ecologically strange situation. It says that if we perturb the system from equilibrium at some time, the populations will then regularly cycle thereafter, the amplitude of the cycle depending on the size of this one perturbation. Of course, real populations are being "perturbed" all the time, so real cycles cannot be represented by these initial-condition-dependent oscillations. Equally important, the model is "structurally unstable"; that is, if we make almost any change to its structure, its dynamics will change. This, commonly argued to be a fatal flaw, is a useful feature: the model acts as a kind of balance or reference point. Any changed assumption will make the equilibrium either stable or unstable, so we can use the model to get insight into whether a particular ecological process is likely to stabilize or destabilize real systems.

Predator-Prey Cycles: Causes, Period in Relation to Predator and Prey Biology, and Non-Linear Effects

Although the Lotka-Volterra is the simplest possible consumer-resource model, its tendency to cycle is characteristic of consumer-resource models in general. The underlying cause of consumer-resource cycles is present in this simplest case. Cycles occur because each population takes time to respond to changes in the density of the other, even though the per-head responses are instantaneous.

There are two linked aspects to the creation of cycles. (1) If somehow the prey increases in density above its equilibrium, there will be a delay before the predator population increases enough to reduce prey density. This phase is *prey escape* (de Roos et al. 1990). (2) Prey are driven below their equilibrium density because there is a delay before predator density declines to the level that allows prey increase. This is *overexploitation*.

From Box 3.1 we can see that the period of small-amplitude Lotka-Volterra cycles is approximately $2\pi/\sqrt{rd}$. This means that a complete cycle (e.g., from one prey peak to the next) will be swept out every $2\pi/\sqrt{rd}$ time units. The formula for the period makes intuitive sense. Population densities will cycle faster (shorter period) as the per-head rate of increase of the prey population (r) increases and as the per-head death rate of the predator (d) increases, both of which induce faster turnover of individuals in the population. These parameters are typically large in small organisms, so protozoa and bacteria have short-period cycles, and small in large organisms, so lynx and hares have long-period cycles.

In fact, when we simulate the Lotka-Volterra model, we see that the observed period (not just the amplitude) depends on the size of the initial starting densities (Fig. 3.1b), and it does not have exactly the value $2\pi/\sqrt{rd}$. This is because that precise value obtains only near the equilibrium. Away from equilibrium, the model's non-linearities, ignored in local stability analysis, affect the period. However, the value calculated from local stability analysis is a good approximation, except when the cycle amplitude is very large.

Extension: Prey Growth is Logistic

In the next section we add, one at a time, processes to the Lotka-Volterra model that are destabilizing and then those that are stabilizing. To illustrate the procedure, however, we first look at a stabilizing mechanism whose effect is easy to determine.

The Lotka-Volterra model assumes the prey population can increase exponentially. This is reasonable for a population at low density, but the per-head rate of increase will decline when density becomes high enough. In real populations, higher prey density will often depress the prey's resource population, decrease the prey's feeding rate, and hence decrease the prey's per-head birth rate or increase its per-head death rate. Adding density dependence directly to the prey is a cheap way to model this process without modeling the dynamics of the prey's resources. The logistic model is a simple form for such density dependence (where K is carrying capacity), and adding it to the Lotka-Volterra model gives

$$\frac{dH}{dt} = rH\left(1 - \frac{H}{K}\right) - aHP$$

$$\frac{dP}{dt} = eaHP - dP.$$

(3.3)

Numerical simulations of equation 3.3 suggest that the logistic term has a stabilizing effect (Fig. 3.2a). The populations are "attracted" to the equilibrium, and the cycles are "damped." To prove this mathematically, we first calculate the new equilibria, which are

$$H^* = \frac{d}{ea}$$

$$P^* = \frac{r}{a}\left[1 - \frac{H^*}{K}\right].$$

(3.4)

These formulas emphasize the consistent behavior of Lotka-Volterra-like equilibria, summarized in Table 3.1. The equilibrium of the prey population is still determined only by the predator parameters and not by its own carrying capacity. The prey equilibrium in the presence of the predator, however, must be less than K because the predator cannot persist if the prey has zero or negative population growth. (If the value of H^* calculated from equation 3.4 turns out to be greater than K, the predator goes extinct in all simulations.) As in the Lotka-Volterra model, the predator equilibrium increases as we enrich the prey resources, i.e., as either r or K increases.

Local stability analysis is easy. All the partials have the same value as in the Lotka-Volterra model except for A_{11}. Since there is self-damping in the prey, the key partial is A_{11}, which can be evaluated using the recipe in the appendix and shown to have the value $-rd/(eaK)$.

Thus our matrix has the sign structure

$$\begin{bmatrix} - & - \\ + & 0 \end{bmatrix}$$

and the equilibrium is locally stable.

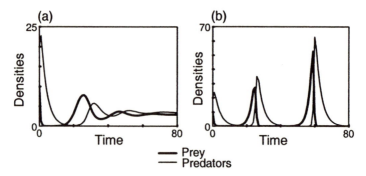

FIGURE 3.2. (a) Simulation of equation 3.3, the Lotka-Volterra model plus logistic prey, $K = 10$. (b) Simulation of Lotka-Volterra model with type 2 functional response added (equation 3.6); handling time $T_h = 0.2$. As in Fig. 3.1, $r = 0.5, a = 0.1, e = 1, d = 0.3$.

EFFECTS OF STABILIZING AND DESTABILIZING PROCESSES: A SURVEY

Although the Lotka-Volterra model captures the tendency of consumer-resource models to cycle, it is too simple to answer one of our central questions: why do some consumer-resource interactions lead to systems with a stable equilibrium whereas other consumer-resource systems exhibit unstable dynamics (chapter 2)? To begin to get at this issue, we need to modify the model in various ways to look at ecological processes that may stabilize or destabilize the dynamics.

First we look at two processes that destabilize the equilibrium: a type 2 functional response (defined later) and developmental time lags in the predator. We do these first because satiation and delays are ineluctable properties of individuals and so confront us with the fact that a consumer-resource model with even minimal realism tends to be unstable.

We then examine four stabilizing processes other than logistic prey, which was analyzed above. These are a type 3 functional response, a prey refuge, density dependence in the predator per-head death rate, and density dependence in the searching rate. (We examine additional stabilizing mechanisms in a later section.)

The resulting models help us understand how the processes have their effect. *Several also demonstrate that stabilizing processes tend to increase the prey equilibrium density.* This turns out to be a quite general rule. In Table 3.2 we collect most of the models, their equilibria, and the key partials determining stability or instability.

Ecological Processes Inducing Instability

All real predators share two features missing in the Lotka-Volterra model. They cannot eat an unlimited amount of prey per unit time, and their responses to changes in prey density involve time delays of one sort or another. Here we show that both features destabilize the Lotka-Volterra equilibrium. Ecologists therefore need to explain how it is that many real predator-prey interactions are stable, whereas models with an irreducible minimum of reality predict instability.

PREDATOR SATIATION: TYPE 2 FUNCTIONAL RESPONSE
The *functional response* describes how feeding rate per consumer (i.e., number or biomass of food items consumed per consumer per unit time) varies with the density of food in its environment. In equation 3.1, the functional response increases linearly with prey density—a type 1 response (Fig. 3.3a). This is clearly infeasible when prey density is high. The predator will run out of time in which to search and find new prey (because it is busy dealing with prey already encountered), its gut will fill, or if it is a parasitoid, it will run out of eggs.

Suppose it takes a time, T_h, to "handle" each prey attacked and killed (Holling 1959), e.g., to subdue, probe, paralyze, oviposit, and move a new egg into position for the next encounter if the consumer is a parasitoid. Although the number encountered per unit of search time is still aH, the fraction of time spent searching will decrease as H increases and more prey are attacked. This yields a type 2 response, with the maximum predation rate being $1/T_h$ prey per day (Fig. 3.3a). In predators and herbivores, true satiation may be the cause of the type 2 response. We can still think of this as being a result of a handling time per prey, but in this instance handling time includes the latent period while the gut is being cleared. A common form for the type 2 functional response, $g(H)$, which is substituted for aH in equation 3.1, is

$$g(H) = \frac{aH}{1 + aT_h H}. \tag{3.5}$$

TABLE 3.2. Simple variants of the Lotka-Volterra model

Model	Equilibria	Local Stability
Lotka-Volterra		
$\dfrac{dH}{dt} = rH - aHP$	$H^* = \dfrac{d}{ea}$	Neutrally stable because both $A_{11} = 0$ and $A_{22} = 0$
$\dfrac{dP}{dt} = eaHP - dP$	$P^* = \dfrac{r}{a}$	
Logistic prey		
$\dfrac{dH}{dt} = rH\left(1 - \dfrac{H}{K}\right) - aHP$	$H^* = \dfrac{d}{ea}$	Stable because
$\dfrac{dP}{dt} = eaHP - dP$	$P^* = \dfrac{r}{a}\left(1 - \dfrac{H^*}{K}\right)$	$A_{11} = \dfrac{-rd}{eaK} < 0$
Type 2 functional response		
$\dfrac{dH}{dt} = rH - \dfrac{aHP}{1 + aT_hH}$	$H^* = \dfrac{d}{ea - daT_h}$	Unstable because
$\dfrac{dP}{dt} = e\dfrac{aHP}{1 + aT_hH} - dP$	$P^* = \dfrac{r(1 + aT_hH^*)}{a}$	$A_{11} = \dfrac{rdT_h}{e} > 0$
Density-dependent predator deaths		
$\dfrac{dH}{dt} = rH - aHP$	$H^* = \dfrac{d}{ea}\left(1 + \dfrac{zr}{ad}\right)$	Stable because
$\dfrac{dP}{dt} = eaHP - dP - zP^2$	$P^* = \dfrac{r}{a}$	$A_{22} = \dfrac{-rz}{a} < 0$
Predator-dependent functional response		
$\dfrac{dH}{dt} = rH - \dfrac{aHP}{1 + zP}$	$H^* = \dfrac{d}{ea}(1 + zP^*)$	Stable because
$\dfrac{dP}{dt} = e\dfrac{aHP}{1 + zP} - dP$	$P^* = \dfrac{r}{a - rz}$	$A_{22} = \dfrac{-dzP^*}{1 + zP^*} < 0$

TABLE 3.2. *continued*

Type 3 functional response

$$\frac{dH}{dt} = rH - \frac{aH^2P}{H^2 + z}$$

$$H^* = \sqrt{\frac{dz}{ea - d}}$$

Stable if $A_{11} =$

$$r\left(1 - \frac{2z}{(H^*)^2 + z}\right) < 0$$

$$\frac{dP}{dt} = e\frac{aH^2P}{H^2 + z} - dP$$

$$P^* = \frac{r((H^*)^2 + z)}{aH^*}$$

Prey has refuge

$$\frac{dV}{dt} = rV - aVP + rR$$

$$V^* = \frac{d}{ea}$$

Stable because

$$A_{11} = r - aP^* < 0$$

$$\frac{dP}{dt} = eaVP - dP$$

$$P^* = \frac{r}{a}\left(1 + \frac{R}{V^*}\right)$$

Note: $H^* = V^* + R$

External source of predator recruits

$$\frac{dH}{dt} = rH - aHP$$

$$H^* = \frac{d}{ea} - \frac{R}{eaP^*}$$

Stable if

$A_{22} = eaH^* - d < 0$

$H^* > 0$

$$\frac{dP}{dt} = eaHP - dP + R$$

$$P^* = \frac{r}{a}$$

Rosenzweig-MacArthur model

$$\frac{dH}{dt} = rH\left(1 - \frac{H}{K}\right)$$
$$- \frac{aHP}{1 + aT_hH}$$

$$H^* = \frac{d}{a(e - dT_h)}$$

Stable if

$$K < \frac{1}{aT_h}\left[\frac{e + dT_h}{e - dT_h}\right]$$

$$\frac{dP}{dt} = e\frac{aHP}{1 + aT_hH} - dP$$

$$P^* = \frac{r}{a}\left(1 - \frac{H^*}{K}\right)$$
$$\times (1 + aT_hH^*)$$

The column "Local Stability" contains the key item of information that determines stability or instability.

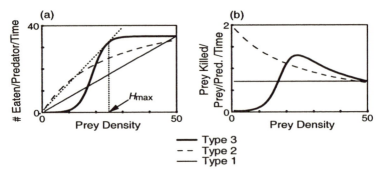

FIGURE 3.3. Three types of functional responses. (a) Number of prey eaten per predator per unit time as a function of prey density. Solid straight line is type 1, saturating dashed curve type 2, sigmoid curve type 3. Fine dotted line tangent to the type 3 curve defines H_{max} and hence the largest stable equilibrium prey density with that type 3 response (see Box 3.2). (b) Risk the average prey runs of being attacked by a given predator per unit time as a function of prey density. Risk is functional response divided by prey density.

The dynamical effect of the functional response can be understood by considering the risk run by the average prey individual from the average predator individual (Fig. 3.3b). In type 1, the risk is constant with prey density, i.e., the per-head death rate induced per predator is density independent. In type 2, risk declines with prey density (and is sometimes described as "inversely density dependent") and thereby reinforces the two processes causing cycles. (1) It facilitates prey escape, a phase of the cycle where the prey population grows faster than the lagging predator population, because now each predator is less and less effective as prey density increases. (2) Similarly, it exacerbates overexploitation because each predator becomes more and more effective as prey density decreases.

The above intuition is confirmed by adding a type 2 response to the Lotka-Volterra model, yielding

$$\frac{dH}{dt} = rH - Pg(H)$$
$$\frac{dP}{dt} = ePg(H) - dP$$

(3.6)

with

$$g(H) = \frac{aH}{1 + aT_hH}.$$

The equilibria are

$$H^* = \frac{d}{ea - daT_h} \quad \text{and} \quad P^* = \frac{r(1 + aT_h H^*)}{a}.$$

Now, following a disturbance, the oscillations get larger with time (Fig. 3.2b). Although extinction is not possible in the model, such oscillations imply certain extinction in real populations.

Following the steps in the appendix establishes that all the partials have the same sign as in the Lotka-Volterra model except for A_{11}. That is, the type 2 functional response alters the way prey density affects the prey's rate of change. We find that

$$A_{11} = \frac{rdT_H}{e} > 0.$$

Thus our Jacobian matrix has the sign pattern

$$\begin{bmatrix} + & - \\ + & 0 \end{bmatrix},$$

confirming that the equilibrium is unstable.

Oaten and Murdoch (1975) described a graphical method for analyzing the functional response and showed that all type 2 responses are destabilizing regardless of the particular function.

TIME LAGS IN DENSITY-DEPENDENT PROCESSES

Time lags are universal in biological systems. Some, including many physiological or behavioral time lags, are so short we can ignore their dynamical effect. Lags induced by developmental delays, however, may have profound dynamical effects. For example, the number of adults now feeding and giving birth is a result of reproduction by adults that were present at least one development time previously. This is one reason we move to age-structured models in chapter 5.

A simple way to incorporate a developmental time lag in the predator is to let P represent the density of adult predators and insert a fixed delay between the act of consuming and the emergence of a new adult predator:

$$\frac{dH}{dt} = rH - aHP$$

$$\frac{dP}{dt} = eaH(t - T)P(t - T) - dP$$

(3.7)

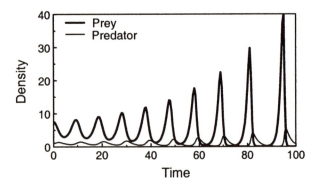

FIGURE 3.4. Simulation of equation 3.7. $r = 1$, $a = 1$, $e = 0.1$, $d = 0.5$, $T = 0.1$.

where T is the delay between killing prey and producing a new adult predator (Bartlett 1957) and $H(t - T)$ is the value of H at time $t - T$. The equilibria have the same values as in the Lotka-Volterra model. Fig. 3.4 shows how the resulting unstable equilibrium is associated with fluctuations of ever-increasing amplitude, implying certain extinction. We omit formal analysis of this model since chapters 5 and 6 deal with delays in development more realistically. However, stability analysis is possible using the methods detailed in the appendix to chapter 5.

Ecological Processes Inducing Stability: Direct Density Dependence

Here we look at processes that involve the classical source of stability: explicit density dependence in the prey or predator vital rates. The first three cases below share the Lotka-Volterra property that predator equilibrium depends on prey parameters only (Table 3.2). This changes in the last two cases.

CASE 1. DENSITY-DEPENDENT PREY GROWTH: LOGISTIC PREY
We analyzed this mechanism above (equation 3.3).

CASE 2. TYPE 3 FUNCTIONAL RESPONSE
An example of a type 3 functional response is shown by the thick line in Fig. 3.3a. The potential stabilizing effect of a type 3 response is

intuitively obvious from Fig. 3.3b. Risk to the average prey at first rises with increasing prey density (risk is density dependent) and then declines, becoming inversely density dependent. The equilibrium is stable if the prey equilibrium is low enough (roughly in the region where per-head prey risk increases with prey density in Fig. 3.3b) and unstable if it is high enough (roughly in the region where per-head risk declines with prey density in Fig. 3.3b). The precise value for the largest stable prey equilibrium is indicated in Fig. 3.3b by H_{max}, whose calculation is explained in Box 3.2.

Formal stability analysis shows A_{11} first increasing and then decreasing approximately as risk does in Fig. 3.3b. We use the following response as an example of a type 3 functional response:

$$g(H) = \frac{aH^2}{H^2 + z},$$

where z is a constant. The equilibria are:

$$H^* = \sqrt{\frac{dz}{ea - d}}$$

$$P^* = \frac{r((H^*)^2 + z)}{aH^*}. \tag{3.8}$$

$$A_{11} = r\left[1 - \frac{2z}{(H^*)^2 + z}\right],$$

so stability requires A_{11} to be less than zero, which is true only if $2z$ is greater than $(H^*)^2 + z$, which is true only if $(H^*)^2$ is less than z. This serves to emphasize the point made by Fig. 3.3a (and in Box 3.2) that the equilibrium becomes unstable once H^* exceeds a critical value (which is H_{max} in Fig. 3.3a).

Substituting for H^* in equation 3.8 establishes that stability requires $d < ea/2$. Thus, the equilibrium is stable for $0 < d < ea/2$ (obviously prey existence [$H^* > 0$] requires $d < ea$). Since H^* increases with predator death rate, d, the inequality again emphasizes that for H^* above some threshold value, the equilibrium is unstable.

This model is thus the first case in which the equilibrium can be stable or unstable, depending on parameter values.

Oaten and Murdoch (1975) showed that a stabilizing tendency (for appropriately low values of H^*) is a general property of a type 3

BOX 3.2.

DETERMINING THE EFFECT OF A FUNCTIONAL RESPONSE ON STABILITY

For a Lotka-Volterra model with a non-linear functional response, $g(H)$, where F defines the rate of change in prey density, the effect of a change in prey density on its own rate of change is

$$A_{11} = \left.\frac{\partial F}{\partial H}\right|_* = r - g'(H^*)P^*,$$

where $g'(H^*) = \left.\dfrac{\partial g}{\partial H}\right|_*$.

Setting dH/dt equal to zero gives (equation 3.6)

$$P^* = \frac{rH^*}{g(H^*)}, \qquad (B3.7)$$

so

$$A_{11} = r\left(1 - \frac{g'(H^*)H^*}{g(H^*)}\right), \qquad (B3.8)$$

which is less than zero if

$$\frac{g'(H^*)H^*}{g(H^*)} \text{ is greater than 1,} \qquad (B3.9)$$

which is true if

$$g'(H^*) \text{ is greater than } \frac{g(H^*)}{H^*}. \qquad (B3.10)$$

Equation B3.9 yields a nice graphical interpretation. The equilibrium is stable for all prey equilibria below H^* for which $g'(H^*)$ is equal to $g(H^*)/H^*$, which is the value at which the tangent to the functional response has the same slope as a line drawn from the origin to the curve at the value $g(H^*)$. Up to that point, the tangent has a greater slope. The value of H at which this equality holds is indicated as H_{max} in Fig. 3.3a.

---

response, regardless of the particular function. Box 3.2 lays out the Oaten and Murdoch graphical criterion since it reappears in chapter 10.

CASE 3. A PHYSICAL REFUGE FOR THE PREY

Suppose that in addition to the standard (vulnerable) prey population, now designated V, there is also a fixed number of prey, R, in a physical refuge. If all prey reproduce at the same rate, the equation is

$$\frac{dV}{dt} = rV - aVP + rR$$

$$\frac{dP}{dt} = eaVP - dP.$$

(3.9)

The equilibria behave in the standard Lotka-Volterra way (Table 3.2). The vulnerable prey equilibrium is the same as for equation 3.1. The predator equilibrium is higher because prey productivity $(rV + rR)$ is higher. The equilibrium is always stable (Table 3.2). The total prey equilibrium is of course $V^* + R$, so stability is associated with an increase above the Lotka-Volterra prey equilibrium.

We can rewrite the equation to show that the refuge acts as a special case of a stabilizing functional response. Let H be the total prey population, i.e., $H = V + R$. Then we have

$$\frac{dH}{dt} = rH - a(H - R)P.$$

(3.10)

Now the attack rate on prey is zero up to R, and thereafter increases linearly (of course in reality there would need to be an upper limit to the attack rate). Thus risk to the average prey is density dependent. Per-head prey mortality from predation is thus explicitly density dependent.

Notice that a refuge that holds a fixed fraction of the population has no effect on stability (though it does affect predator equilibrium density). Thus, if a fraction, s, of the population is safe and $1 - s$ is available to the predator, the prey equation becomes

$$\frac{dH}{dt} = rH - a(1 - s)HP.$$

(3.11)

We can simply define a new $a' = a(1 - s)$ and regain the basic model.

CASE 4. DENSITY-DEPENDENT PREDATOR DEATH RATE
The predator's vital rates may be a function of predator density. Searching predators may fight and kill each other, or those lacking territories may suffer higher death rates. If the per-head death rate increases

linearly with predator density, at rate z, then the predator equation becomes

$$\frac{dP}{dt} = eaHP - (d + zP)P. \tag{3.12}$$

We now have a negative squared term in the predator population, just as the logistic model introduces such a term to the prey population, and the equilibrium is stable (Table 3.2).

Density dependence in the predator death rate has two major effects on the prey equilibrium. First, *it causes the prey equilibrium to depend on prey parameters* in addition to predator parameters. While the predator equilibrium (r/a) equals that in the Lotka-Volterra model, the prey equilibrium, $H^* = d/ea + rz/ea^2$, now increases if prey rate of increase, r, increases (Table 3.2). So the basic Lotka-Volterra pattern—prey equilibrium depends only on predator parameters—no longer holds. In particular, *prey equilibrium density increases as prey productivity (r) increases.*

Second, the prey equilibrium is again higher than the Lotka-Volterra equilibrium (d/ea). As in the previous model, a gain in stability is associated with an increase in prey equilibrium density.

CASE 5. PREDATOR-DEPENDENT ATTACK RATE

Interference between searching predators could reduce the time available for search, or the per-head attack rate per unit of search time (Hassell 1978). Beddington (1975) suggested a predator-dependent attack rate that causes a type 2 functional response to decrease with predator density:

$$g(H) = \frac{aH}{1 + aT_hH + zP}. \tag{3.13}$$

Ruxton et al. (1992) showed that this model could be derived, albeit with quite strong assumptions, as the limit when predators interfere with each other as they search and handling time is very short. If we retain a Lotka-Volterra linear (type 1) functional response, we get

$$\frac{dH}{dt} = rH - \frac{aHP}{1 + zP}$$
$$\frac{dP}{dt} = e\frac{aHP}{1 + zP} - dP. \tag{3.14}$$

The predator population attack rate decelerates with predator density, P. As P approaches zero, the rate approaches aHP, the Lotka-Volterra rate. The equilibrium in this model, if it exists, is always stable. As in the previous example, this type of density dependence in the predator causes the prey equilibrium, as well as the predator equilibrium, to increase with the prey's rate of increase. In addition, stability implies an increase in prey density, which is always greater than in the Lotka-Volterra model (Table 3.2).

The models to this point contain only a single "added" process that is either stabilizing (e.g., logistic prey) or destabilizing (e.g., a type 2 functional response). In real populations more than one of these processes is likely to operate concurrently. We obviously need to investigate the interaction between such opposing forces. We do this next.

COMBINING STABILIZING AND DESTABILIZING PROCESSES: FROM NEUTRAL STABILITY TO LIMIT CYCLES

Chapter 2 illustrates two types of dynamics seen in real systems that our models need to be able to account for. The first is populations that persistently cycle with approximately constant period and amplitude, in spite of being continually perturbed by a variable environment (Fig. 2.2a). The second is species whose dynamics change from stable to cyclic more or less smoothly as the environment does (Fig. 2.2b–d). None of the models presented so far can explain these two phenomena.

Both phenomena, however, can arise in a model with (1) a destabilizing process that can create an unstable equilibrium and generate oscillations of increasing amplitude and (2) a stabilizing process that can ultimately constrain the amplitude. This combination, first, can lead to a stable limit cycle with appropriate parameter values. Second, typically it can generate a stable equilibrium with one set of parameter values and a stable limit cycle with a different set.

To illustrate these results, we combine logistic prey growth (equation 3.3) and a type 2 functional response (equation 3.5) to give a model commonly attributed to Rosenzweig and MacArthur (1963),

$$\frac{dH}{dt} = rH\left(1 - \frac{H}{K}\right) - \frac{aHP}{1 + aT_hH}$$

$$\frac{dP}{dt} = e\frac{aHP}{1 + aT_hH} - dP.$$

(3.15)

Figs. 3.5a and b illustrate that one set of parameter values (with $K = 100$) gives a stable equilibrium, so an initial perturbation dies away through time, whereas a different set (with $K = 150$) gives perturbations that initially grow with time but are eventually captured in a limit cycle. These cycles eventually have a fixed period and amplitude and are repeated endlessly if the populations are not perturbed again.

Fig. 3.5c demonstrates that the cycle in Fig. 3.5b is a stable limit cycle. On the y-axis we plot predator density at any time, t, against prey density at precisely the same time on the x-axis. The actual limit cycle is the heavy closed loop in this "phase plane" plot. Once on this closed loop, the populations will perpetually cycle around in a counter-clockwise direction, tracing out the densities in Fig. 3.5b. Fig. 3.5c shows that if we perturb the populations, either above or below the limit cycle, they return to the cycle, just as perturbations from a stable equilibrium return toward the equilibrium. The limit cycle is thus stable (Fig. 3.5c).

Stable limit cycles are crucially different from the cycles seen in the Lotka-Volterra model with its neutrally stable equilibrium. In the latter, each perturbation moves the system to a new neutrally stable cycle whose amplitude is determined by the size of the perturbation (Fig. 3.1b). By contrast, the amplitude of a limit cycle is set by the model's parameters and is unaffected by the size of any perturbation.

Stable limit cycles are a general feature of two-dimensional models of the sort described here (the two dimensions being defined by H and P). A theorem by A. N. Kolmogorov, and its extensions (May 1972; Brauer 1979), proves that a large class of two-dimensional predator-prey systems lacking time lags, and having smooth functions, have either a single stable equilibrium, or if the equilibrium is unstable, they also have a stable limit cycle (Fig. 3.5c). We refer throughout the book to cycles of this sort as consumer-resource, predator-prey, or parasitoid-host cycles. Circumstances where the assumptions of the Kolomogorov,

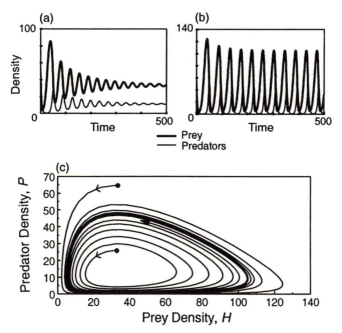

FIGURE 3.5. Simulations of equation 3.15 with $r = 0.2, a = 0.02, e = 1, T_h = 1, d = 0.4$. (a) $K = 100$, implying a relatively nutrient-poor environment for prey. (b) $K = 150$, implying a richer prey environment. (c) Phase plane plot of prey and predator densities arising from simulations with parameters as in (b). One simulation was started outside the limit cycle; the trajectory moves into the limit cycle. The other simulation was started inside the limit cycle; the trajectory spirals out to the limit cycle.

May, or Brauer results do not hold can arise, however, where there is both a locally stable equilibrium and a stable limit cycle (i.e., there are multiple attractors) and the size of the perturbation determines which attracts the population trajectory. Unfortunately, the general results do not apply when we have more than two dimensions, as occurs if there are, for example, age classes or additional populations, though it turns out in practice that such systems frequently show bounded fluctuations.

Cyclic trajectories can also arise in the following way. If the deterministic dynamics are actually like those in Fig. 3.5a (damped cycles around a stable equilibrium), random environmental variation, or some other type of external forcing, can create persistent cycles called *quasi-cycles*. Since real environments are stochastic, it is often difficult in practice to tell whether an observed time series actually arises from a stable limit cycle or a quasi-cycle. If we are concerned about mechanism, the difference may not be important.

Suppression-Stability Trade-off: A Simple Model
Applied in the Field

Equation 3.15 is the classical representation of the paradox of enrichment. The so-called paradox is that in many predator-prey models, enrichment of the prey's environment (an increase in K) tends to destabilize a stable equilibrium. This relationship is actually not paradoxical, as we shall see.

The emphasis on enrichment as the source of instability distracts from what is perhaps a more relevant consumer-resource issue: the further the predator suppresses the prey equilibrium below the limit set by the prey's resources, the more likely is the system's equilibrium to be unstable. That is, there is a trade-off between degree of suppression of the prey and stability. We therefore refer throughout the book to this suppression-stability trade-off, rather than the paradox of enrichment.

The trade-off has fascinated ecologists since it was first posed by Rosenzweig (1971). Since these aspects were discussed by Rosenzweig and MacArthur (1963), it has become common practice, which we follow, to refer to equation 3.15 as the Rosenzweig-MacArthur model. Its equilibria are

$$H^* = \frac{d}{a(e - dT_h)}$$

$$P^* = \frac{r}{a} \left(1 - \frac{H^*}{K} \right) (1 + aT_h H^*). \tag{3.16}$$

Notice that this is still solidly a part of Lotka-Volterra-type theory: the prey equilibrium does not depend on its own parameters, and only the predator equilibrium increases with a measure of the "richness" of the environment for the prey—the carrying capacity, K, and/or r.

Local stability analysis (Murdoch et al. 1998b) shows that the equilibrium is stable if

$$K < \frac{1}{aT_h} \left[\frac{e + dT_h}{e - dT_h} \right], \qquad (3.17)$$

which clearly establishes the trade-off. Thus, when K is small, the stability criterion is met (Fig. 3.5a). Now suppose all the parameters in inequality 3.17 have values such that K is just barely smaller than the right-hand side, so the equilibrium is stable. A small increase in the value of K will now reverse the inequality, giving an unstable equilibrium (Fig. 3.5b). In addition, the amplitude of prey and predator fluctuations also increases with environmental enrichment (i.e., as K increases). These results exemplify the trade-off.

In this model, unlike those discussed to this point, reductions in predator "efficiency" are stabilizing. Thus, inequality 3.17 shows that an unstable equilibrium can be stabilized by an increase in the predator's death rate, d, or by a decrease in the individual predator's efficiency. Efficiency can be reduced by a decrease in predator attack rate, a, by an increase in its handling time, T_h, or by a decrease in the conversion efficiency, e. Therefore, as in earlier examples, gains in stability from a decrease in predator efficiency cause H^* to increase (equation 3.16).

Like the earlier model containing a type 3 functional response, stability of the equilibrium depends on the parameter values. Such models thus provide a potential explanation for shifts in stability properties along an environmental gradient marked, for example, by gradual changes in temperature (e.g., Figs. 2.2b–d).

Equation 3.15 has been used as the starting point for a variety of purely theoretical studies (for example, several spatial models discussed in chapter 10). We now show it can be used to test hypotheses in the field, demonstrating that simple models can sometimes serve an empirical end. The field problem has two parts.

First, for decades ecologists have argued about whether resource populations remain approximately constant, or increase, as their environment is enriched (Oksanen et al. 1981; Sarnelle 1992). In a two-trophic-level system, Lotka-Volterra-type models predict resource abundance will remain constant whereas consumer abundance will increase under enrichment. In lakes, where the basic resource is algae, there is some, but a quite small, increase in the density of edible algae with enrichment, i.e., with an increase in phosphorus loading (Watson et al. 1992). In particular, in lakes where the main algal consumer is the zooplankter *Daphnia*, edible algal abundance changes little over a wide range of phosphorus levels and algal productivity.

Second, in *Daphnia*-algal populations in temperate lakes and experimental microcosms, cycles do not occur more frequently, nor does their amplitude increase, in richer environments (McCauley and Murdoch 1990; Murdoch et al. 1998b). Clearly, this observation does not match theory, which predicts increasing instability with enrichment.

Equation 3.15 can be used to study the problem because the following are quite good approximations to reality (Nisbet et al. 1991). The biomass of algae grows logistically in the absence of *Daphnia*. A *Daphnia*'s rate of feeding and its metabolic costs increase in proportion to its biomass. Finally, it requires the same amount of assimilate to produce a given mass of egg or of new body tissue. Thus we can think of H and P as the biomass of algae and *Daphnia*, respectively. The per-head vital rates are then in terms of per milligram of biomass. All retain their original meaning except that the per-head "death" rate of *Daphnia* now combines losses from both death and respiration. The model does very well at predicting the biomass dynamics of *Daphnia* and algae in laboratory populations (Nisbet et al. 1997b).

The field problem is nicely summarized in Fig. 3.6, which portrays stability as a function of algal carrying capacity, K, and *Daphnia*'s death rate, d. Below the lower curve, algal productivity is too low to allow *Daphnia* to persist. The higher curve is the set of values of K and d at which the two sides of inequality 3.17 are exactly equal. Above this curve, large-amplitude consumer-resource cycles are predicted, yet many stable *Daphnia*-algal interactions in nature exist in this theoretically unstable parameter space (Murdoch and McCauley 1985; McCauley and Murdoch 1987). Fig. 11.2 shows qualitatively similar dynamical behavior in a stage-structured version of equation 3.15.

Four hypotheses have been proposed, mainly to account for an increase in algal biomass with enrichment, but these mechanisms are also in theory stabilizing. They all assume that *Daphnia*'s efficiency decreases as nutrient level, or *Daphnia* density, increases. The

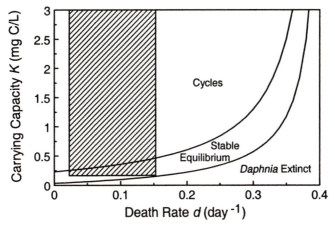

FIGURE 3.6. Stability properties of equilibria of equation 3.15 (Rosenzweig-MacArthur model) as a function of lake nutrient status, K, and *Daphnia* mortality rate d. Real, stable populations exist in the shaded area, but in most of this parameter region equilibria are predicted to be unstable. Modified from Murdoch et al. 1998b with permission of the Ecological Society of America.

four hypotheses are that *Daphnia*'s (1) filtering rate, a, decreases or (2) its death rate, d, increases with enrichment, K, or (3) its death rate or its (4) functional response decreases as *Daphnia* density increases. Murdoch et al. (1998b) incorporated each of the four hypotheses into equation 3.15 and then calculated the lowest H^* that was consistent with stability for all levels of K up to the most nutrient-rich environment for which data were available. All of the H^* values predicted by the four hypotheses substantially exceed those observed at high nutrient levels, so all of the hypotheses can be rejected (Fig. 3.7).

An alternative hypothesis is that inedible algal species, which increase in abundance with lake nutrient status, take up nutrients and therefore lower the effective K of edible algae. Microcosm experiments showed that indeed predator-prey limit cycles occur in *Daphnia* and edible algal populations in the absence, but not in the presence, of inedible algae (McCauley et al. 1999). The cycles, however, were of quite low amplitude, so the story is not yet complete.

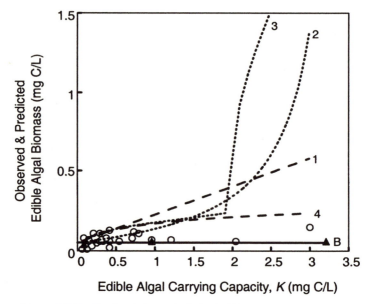

FIGURE 3.7. Predicted and observed edible algal biomass values over a wide range of nutrient levels (K). The various curves are predicted equilibrial values; points are observed values. Circles are data from lakes and a reservoir, triangles from experimental stock tanks, and were compared with predicted curves to test four hypotheses. Line B is the MacArthur-Rosenzweig model and predicts well the equilibrium values across the entire range of nutrient levels but fails to predict the observed stability. The four hypotheses to explain stability (see text) all predict algal equilibrium values far higher than those observed at high nutrient levels. Numbered lines and curves correspond with hypotheses that assume (1) *Daphnia* filtering rate decreases with K, (2) *Daphnia* mortality increases with K, (3) *Daphnia* mortality increases with *Daphnia* density, and (4) functional response declines with *Daphnia* density. Modified from Murdoch et al. 1998b with permission of the Ecological Society of America.

SIMPLE MODELS OF STAGE AND SPATIAL STRUCTURE: THE CREATION OF INDIRECT DENSITY DEPENDENCE

Up to this point, we have explored ecological processes that affect stability in a direct, intuitively obvious, and straightforward way. In each case, stability arose because some per-head rate was density dependent; we called this direct density dependence.

In this section we establish two key results. First, we show that stability can be influenced by a more subtle process involving indirect density dependence: the equilibrium can be stabilized even though all explicit vital rates are density independent. Second, *we show that the underlying mechanism is the same in simple stage-structured models and simple spatially structured models*, a result that may have deep implications for the dynamics of real populations. The second result arises because stage structure and spatial structure take the same form in simple models.

The stabilizing mechanism is indirect-density-dependent recruitment to some ageclass or spatially distinct subpopulation. It is created by *asynchronous trajectories* in different components of the model. By this we mean that the fluctuations in density of one component are asynchronous with those of another component. (The initial trajectories following a perturbation from equilibrium also often have different shapes, but these shapes approach each other as the populations move toward equilibrium, so we focus only on the asynchrony.)

We begin, however, with three models that actually contain direct dependence in a vital rate. These models also exhibit asynchronous trajectories and thus provide a link between the models and mechanisms discussed in the previous sections and the last two models in this section.

CASE 1. A PHYSICAL REFUGE FOR THE PREY

This is equation 3.9,

$$\frac{dV}{dt} = rV - aVP + rR$$

$$\frac{dP}{dt} = eaVP - dP,$$

which was analyzed above. It illustrates the role of asynchronous trajectories in a straightforward way. Abundance in the vulnerable prey component varies (as a result of births, predation, and recruitment from the refuge), whereas abundance in the refuge is constant. This leads to density dependence in per-head rate of prey recruitment from the refuge: since V is fluctuating but rR is constant, the per-head recruitment rate to the vulnerable population, rR/V, decreases as V increases.

This is a key observation. This type of density dependence turns out to be fundamental to many models that are stabilized by asynchronous trajectories. It is starkly obvious in this case, but less so in the more complex models we will see later.

CASE 2. AN EXTERNAL SOURCE OF PREY RECRUITS

Now consider a local population of prey, L, that receives R recruits per unit time from "elsewhere." Such recruitment is a central concern in marine ecology, where larvae that land on a coral reef or some section of the rocky seashore may have been produced by adults living many kilometers distant. In fact, dispersal occurs in every species, and frequently at least some dispersers move great distances, so this is a widespread feature in nature.

The number of recruits is of course likely to vary through time, but for simplicity assume R is fixed. The equation for local prey dynamics is now

$$\frac{dL}{dt} = rL - aLP + R. \tag{3.18}$$

This is effectively the same as the prey-refuge model. Once again the equilibria behave in the standard way and the equilibrium is stable. The predator and prey equilibria are analogous to the refuge case (Table 3.2).

An external source of recruits is clearly just like a refuge: if R in the former equals rR in the latter, it is the same model. Once again, there are asynchronous trajectories in the two subpopulations of prey: away from equilibrium the component L fluctuates whereas R does not. The per-head recruitment rate to the prey population from "elsewhere" is now density dependent (R/L decreases with L).

CASE 3. AN EXTERNAL SOURCE OF PREDATOR RECRUITS

In the predator population a constant number of recruits simply adds R individuals per unit time to the predator equation

$$\frac{dP}{dt} = eaHP - dP + R. \tag{3.19}$$

The predator equilibrium is as in the Lotka-Volterra model. The prey equilibrium is

$$H^* = \frac{d}{ea} - \frac{R}{eaP^*}. \tag{3.20}$$

If R is large enough ($R > dP^*$), the prey are driven extinct (Nisbet et al. 1997a). If H^* is greater than zero, however, external predator recruitment is stabilizing, and for the same reason we discussed above: the total recruitment rate of the predator is now density dependent (Table 3.2) because P fluctuates whereas R is constant.

The reduction in prey equilibrium by an external source of predator recruits provides a simple theoretical justification for the common practice of continually releasing insect parasitoids for biological control of pests. This mechanism is therefore an exception to the generalization that stability is associated with an increase in prey equilibrium density. However, since there is density dependence in the predator recruitment rate, prey density now increases with per-head prey growth rate, r (Table 3.2).

CASE 4. PREY LIFE-STAGES DIFFER IN VULNERABILITY

Now consider a prey with a juvenile and adult stage, both of which are attacked by the predator but differ in their vulnerability. Although this is a biologically quite different situation from the three previous models, we will see how the key to stability is again asynchronous trajectories.

We now need an equation for the rate of change of the adult prey population, A, and the juvenile prey population, I. McNair (1987) wrote a model for this situation. A simplified version is

$$\frac{dI}{dt} = rA - mI - a_I I P$$

$$\frac{dA}{dt} = mI - a_A A P \qquad (3.21)$$

$$\frac{dP}{dt} = eP(a_A A + a_I I) - dP,$$

where r is the prey per-head rate of increase in the absence of the predator, m is the per-head rate of maturation of juvenile prey, and a_A and a_I are the per-head attack rates on adults and juveniles, respectively. The predator is assumed to convert juvenile and adult meals into predators with the same efficiency, e.

McNair showed that the equilibrium is more likely to be stable as the difference in vulnerability increases, especially when the juveniles are more vulnerable ($a_I > a_A$). Fig. 3.8 gives insight into the

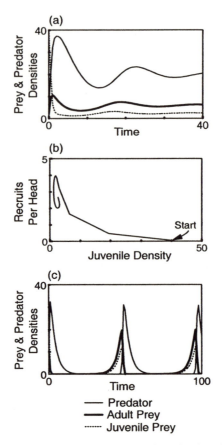

FIGURE 3.8. (a) Simulation of equation 3.21, in which juvenile and adult prey have different levels of vulnerability to predation. Parameter values are $m = 0.5, a_A = 0.01, r = 1, a_I = 0.1, e = 1, d = 0.3$. (b) Density-dependent juvenile recruitment (number of juvenile recruits per juvenile in the population vs. density of juveniles) over the first 20 time units of the simulation, beginning at the indicated starting point. (c) When attack rates are equal ($a_A = a_I = 0.1$), dynamics collapse to neutrally stable cycles in which adult and juvenile prey populations are exactly in phase.

stabilizing mechanism. Fig. 3.8a shows the population densities following a perturbation. The densities rapidly approach equilibrium, where the juvenile-to-adult ratio is of course constant. However, the initial adult and juvenile prey trajectories are not in phase and have slightly different shapes.

Fig. 3.8b shows how this initial difference between the juvenile and adult trajectories affects per-head recruitment to the juvenile prey class (i.e., number of recruits per juvenile). We see that even though none of the vital rates in the model is density dependent, per-head recruitment of juveniles declines as juvenile density increases, creating a density-dependent component (though the relationship is also affected by the cyclic tendency of the initial trajectories). This indirect density dependence arises because A does not track I closely, so the number of recruits produced per unit time (rA) is largely independent of the number of juveniles; this is analogous to the stabilizing mechanisms in cases 1–3 in this group. (A similar plot to Fig. 3.8b shows that per-head recruitment to the adult stage is not obviously dependent on adult density, but the single density-dependent component is enough to stabilize the interaction.)

When the two prey stages are equally vulnerable ($a_A = a_I = a$), the equilibrium is neutrally stable and we regain essentially the Lotka-Volterra model (Fig. 3.8c). Thus, the mere existence of different classes does not give stability.

Although this model illustrates the stabilizing process well, it makes a peculiar assumption about development. At time t, we can think of a constant fraction of the juvenile population pouring via maturation into a bucket, where they mix with adults already present. Adults die as a constant fraction leak out of a hole in the bucket. Since they are well mixed, however, all adults have the same probability of dying, regardless of how long they have been in the bucket—i.e., regardless of their age. Similarly, adults produce juveniles at a constant per-head rate, and at any time t every juvenile has the same (constant) probability of maturing to an adult, regardless of its age. The stage-structured models later in the book deal more realistically with maturation.

CASE 5. SPATIAL STRUCTURE AND HETEROGENEITY

The key point we want to make here is that, in simple cases, models of patches linked by individual movement look very like models with stage structure, and can be stabilized through the same generic process. For example, with little effort we can turn the age-structured equation 3.21 into a model with two patches but no age structure.

Suppose, now, that individual prey are all the same—there are only adults, but there are two patches in the environment. Suppose there is more cover in one patch than in the other, so the per-head attack rate of the predator is lower in one of the patches. To keep the notation as in equation 3.21, and to stress that the models are essentially the same, call the number of prey in the patch with a high attack rate I, and the number in the patch with a lower attack rate A; so their relative vulnerabilities remain as in the example above. In each patch a constant fraction of the population, which may be the same or different in the two patches, is continually diffusing to the other patch. Finally, suppose the predator distributes itself evenly between the patches.

Prey can now reproduce in each patch and move between patches in both directions. Equation 3.21 is modified to give

$$\frac{dI}{dt} = rI + mA - mI - a_I I P$$

$$\frac{dA}{dt} = rA + mI - mA - a_A A P \tag{3.22}$$

$$\frac{dP}{dt} = eP(a_A A + a_I I) - dP.$$

In equation 3.21, each component contributed a constant fraction to the other component, via reproduction (rA) and maturation (mI); now the constant fractions are contributed through movement (mA and mI). So the model is effectively unchanged except that both components reproduce. Sabelis et al. (1991) analyzed a similar model but interpreted A and I as the number of prey patches suffering different levels of risk, with P being the number of patches with both predators and prey.

Unequal attack rates in the two patches ($a_A \neq a_I$) can again stabilize the equilibrium. Prey densities in the two patches initially oscillate out of synchrony and with a slightly different frequency, but move toward the stable equilibrium densities (Fig. 3.9a). Now per-head immigration to patch A from patch I has a density-dependent component (Fig. 3.9b) and is the stabilizing process. As in equation 3.21, the process is asymmetrical: there is little evidence for density dependence in the immigration rate to patch I from patch A.

Equation 3.22 is a simplified version of a model, first proposed by Murdoch and Oaten (1975), of Lotka-Volterra predator-prey populations in two patches that are heterogeneous and linked by random

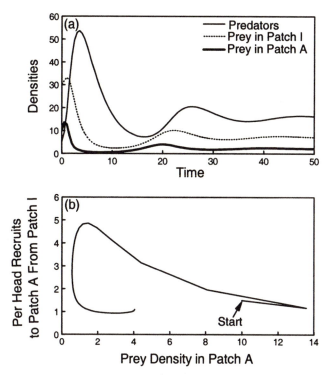

FIGURE 3.9. (a) Simulation of equation 3.22, in which spatial structure has been substituted for age structure. Parameter values are $r = 0.5$, $m = 0.5, a_I = 0.1, a_A = 0.01, e = 1, d = 0.3$. (b) Density-dependent per-head recruitment of prey, to patch A from patch I, versus prey density in patch A over the first 20 time units of the simulation, beginning at the indicated starting point.

migration of both prey and predator. The stabilizing role of density-dependent immigration in that model is demonstrated in Murdoch et al. (1992a), and a simpler "master-slave" version is analyzed by Nisbet et al. (1993).

Equations 3.21 and 3.22 are very simple stage- and spatially structured models. Asynchronous trajectories, however, can still play a stabilizing role in our more realistic stage-structured models (chapter 5) and in some more realistic spatial models (chapter 10). Spatially and stage-structured models of course are not always good analogs. For example, there can be diffusive movement among all patches in a population, but individuals cannot diffuse backward from the pupal stage to the larval stage of an insect. In both cases, however, asynchronous trajectories of different components may still be stabilizing.

Limitations of Exponentially Distributed Development Times

We pointed out at the end of case 4, above, that this form of stage structure makes a very odd assumption about development: a constant fraction of the juveniles mature to the adult stage, regardless of their age. This says each individual juvenile has an exponentially distributed waiting time in the stage. This is not realistic. In spite of this, some qualitative results of such models persist into the properly stage-structured models of later chapters. For example, an invulnerable stage still tends to be stabilizing.

Such models, however, are not reliable predictors of quantitative dynamics. For example, the equilibrium of equation 3.21 is always stable if one of the prey stages is totally invulnerable, which is not true in the analogous stage-structured model formulated with explicit developmental delays (chapter 5). Models with exponentially distributed waiting times also do not reliably predict cycle periods, an important fact for applications to real populations. A result from a single-species model of this type analyzed by Thieme (in press) shows that the cycle period is substantially different from that predicted by its stage-structured analog (Gurney et al. 1980; Gurney and Nisbet 1985). For this reason, we suspect that fits to cyclic data, using models with exponential waiting times, may often be spurious.

Generality of Stabilization via Decoupling and Asymmetry

Equations 3.21 and 3.22 show in slightly more complicated form the mechanism revealed by the ultra-simple models of a prey refuge and an external source of prey recruits. We see the same process occurring in more complex stage-structured models (chapter 5) and in some more realistic spatial models (chapter 10).

Since all real populations have age structure and are spatially distributed, we may expect asynchrony to be an important stabilizing property. In general, this may explain many cases in the field where stability has been observed but no direct density dependence has been detected in the vital rates.

Density Dependence and Stability

The ecological literature is replete with statements that stability can only arise from density-dependent processes. Typically what is meant is that some vital rate is related to population density. Most of the stabilizing mechanisms analyzed in this chapter exemplify this type of density dependence. Yet many analyses of population time series have failed to find density dependence. Models of the type discussed in the previous section provide a potential explanation: asynchronous trajectories in different components of the population induce indirect density dependence.

Density dependence defined as "a vital rate dependent on density" makes perfectly good sense in two-dimensional interactions, and corresponds with the fact that one or other of the "partials" discussed above is negative. However, whereas density dependence is also needed for stability in more complex situations, the fact is that stability can be induced in high-dimensional interactions by processes whose interpretation as "density dependent" is far from immediately obvious. By high dimensional we mean there are several to many dynamic components, such as life stages, species, or spatial subpopulations. Such models are common in later chapters.

The point is made relatively simply by equation 3.22. The equilibrium of this model is always stable. Yet the vital rates are explicitly density independent. Fig. 3.9b shows that we can detect density dependence when we know where to look, but we would not know where to look if we did not understand the underlying dynamical processes, even in this simple model.

BASIC AND POTENTIAL GENERAL PROPERTIES OF PREDATOR-PREY SYSTEMS

Here we step back from the details of the various models to ask: what fundamental properties have we seen that seem to characterize continuous-time predator-prey models in general? If these properties

TABLE 3.3. Fundamental properties of predator-prey models

1. Interacting predator and prey populations tend to oscillate.
2. Satiation in predators, and time lags in predator response, increase the tendency to oscillate.
3. Direct density dependence in vital rates tends to dampen oscillations.
4. Combination of factors in # 2 and direct density dependence in prey tends to produce limit cycles.
5. Most processes suppressing oscillations also increase prey equilibrium density.
6. Spatial heterogeneity and/or age structure can lead, via asynchronous trajectories of component populations, to indirect density dependence which may stabilize populations.

turn out to be general in models, they may provide fundamental insights into the behavior of real systems. They are listed in Table 3.3. We return to the question of fundamental properties in the final chapter, to check whether our chosen properties are indeed shared by the range of models explored in the book.

There is an inherent tendency for model predator-prey systems to oscillate. This contrasts with, for example, species that compete with each other. Furthermore, adding two of the most basic features of biology—time lags and satiation in predators—always tends to destabilize the equilibrium and enhance the tendency to oscillate. So we take this tendency to oscillation to be an inherent property of real predator-prey systems.

Two kinds of processes can counter the tendency to instability, or can constrain oscillations to stable limit cycles or other bounded dynamics. The first is density dependence in vital rates, as is induced, for example, by limited prey resources, antagonistic encounters among predators, refuges, and so on. The second is indirect density dependence induced by asynchronous trajectories in different component populations. We demonstrated that spatial processes or age-structure can act in dynamically equivalent ways to create such indirect density dependence.

One or another of these stabilizing features is probably also universal in real systems. Even when they occur, however, there is frequently a large area of potentially realistic parameter space in which models predict cyclic dynamics. The Rosenzweig-MacArthur model is a good example. The fundamental problem posed by these simple models is therefore why, in nature, unstable equilibria—and hence cyclic or chaotic behavior—seem to be relatively uncommon (chapter 2).

Finally, in most models in this chapter, changes that result in stability increase prey equilibrium density. This theoretical result is quite general in models with self-limiting prey (May 1976b). Stability is more likely the closer the prey equilibrium is to its own carrying capacity, because at such densities the prey is more subject to its own density-dependent processes. When the predator suppresses the prey far below the prey's carrying capacity, the instability inherent in the predator-prey interaction dominates the dynamics. The trade-off between prey suppression and stability is especially marked when stability is induced by density dependence in the predator population (which implies decreasing predator efficiency as predator density increases). In this case, prey density increases as the prey's environment is enriched. Thus, an additional question naturally arises: how do we account for the apparent stability of real systems in which predators routinely suppress the prey far below prey carrying capacity in productive environments? This question is nowhere more acute than in apparently stable cases of biological control of pest insects (chapter 9).

Local Stability Conditions, Computation of Stability Boundaries, and Cycle Periods for Systems of Differential Equations

The mathematical technicalities for local stability analyses of systems of ordinary differential equations are covered in many texts, including several targeted at biologists and ecologists (e.g., Edelstein-Keshet 1988; Murray 1993; Hastings 1997; Adler 1998; Gurney and Nisbet 1998; Case 2000; Kot 2001). This appendix has two aims. First, we summarize the key steps for analyzing the stability of continuous-time systems with two and three variables, with particular emphasis on the ecological interpretation of the eigenvalues of the stability matrix. We follow the approach in Gurney and Nisbet 1998, but the summary here may suffice for readers who have previously encountered, but forgotten, the details of local stability analysis. Second, we describe step by step how to compute stability and/or extinction boundaries numerically, and also to estimate the period of cycles near these boundaries. Similar computations are required in subsequent chapters for discrete-time models and for continuous-time systems with time delays. The computations are straightforward for the mathematically adept, many of whom will have their own preferred approach. They are seldom spelled out in the standard texts, however, and we hope the examples here will help the inexperienced make a start.

EIGENVALUES AND THEIR INTERPRETATION

We first describe the procedures for linearizing a system of two equations describing interacting populations. If we write the equations in the form

$$\frac{dH}{dt} = F(H, P), \quad \frac{dP}{dt} = G(H, P), \quad \text{(A3.1)}$$

then the equilibria are obtained by solving simultaneously the algebraic equations

$$F(H^*, P^*) = 0, \quad G(H^*, P^*) = 0. \quad \text{(A3.2)}$$

Small deviations, h and p, from equilibrium follow the differential equations

$$\frac{dh}{dt} = A_{11}h + A_{12}p$$
$$\frac{dp}{dt} = A_{21}h + A_{22}p \quad \text{(A3.3)}$$

where the recipe for calculating the four quantities A_{11}, A_{12}, A_{21}, and A_{22} is

$$A_{11} = \left.\frac{\partial F}{\partial H}\right|_*, \quad A_{12} = \left.\frac{\partial F}{\partial P}\right|_*, \quad A_{21} = \left.\frac{\partial G}{\partial H}\right|_*, \quad A_{22} = \left.\frac{\partial G}{\partial P}\right|_*$$

$$\text{(A3.4)}$$

where the asterisk implies that the partial derivatives are evaluated with H and P set equal to their equilibrium values. In some models there may be more than one equilibrium and the As must be calculated separately for each. Note that for any particular equilibrium the As have constant values that are determined by the model parameters and do not vary as h and p vary. They can be arranged as the elements of a 2×2 matrix called the Jacobian matrix.

The system of equations A3.3 is linear, meaning that the right-hand side of the equations involves only terms proportional to h and p and not terms such as p^2, hp, e^p. It is shown in the texts cited above, and in countless other texts on differential equations, that the solution to equations A3.3 is of the form

$$h(t) = C_1 e^{\lambda_1 t} + C_2 e^{\lambda_2 t}$$
$$p(t) = D_1 e^{\lambda_1 t} + D_2 e^{\lambda_2 t}$$

(A3.5)

where C_1, C_2, D_1, and D_2 are constants whose values are determined by the initial conditions (i.e., the values of h and p at time $t = 0$). The quantities λ_1 and λ_2 are the two eigenvalues of the Jacobian matrix; their values are unaffected by the initial conditions and thus depend only on the elements of the Jacobian matrix.

The eigenvalues of a 2×2 matrix are the two roots of the *characteristic equation*

$$\lambda^2 + B_1 \lambda + B_2 = 0$$

(A3.6)

where

$$B_1 = -(A_{11} + A_{22}) \quad \text{and} \quad B_2 = A_{11} A_{22} - A_{12} A_{21}.$$

(A3.7)

Equation A3.6 is quadratic and has two roots which can be written in the form

$$\lambda_1 = \frac{1}{2}\left(-B_1 + \sqrt{\Delta}\right); \quad \lambda_2 = \frac{1}{2}\left(-B_1 - \sqrt{\Delta}\right)$$

(A3.8)

where

$$\Delta = B_1^2 - 4B_2.$$

(A3.9)

If Δ is greater than zero, then both roots are real. Equation A3.5 then shows that deviations from equilibrium will ultimately decay to zero if both roots are negative, and will (except for special initial conditions) ultimately grow exponentially if one or both roots are positive. The case where both roots are negative then corresponds to a locally stable equilibrium. The magnitude of the eigenvalue giving the slowest decay rate (i.e., the least negative eigenvalue, which is the one with the smallest magnitude, λ_1) determines the rate of return to equilibrium.

If Δ is less than zero, then equation A3.3 has a complex conjugate pair of roots which we can write in the form

$$\lambda_1 = \xi + i\omega, \quad \lambda_2 = \xi - i\omega, \qquad (A3.10)$$

with

$$i = \sqrt{-1}, \quad \xi = -\frac{B_1}{2}, \quad \omega = \frac{\sqrt{-\Delta}}{2}. \qquad (A3.11)$$

Solutions involving complex numbers may appear ecologically absurd, but exponentials of complex numbers are related to real exponentials and to the standard trigonometric functions; thus it can be shown with a little algebra that the solution for $h(t)$ can be written in the form

$$h(t) = C \exp(\xi\, t) \cos(\omega\, t - \phi) \qquad (A3.12)$$

where C and ϕ are new constants related to the initial conditions. A similar expression can be written for $p(t)$. Mathematically, the quantities ξ and ω are just the real and imaginary parts of the (complex) eigenvalues, but they also have simple ecological interpretations. Equation A3.12 describes oscillations, with a period equal to $2\pi/\omega$, whose amplitude grows exponentially if ξ is greater than zero and declines exponentially if ξ is less than zero. For stability we thus require ξ to be less than zero, and the magnitude of ξ is a measure of the rate of approach to equilibrium.

The above theory generalizes to systems of M differential equations, the end result being that local stability requires that all the eigenvalues of the Jacobian matrix are either real and negative, or complex with negative real parts.

DETERMINING STABILITY WITHOUT EVALUATING EIGENVALUES

Much of chapter 3 is devoted to general statements about stability in particular models. For these, it is convenient to have stability conditions in a form that do not require explicit computation of eigenvalues. For example, with the 2×2 Jacobian matrix above, it is found that necessary and sufficient conditions for all the eigenvalues to be either real and negative or complex with negative real parts are:

$$B_1 > 0 \quad \text{and} \quad B_2 > 0. \tag{A3.13}$$

With three variables, the characteristic equation takes the form

$$\lambda^3 + B_1\lambda^2 + B_2\lambda + B_3 = 0. \tag{A3.14}$$

The necessary and sufficient conditions for local stability are then

$$B_1 > 0, \quad B_3 > 0, \quad B_1B_2 - B_3 > 0. \tag{A3.15}$$

There are corresponding inequalities (known as the Routh-Hurwitz criteria) for systems with arbitrary numbers of variables, but the general forms are not very instructive for our purposes. Also, with larger numbers of variables the number of inequalities grows rapidly, leading May (1973a) to remark that "no-one in their right mind is going to use these criteria on $M > 5$."

COMPUTATION OF STABILITY AND EXTINCTION BOUNDARIES

Chapter 3 contains several examples of models where changes in the value(s) of one or more model parameters can lead to a change in stability or to extinction of the consumer. For these, we frequently need to calculate *stability* or *extinction boundaries*, like those drawn in Fig. 3.6 for the Rosenzweig-MacArthur model.

At the critical parameter values where a transition from stability to instability occurs, the characteristic equation either has a real root equal to zero, or there is a complex root with zero real part. In consumer-resource systems, the latter is frequently the focus of interest, since a complex root with zero real part (which we will write as $\lambda = i\omega$ with ω a non-zero real number) is associated with a transition from damped oscillations to oscillations whose amplitude grows exponentially. For a two-variable system, substituting this form into equation A3.6 leads to

$$-\omega^2 + iB_1\omega + B_2 = 0. \tag{A3.16}$$

Since both the real and imaginary parts of this equation must be zero, and we have already assumed that ω does not equal zero, equation A3.16 implies that

$$\omega^2 = B_2 \quad \text{and} \quad B_1 = 0. \tag{A3.17}$$

It follows that at the transition from damped to growing oscillations, i.e., on the stability boundary, B_1 is equal to zero. The period of the oscillations on the boundary is equal to $2\pi/|\omega| = 2\pi/\sqrt{B_2}$. The corresponding results for a system with three variables are that on the stability boundary

$$\omega^2 = B_3/B_1 \quad \text{and} \quad B_1 B_2 - B_3 = 0. \tag{A3.18}$$

The mechanics for computing a stability boundary and periods on this boundary can be summarized as follows.

1. If the equilibrium can be calculated analytically, do so, and then evaluate the Jacobian matrix. Calculate the coefficients in the characteristic equation and compute the stability boundary and periods using A3.17 or A3.18.

2. If the equilibrium cannot be calculated analytically, evaluate the Jacobian matrix and the coefficients of the characteristic equation in terms of the equilibrium values (e.g., H^* and P^*). Then work with three simultaneous algebraic equations (both equilibrium conditions [A3.2] and the right-hand equations of A3.17 or A3.18). If the stability boundary is to be drawn in a plane whose axes represent two model parameters, we now have three equations in four unknowns (H^*, P^*, and the two parameters). These define a curve in the relevant plane, which can be plotted. To determine cycle periods on the stability boundary, compute ω (and hence the cycle periods) using the left-hand equations of A3.17 or A3.18.

Computation of extinction boundaries is very similar. Consumer extinction occurs at the value of H^* for which $P^{-1}(dP/dt)$ is equal to zero when P approaches zero. Thus we solve simultaneously

$$F(H^*, 0) = 0 \quad \text{and} \quad \lim_{P \to 0}[P^{-1}G(H^*, P^*)] = 0, \tag{A3.19}$$

which we treat as two equations relating H^* and any two model parameters.

EXAMPLES

Example 1

The Rosenzweig-MacArthur model (see Table 3.2) provides an example of the first (simpler) methodology. The functions $F(H, P)$ and $G(H, P)$ in equation A3.1 are

$$F(H, P) = rH\left(1 - \frac{H}{K}\right) - \frac{aHP}{1 + aT_hH};$$

$$G(H, P) = e\frac{aHP}{1 + aT_hH} - dP.$$

(A3.20)

The non-zero equilibrium densities are

$$H^* = \frac{d}{a(e - dT_h)}; \quad P^* = \frac{r}{a}\left(1 - \frac{H^*}{K}\right)(1 + aT_hH^*).$$

(A3.21)

The Jacobian matrix elements are obtained from equation A3.4; for example,

$$A_{11} = \frac{\partial F}{\partial H}\bigg|_* = \frac{\partial}{\partial H}\left(rH\left(1 - \frac{H}{k}\right) - \frac{aHP}{1 + aT_hH}\right)\bigg|_*.$$

(A3.22)

The differentiation and substitution can be performed analytically with the result

$$A_{11} = \frac{dr(e - aeKT_h + dT_h(1 + aKT_h))}{aeK(-e + dT_h)};$$

(A3.23a)

$$A_{22} = \frac{\partial G}{\partial P}\bigg|_* = e\frac{aH^*}{1 + aT_hH^*} - d = 0.$$

(A3.23b)

The stability boundary is the line on which $B_1 = -(A_{11} + A_{22}) = 0$. In this example this reduces to $A_{11} = 0$. The stability condition in Table 3.2 is then obtained by solving the equation $A_{11} = 0$ for K.

For the lazier reader, who is familiar with computer algebra software such as Mathematica or Maple, the calculation can be automated. We illustrate this for Mathematica in Box A3.1.

BOX A3.1.

MATHEMATICA NOTEBOOK FOR ANALYTIC COMPUTATION OF THE
STABILITY BOUNDARY IN THE ROSENZWEIG-MACARTHUR MODEL

Clear[F, a, Th, H, P, Hstar, Pstar, A11]
F:=r H (1-H/K) - a P H/(1+a Th H)
Hstar:=d/(a (e - d Th))
Pstar:=r/a (1- Hstar/K) (1+a Th Hstar)
A11:=D[F,H] /.{H->Hstar, P->Pstar}
Solve[A11==0, K]

This produces the output

$$\left\{\left\{ K-> \frac{e+dTh}{a\ Th\ (e-d\ Th)} \right\}\right\}$$

which is the stability condition in Table 3.2.

Example 2

We illustrate the numerical computation of the stability boundary, extinction boundary, and cycle period without explicit computation of equilibria using a modification of the Rosenzweig-MacArthur model with predator-dependent attack rate given by equation 3.13. There are several practical ways of implementing the numerical computation of stability boundaries. For many of the calculations presented in this book we used the program Contour, which is part of a software package known as Solver, written by W. S. C. Gurney and colleagues at the University of Strathclyde and available from http:/www.stams.strath.ac.uk/external/solver. The reader already familiar with alternative software such as Mathematica will certainly prefer to use that software. To illustrate the mechanics of a more complex stability calculation using Mathematica, Box A3.2 contains a notebook for calculating the stability boundary, extinction boundary, and cycle period in a modification of the Rosenzweig-MacArthur model with predator-dependent attack rate given by equation 3.13.

BOX A3.2.

MATHEMATICA NOTEBOOK FOR NUMERICAL COMPUTATION OF THE
STABILITY BOUNDARY, EXTINCTION BOUNDARY, AND CYCLE PERIOD IN
A MODIFICATION OF THE ROSENZWEIG-MACARTHUR MODEL WITH
PREDATOR-DEPENDENT ATTACK RATE GIVEN BY EQUATION 3.13

Clear[f, g, H, P, A11, A12, A21, A22, B1, A2, r, K, a, b, d, Th, ee, LSB,
EXT, Period, w]

(* Define functions f and g *)

f[H_, P_] := r (1 - H/K) - a P /(1 + a Th H + b P);

g[H_, P_] := ee a H /(1 + a Th H + b P) - d ;

(* Calculate elements of Jacobian matrix *)

A11:= H D[f[H, P], H]; A12 := H D[f[H, P], P];

A21:= P D[g[H, P], H]; A22 := P D[g[H, P], P];

(* Calculate coefficients in characteristic equation *)

B1:= -(A11 + A22); B2 := A11 A22 - A21 A12;

(* Supply parameter values *)

a = 1; Th = 1; r = 1; ee = 0.5; b = 0.2;

(* Define ranges and initial values for numerical root finding *)
(* Some trial and error required here *)

Hmin = 0.001; Pmin = 0.001; Kmin = 0.001; Kmax = 20; Hmax = Kmax;
Pmax = Kmax; wmin = 0.01; wmax = 10; Hstart = 0.2; Pstart = 0.1;
    Kstart = 1.0;

wstart = Sqrt[r d]; (* This is the value for the Lotka-Volterra model *)

(* Rules for calculating the local stability boundary [LSB], consumer extinction boundary [EXT], and period on the local stability
boundary *)

(Box A3.2. continued)

LSB[d_] :=
  K /. FindRoot[{f[H, P] == 0, g[H, P] == 0, B1 == 0},
    {H, Hstart, Hmin, Hmax}, {P, Pstart, Pmin, Pmax}, {K, Kstart,
    Kmin, Kmax}, MaxIterations -> 50];

EXT[d_] :=
  K /. FindRoot[{H == K, g[H, 0] == 0}, {H, Hstart, Hmin, Hmax},
    {K, 2.0, Kmin, Kmax}, MaxIterations -> 50];
Period[d_] := (2 Pi/Abs[w]) /.
  FindRoot[{f[H, P] == 0, g[H, P] == 0, B1 == 0, B2 == w^2},
    {H, Hstart, Hmin, Hmax}, {P, Pstart, Pmin, Pmax},
    {K, Kstart, Kmin, Kmax}, {w, wstart, wmin, wmax},
    MaxIterations -> 50];

(* **Switch off warning messages** –
USE THIS OPTION WITH CARE ** AFTER
** DEBUGGING!! *)
Off[FindRoot::frnum]
Off[ReplaceAll::reps]
Off[FindRoot::regex]
Off[FindRoot::frns]

(* **Specification of plots**
Local stability boundary is gray, extinction boundary
black in first plot *)
Plot[{LSB[d], EXT[d]}, {d, 0.01, 0.39}, PlotStyle -> {GrayLevel[0.5],
  GrayLevel[0]}, AxesLabel -> {d, K}, PlotRange -> {0, 5}]
Plot[Period[d], {d, 0.01, 0.39}, AxesLabel -> {d, Period},
  PlotRange -> {0, 100}]

(Box A3.2. continued)

This code produces the two plots shown in Fig. A3.1.

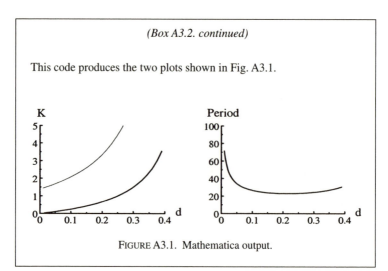

FIGURE A3.1. Mathematica output.

Simple Models in Discrete Time

Most of the models in this book are framed in continuous time, which is a common framework for models of true predators and prey. Because they assume that dynamical processes such as births and deaths occur more or less all the time, however, they can mislead when applied to organisms in which reproduction occurs in a discrete pulse determined by season. Seasonal breeding can be incorporated into continuous-time models, for example by making the per-head birth rate a function of Julian day. The usual practice, though, is to turn to discrete-time models, i.e., difference equations.

Discrete-time models describe changes in density at discrete points in time, for example each year. They must be used with care as life processes are continuous. Consumers feed and grow, and hence change the resource density continuously. Individuals are likely to be subject to mortality all the time—though each birth and death is of course a discrete event. Thus discrete-time models must be formulated with an eye on the sequence of events assumed to occur between updates of the state variables. This process is sometimes made explicit, a point we emphasize when formulating a host-parasitoid model in Box 4.1, and later in the chapter where we explore briefly models that combine continuous growth within a season and pulsed breeding at the start of the next season.

There are two main classes of discrete breeders. The first have non-overlapping generations and include, for example, the many temperate insects that emerge in spring and lay their eggs over a few weeks or so. The eggs then develop through a series of larval stages, overwinter in the pupal stage, and emerge as adults again the following spring. There is typically little overlap of life stages and no overlap of generations; the adults die after laying their eggs and are not present when next year's adults emerge. The second class breeds in pulses but has overlapping generations. Many vertebrates are in this class. Birds and fish, for example, typically take one or more years to mature and then breed each spring thereafter, often living for a decade or more. Thus generations (or "year classes") overlap in these species.

Discrete-time models are traditional in parasitoid-host theory. This is partly historical artifact, stemming from the seminal work of Nicholson and Bailey (Nicholson and Bailey 1935; Bailey et al. 1962), but mainly is a response to the life-history features of many temperate-region insects.

As in chapters 2 and 3, we begin with single-species models and then go to consumer-resource interactions.

SINGLE-SPECIES MODELS IN DISCRETE TIME

The obvious starting point is a discrete-time analog of the logistic model. This can take a variety of forms, but a common one assumes one generation per year and defines the density in generation $t + 1$, given the density in generation t, as

$$N_{t+1} = N_t e^{r\left(1 - \frac{N_t}{K}\right)}. \tag{4.1}$$

The model is equivalent to the well-known Ricker model for a population with non-overlapping generations. Recall that $e^r = R$ is the multiplication factor of the population in each generation in the absence of density dependence. Then substituting $c = r/K$, we get the Ricker model, namely

$$N_{t+1} = R N_t e^{-c N_t}. \tag{4.2}$$

In this form, the parameter c determines the strength of density dependence, i.e., the speed at which the per-head rate of increase decreases with increasing density. When there is no density dependence, we have

$$N_{t+1} = R N_t, \tag{4.3}$$

the equation for geometric growth in which the population multiplies by a factor of R each generation. For the population to persist, we need R greater than 1.

Equation 4.2 is often used to describe the dynamics of fish or insects that develop from egg to adult in 1 year, reproduce, and then die. If we model only females, R is the number of females surviving per female in the absence of any density dependence. For example, it can be the maximum number of female eggs per female (which occurs at a female density of effectively zero) times the fraction that survive density-independent mortality through the adult stage. The term e^{-cN_t} can be interpreted as the fractional decrease in fecundity, or in subsequent survivorship (e.g., via cannibalism), in response to an increase in adult female density. Fig. 5.1b gives an example, in blowflies, of fecundity declining roughly exponentially with adult density. Other regulatory processes lead to different models; for example, a form of density dependence in which mortality of immatures at any given time depends on their current density leads to a very different model—the Beverton-Holt model (Gurney and Nisbet 1998, chapter 5).

The model's equilibrium is obtained by setting $N_{t+1} = N_t = N^*$. Both sides of equation 4.2 are equal either if N^* is equal to zero (the population is extinct) or (dividing both sides by N_t and taking the natural logarithm of both sides) if

$$N^* = \frac{\ln R}{c}, \qquad (4.4)$$

where $\ln R = r$ and $c = r/K$, so $N^* = K$, as expected.

Local stability analysis, which describes the behavior of a small perturbation from equilibrium, follows steps analogous to those for differential-equation models described in chapters 2 and 3. The methodology is outlined in the appendix to this chapter. As we might expect, if R is less than 1, the "extinction" equilibrium ($N^* = 0$) is stable (the population always declines toward zero from any initial density), and there is no positive equilibrium (the natural logarithm in equation 4.4 is negative). For R greater than 1, stability of the non-zero equilibrium requires that the value of the eigenvalue, μ (defined in the appendix to this chapter), be between -1 and 1. In the appendix, we show that for the Ricker model

$$\mu = 1 - \ln R = 1 - r. \qquad (4.5)$$

Stability of the equilibrium thus requires

$$0 < r < 2. \tag{4.6}$$

This finding contrasts starkly with our previous analysis of the continuous-time logistic model (chapter 2) where the equilibrium is invariably stable. Only a narrow range of values of r is consistent with stability in the Ricker model. This exemplifies a general point we explore in detail in chapter 5: delayed density dependence, although it may stabilize an otherwise unstable equilibrium, may also induce instability. Here the delay is a consequence of converting to discrete time, where the effect of density, for example current adult density suppressing fecundity, is not expressed until adult recruitment at the next time step or generation.

We illustrate the discrete-time logistic model's behavior by explicit simulations (Fig. 4.1). Fig. 4.1a plots the relationship between N_{t+1} and N_t for three values of r. The equilibrium (in this model, the carrying capacity K) occurs where each curve meets the dashed 45-degree line, since at this point N_{t+1} equals N_t. The eigenvalue, introduced above, is the slope of the curve as it passes through this point. For a given value of K, larger values of r in equation 4.1 imply stronger density dependence, and a higher hump in the relationship between N_{t+1} and N_t (Fig. 4.1a). In conformity with the analytical results above, the equilibrium is a stable point if r is between zero and 2 ($R < 7.39$) (e.g., Fig. 4.1b). Local stability analysis tells us nothing about what happens with r greater than 2, but for this model, these dynamics have been investigated by many authors (e.g., May 1976c; May and Oster 1976) and are well described in text books (e.g., Kot 2001). If $2 < r < 2.69$ ($7.39 < R < 14.73$), the equilibrium is unstable but there is a stable limit cycle (e.g., Fig. 4.1c). Chaotic fluctuations, or cycles with a diverse range of periods (famously 3), arise if r is greater than 2.69 ($R > 14.73$) (Fig. 4.1d).

The essential feature that leads to the exotic dynamics of Ricker models with a "hump," as in two of the cases in Fig. 4.1a, is *overcompensation*: beyond a certain adult density, further increases in current adult density lead to a decrease in the total number of adults produced in the next generation. Thus, decreases in population size can actually increase the number of recruits.

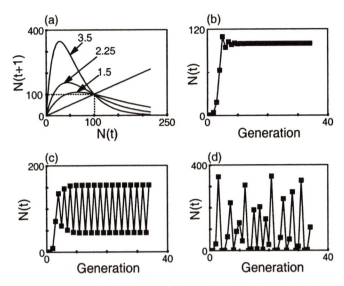

FIGURE 4.1. (a) Relationship between N_{t+1} and N_t for a discrete-time logistic model (equation 4.1) with three different values for r. The number for each curve is the value of r. The equilibrium (K), where the curves meet the 45-degree line, has the value 100. (b) Population trajectory from a simulation of the model with $r = 1.5$. (c) $r = 2.25$ gives a stable limit cycle. (d) $r = 3.5$ gives chaotic dynamics (May 1976a).

Likely Range of R in Natural Populations

R is the factor by which a population would increase each time step if there were no density-dependent factors operating. We stressed above that, like the instantaneous rate of increase, r, the value of R is the end result after the occurrence of all density-independent births and deaths. Its value is therefore typically much lower than the number of female eggs per female. For example, whereas a marine fish might produce hundreds of thousands of eggs in a year, the vast majority of these will die as planktonic organisms, independent of how many parent fish, or eggs, there are.

We suspect that the maximum R for populations in natural and semi-natural environments, and even in many unnatural environments such as orchards, is not much greater than 10. For example, insect pests in forests have been well studied and have among the highest rates of population increase recorded. The larch budmoth may have the largest R of any such insect. We showed in chapter 2 that R for this population is just over 7. Morris (1959) estimated $R = 5$ for an outbreaking budworm population in Canada. In chapter 2 we calculated $R = 3.14$ in a freely expanding population of California red scale.

An analysis by Hassell et al. (1976) gives strong support to this generalization. Using information from life tables, they calculated R for 24 annual insect species; we discovered a few of the estimates are too high. Our revised list shows R is less than 10 for 20 of the 22 species we could check. The two populations with R greater than 10 were grossly unnatural. One was a laboratory population, the other a potato beetle inoculated into isolated experimental plots of potatoes largely lacking natural enemies.

These comments add support to a conjecture by Hassell et al. (1976), who noted that in simple models, chaos only occurs with large values of R. They then argued that R is rarely large enough in real populations, given reasonable levels of density dependence, to produce chaotic dynamics. For example, chaos in the Ricker model requires R to be greater than 15, substantially higher than our upper bound. In chapter 2 we also argued that chaos probably is not common in populations in nature. Some caution is required, however; not all single-species models require large values of R for chaos, and chaos can arise through multispecies interactions.

Demography of a Stable Annual Species

The population trajectory shown in Fig. 4.2 describes the density of a hypothetical population that breeds once a year. Reproduction occurs at the instant in time when the population is at its lowest density and the individuals are all adults. They produce a cohort of young (eggs, say) that then declines at a constant rate over the year as it develops.

In terms of a discrete-time model, this population is at a stable equilibrium. If, year after year, we were to sample the adults at the moment when they breed, we would see the same number each year—the population trajectory would be just a flat line. As May and Oster (1976) pointed out, a discrete-time model is like a stroboscopic photograph taken once per generation, when in fact the real population is a movie.

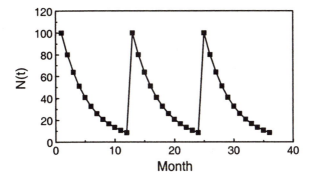

FIGURE 4.2. Numbers present each month in a hypothetically stable population that breeds once per year, producing immature offspring that then decline exponentially until they mature into adults in 12th month and reproduce again. An annual sample from this population would show a stationary population size.

Suppose, however, that the population trajectory in Fig. 4.2 was output from a continuous-time model, in which the population reproduced continuously and in which there was no seasonal forcing. We would infer that we were observing a stable limit cycle. In chapter 5 we find just such fluctuations and call them *single-generation cycles*. Thus, a stable equilibrium in a discrete-time model is the analog of an unstable equilibrium in a continuous-time model that exhibits single-generation cycles (Jansen et al. 1990).

Cycle Period

One reason we have introduced a single-species discrete-time model is to comment on the period of cycles that arise when the equilibrium is unstable. The fundamental period in the above model (equation 4.1) is 2 (the population in successive generations alternates between high and low densities, as in Fig. 4.1c). May and Oster (1976) demonstrated that as r increases in value, a process known as period doubling occurs. This is seen in Figs. 4.3a–c, where we first see the system cycle between two different densities, then four different densities, then eight different densities, so the period doubles as r increases (and eventually the dynamics are chaotic). However, the basic two-year pattern persists in

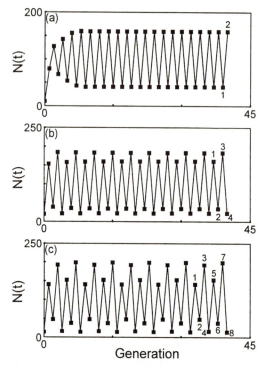

FIGURE 4.3. Illustration of period doubling, as r in equation 4.1 increases
from (a) 2.3 (period = 2), to (b) 2.55 (period = 4) to (c) 2.67 (period = 8).
Notice that the basic period = 2 is evident in all cases.

these trajectories. Different periods arise at some values of r greater
than those marking the transition to chaos (May and Oster 1976; May
1976c), but those typically involve larger values of r (or R) than occur
in nature (see above).

Organisms with Overlapping Generations

Suppose now we have a fish or bird species that matures in one year but
in which a fraction of the adults survives each year to breed again. The
Ricker model can be modified to take this into account simply by adding
the fraction, S, surviving to the next year. Recruitment to the first-year

class of adults is the same as in equation 4.2. The adult population next year is made up of these recruits, together with survivors from this year's adults, i.e.,

$$N_{t+1} = RN_t e^{-cN_t} + SN_t. \tag{4.7}$$

The addition of overlapping generations has a stabilizing effect on the model, as we might expect. For example, when high population density in year t severely suppresses recruitment to the population in year $t + 1$, the survival of some of the existing year classes of adults (SN_t) causes N_{t+1} to be less severely depressed than would occur in the absence of overlapping generations. Multiple year classes thus tend to counter the destabilizing effect of overcompensation in any particular year.

The equilibrium is

$$N^* = \frac{1}{c} \ln \left(\frac{R}{1-S} \right). \tag{4.8}$$

The eigenvalue is

$$\mu = 1 - (1 - S) \ln \left(\frac{R}{1-S} \right). \tag{4.9}$$

Because $0 < (1 - S) < 1$, some algebraic juggling with the second term of equation 4.9 reveals that it is always smaller than the analogous term, $\ln R$, in equation 4.5. Thus overlapping generations reduce the tendency for overcompensating instability, which occurs when μ is less than -1. This effect is illustrated in Fig. 4.4, which shows that as S increases, the equilibrium is stable for a larger and larger range of values of R.

Development to Adulthood Takes More Than One Year, and Generations Overlap

Many species, especially of vertebrates, need more than one year to reach sexual maturity, and some of the adults also typically survive to breed in more than one year.

Let d be the length of the pre-reproductive period. In annual breeders d is measured in years. So the number of new adults recruited in year t is the number of young produced by adults alive in year $t - d$, times the fraction that survive to become adult in year t. As in the previous

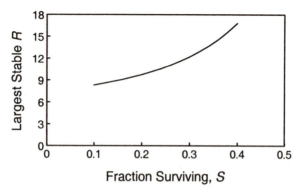

FIGURE 4.4. Effect of increasing the fraction surviving to the next generation, S, on the range of R values consistent with stability in the Ricker model (equation 4.9). The eigenvalue μ was set to -1, and the value of R was calculated from equation 4.9 for a range of values of S that satisfy $\mu = -1$.

section, a fraction, S, of adults survive from one year to the next. The model thus becomes (Higgins et al. 1997a)

$$N_t = RN_{t-d}e^{-cN_{t-d}} + N_{t-1}S. \tag{4.10}$$

Note that the parameter R here is, as previously, the product of fecundity (eggs per adult per year) and the probability of surviving from the egg stage to become an adult in the absence of density dependence. We have written the equation with the left-hand side representing this year's density, rather than next year's as in the equations above, because this simplifies the notation.

Higgins et al. (1997a) analyzed equation 4.10 in detail, by simulating the non-linear dynamics away from the stability boundary. They affirm that, in general, increasing the developmental delay, d, is destabilizing, whereas increasing adult survival is stabilizing in the sense of reducing the amplitude of fluctuations. These authors also extend an interesting result, related to the cycle period, established by Jones et al. (1988). When adults die immediately after breeding, the period is close to $2d$, provided R is not too large. The period increases as adult survival increases, but it is normally never greater than $4d$, though longer periods may be possible in models with extremely large-amplitude cycles (Nisbet and Bence 1989). We encounter similar results in a stage-structured continuous-time model in chapter 5 and use them to some effect in chapter 11.

DISCRETE-GENERATION PARASITOID-HOST MODELS

There is a large literature on simple, discrete-time models of parasitoid-host dynamics. Fortunately, it has recently been well summarized (Hassell 2000), and we discuss it only briefly in this chapter. In addition, Hassell (1978) is still an excellent source of the fundamentals. This theory uses a framework originally attributable to Nicholson and Bailey (Nicholson and Bailey 1935; Bailey et al. 1962) but has also been used to model true predators and prey, and for multi-species interactions (Beddington et al. 1975; Hassell 2000). Here we present no more than an introduction and a few pointers to properties of the models that are relevant later in the book. The formulation of the Nicholson-Bailey model, and of discrete-time models in general, is decribed in Box 4.1.

The Nicholson-Bailey Model

Where H_t and P_t are, respectively, the densities of hosts and parasitoids at the start of year t, R is the host per-head annual rate of increase in the absence of the parasitoid, $f(H_t, P_t)$ is the fraction escaping parasitism in each generation, and each parasitized host gives rise to c adult parasitoids in the next generation, a general model describing host-parasitoid interactions is:

$$H_{t+1} = RH_t f(H_t, P_t)$$

$$P_{t+1} = cH_t[1 - f(H_t, P_t)].$$

(4.11)

A host species usually is vulnerable to a particular parasitoid species at only one stage of the life cycle, say the larval stage (see Fig. 4.5). So the model is commonly taken to describe the dynamics of the vulnerable host stage and the adult parasitoid. Thus, P_t adult parasitoids search and parasitize a fraction, $1 - f$, of the H_t larval hosts. R is the number of new larvae available to be parasitized in the next generation, for each

BOX 4.1.
FORMULATION OF DISCRETE-TIME MODELS, AND
THE CONNECTION BETWEEN THE DISCRETE-TIME
NICHOLSON-BAILEY MODEL AND
CONTINUOUS-TIME LOTKA-VOLTERRA MODEL

FORMULATION OF DISCRETE-TIME MODELS

The updating function $F(N_t)$ in a discrete-time model $N_{t+1} = F(N_t)$ describes all the processes that occur from a fixed date in one time period (1 year) to the next. Discrete-time models are most commonly used in systems such as univoltine insects, in which reproduction occurs during one discrete point during the year. Mortality from parasitism and other causes, however, occurs continuously, at least for some portion of the year. The updating function needs to incorporate the net effect of these sources of mortality.

The best way to formulate the updating function is to write out a continuous-time model that describes the various continuous-mortality sources explicitly, and then integrate the model over the portion of the year in which these processes operate. In some cases, as we will see below, it is possible to obtain an exact function from the integration, which can be used in the models. In other cases it may not be possible to analytically integrate the continuous-time portion of the models, so it is necessary to resort to numerical integration (see "Hybrid Discrete-Time/Continuous-Time Models" below). In those cases it may then be possible to fit a function with an appropriate form to the results of the numerical integration, which can be used in the model.

FORMULATION OF THE DISCRETE-TIME
NICHOLSON-BAILEY MODEL

The Nicholson-Bailey model assumes that hosts and parasitoids have a single generation per year, reproducing at a single point during the year, whereas the Lotka-Volterra model has continuous, overlapping generations, with reproduction occurring at all times during the year.

(Box 4.1. continued)

Unlike the Lotka-Volterra model, where the parasitoids search and attack continuously, the Nicholson-Bailey model has a single function, $f(H_t, P_t)$, that represents the end result of all of the processes that occur during the time the parasitoid is searching for hosts.

We assume the life cycles of host and parasitoid look something like those in Fig. 4.5. Hosts are vulnerable to attack by the parasitoids only during the larval stage, which occurs during the summer. At the end of summer hosts pupate and overwinter in the pupal stage. Adults emerge in the spring, lay eggs, and then die. Adult parasitoids search for hosts during summer and then die. Parasitized hosts enter the parasitoid egg class and mature into parasitoid larvae and then pupae. Parasitoids overwinter in the pupal stage and emerge as adults at the end of the following spring to start the next generation.

We define $H(\tau, t)$ as the density of hosts on day τ within year t. We start our year, $\tau = 0$, at the beginning of the summer, when the larval hosts just enter their vulnerable stage, such that $H(0, t)$ is the density of young host larvae at the start of the summer in year t. $P(\tau, t)$ is the density of parasitoids on day τ within year t, with $P(0, t)$ the density of adult parasitoids at the start of the summer. We assume that the vulnerable portion of the host larval stage during the summer lasts T_S days.

We can model the processes of parasitism and production of new juvenile parasitoids over the summer as a system of differential equations. These differential equations will be integrated, resulting in expressions for the density of hosts and parasitoids at the end of the summer (at $\tau = T_S$) as functions of the densities of hosts and parasitoids at the start of the summer. These functions will then be incorporated into discrete-time models.

We assume that the interaction between the host and parasitoid during the summer is like a Lotka-Volterra interaction, i.e., random encounter between hosts and parasitoids, with an attack rate α. We assume there is no background mortality rate to the hosts or parasitoids during the period of parasitism (these can be included; they just make the integration slightly

(Box 4.1. continued)

more complicated). The continuous-time model is

unparasitized hosts: $\dfrac{dH(\tau, t)}{d\tau} = -\alpha H(\tau, t) P(\tau, t)$ (B4.1a)

adult parasitoids: $\dfrac{dP(\tau, t)}{d\tau} = 0$ (B4.1b)

immature parasitoids: $\dfrac{dI(\tau, t)}{d\tau} = e\alpha H(\tau, t) P(\tau, t)$ (B4.1c)

where $I(\tau, t)$ is the density of immature parasitoids (parasitized hosts) on day τ within year t, and e is the number of parasitoid eggs laid per host.

If we integrate equation B4.1a from day $\tau = 0$ to $\tau = T_S$, with $P(\tau, t) = P(0, t)$, we get $H(T_S, t) = H(0, t)e^{-\alpha P(0,t)T_S}$, and integrating equation B4.1c gives us $I(T_S, t) = eH(0, t)[1 - e^{-\alpha P(0,t)T_S}]$.

The density of hosts at the start of the next year, $H(0, t + 1)$, will be $RH(T_S, t)$, where R includes fecundity and the probability of surviving through all the stages that are not vulnerable to parasitism. Likewise, the density of adult parasitoids at the start of the next year, $P(0, t + 1)$, will be $\sigma I(T_S, t)$ where σ is the survival of the parasitoids through all the juvenile stages. Thus, dropping $\tau = 0$ from the notation and letting a equal αT_S and c equal $e\sigma$ yields the Nicholson-Bailey model:

$$H(t + 1) = RH(t)e^{-aP(t)}$$

$$P(t + 1) = cH(t)(1 - e^{-aP(t)}).$$
     (B4.2)

Finally, we note that by rescaling the parasitoid population density, we can obtain equations that are identical in mathematical form to equation B4.2, but with c equal to 1. This is the form used in the text.

non-parasitized larva present in this generation. R thus combines three components of inter-generational dynamics: density-independent survivorship from larval stage to adult, number of eggs produced per adult, and density-independent survival of these eggs to the larval stage in the next generation, just before parasitism occurs. The model is best

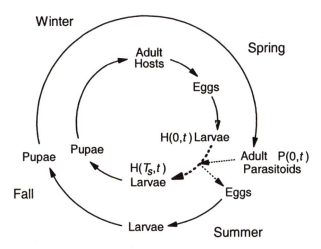

FIGURE 4.5. Typical life cycles of univoltine hosts and parasitoids and their relationship to the models discussed in Box 4.1. Inner ring is the host, outer ring the parasitoid. Heavy dashed line indicates duration of the vulnerable larval stage.

interpreted as following only females, especially since only female parasitoids kill hosts.

This model makes a strong assumption about how events are distributed in time. As we noted for the Ricker model, such discrete-time models take a stroboscopic snapshot at a single point in the generation, whereas the interaction is spread over time within the generation. The approximation is better the more the dynamics throughout the period of interaction are determined mainly by the state of the system at the start.

In the standard Nicholson-Bailey model, the general framework of equation 4.11 is replaced by particular assumptions. As in the Lotka-Volterra model, hosts and parasitoids are assumed to meet at random, and the parasitoid is assumed to have no handling time and a limitless supply of eggs. The final result, derived in Box 4.1, is

$$H_{t+1} = RH_t e^{-aP_t}$$

$$P_{t+1} = cH_t(1 - e^{-aP_t}).$$

(4.12)

EQUILIBRIUM AND STABILITY

We find the equilibria of equation 4.12 by setting H_{t+1} equal to H_t and P_{t+1} equal to P_t to obtain

$$H^* = \frac{R \ln R}{ca(R - 1)} \tag{4.13a}$$

and

$$P^* = \frac{\ln R}{a}. \tag{4.13b}$$

In the appendix, we show that the one-generation time delay causes the Nicholson-Bailey equilibrium to be unstable, in contrast with the neutrally stable equilibrium of the analogous non-delayed Lotka-Volterra model. The populations of host and parasitoid fluctuate through time with cycles of increasing amplitude, implying extinction in the real world (Fig. 4.6a).

STABILIZING PROCESSES

The equilibria in Nicholson-Bailey models can be stabilized by some of the same processes that stabilize the Lotka-Volterra model (chapter 3), but there are some interesting differences. Density-dependent growth in the host has been studied by several authors (e.g., Beddington et al. 1975) by introducing the discrete form of the logistic (equation 4.1) to the host equation and obtaining

$$H_{t+1} = RH_t e^{-(rH_t/K) - aP_t}$$
$$P_{t+1} = H_t(1 - e^{-aP_t}) \tag{4.14}$$

where $r = \ln(R)$ and K is the host carrying capacity.

We show later (in the discussion preceding equation 4.18) that this formulation is problematic and note here just a few points. Host density dependence can induce stable equilibria under some circumstances. Recall that, by contrast, adding the logistic to Lotka-Volterra hosts, which have no time lags, always gives stability. The model also illustrates again the suppression-stability trade-off (May 1976b, chapter 3): for other parameters fixed and r not too high, increasing K increases the difference between H^* and K and hence increases the destabilizing effects of the parasitoid.

The host density dependence can also convert the ever-increasing oscillations of the Nicholson-Bailey model to stable consumer-resource

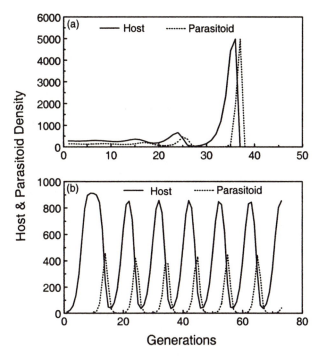

FIGURE 4.6. (a) Basic Nicholson-Bailey model (equation 4.12). A small perturbation from equilibrium at time 0 increases with time until first the host and
then the parasitoid population goes extinct. In this run $R = 2$, $a = 0.005$, and
$c = 1$. (b) Nicholson-Bailey model with Ricker (logistic) density dependence
in the host rate of increase (equation 4.14). Perturbations from equilibrium are
captured by density dependence, and the populations move toward a stable
consumer-resource limit cycle. In this run $R = 2.5$, $a = 0.005$, $r = 0.916$,
and $K = 916$. The period is approximately 10 generations.

limit cycles. Fig. 4.6b gives an example. The limit cycles have a period
much longer than the basic 2-year period in the single-species Ricker
model (compare Figs. 4.3 and 4.6b). We show in the appendix to this
chapter that the period of the cycles is typically close to or greater
than 6 (Lauwerier and Metz 1986; Murdoch et al. 2002). We discuss
the importance of this period for interpreting field data in chapter 11.

Density dependence in the parasitoid is also stabilizing, and
strongly so, as in Lotka-Volterra models. Density dependence in the
parasitoid attack rate is typically introduced by assuming that hosts
vary in their susceptibility to attack; this implies that the constant a

is replaced with a distribution that describes variation in vulnerability among hosts. This idea was first explored by Bailey et al. (1962), who suggested that attacks could have a gamma distribution among individuals. May (1978) noted that vulnerability would have a negative binomial distribution among hosts if they occur in patches, parasitoids within a patch encounter hosts at random (as in the Nicholson-Bailey model), and the number of parasitoids per patch has a gamma distribution. The parameter k describes the shape of the negative binomial distribution by defining the degree to which attacks are concentrated mainly in a small fraction of the population (the degree of aggregation of attacks increases as k decreases). The fraction escaping attack is now the probability of zero attacks in the negative binomial distribution. So the model becomes (May 1978; see also Griffiths 1969)

$$H_{t+1} = RH_t \left(1 + \frac{aP_t}{k}\right)^{-k}$$

$$P_{t+1} = H_t \left[1 - \left(1 + \frac{aP_t}{k}\right)^{-k}\right],$$

(4.15)

which has the nice property that the equilibrium is locally stable if k is less than 1. Interestingly, in this model we either get a stable equilibrium or fluctuations that increase with time—there are never stable limit cycles.

The stability in this, and related models, has been linked to the coefficient of variation (CV) of the distribution of parasitoids among patches (Chesson and Murdoch 1986; Hassell et al. 1991b). The variance of the gamma distribution, with mean equal to 1, is $1/k$, which leads to the result that the CV of the negative binomial distribution is also $1/k$. Since stability requires k less than 1, the equilibrium is stable if CV (or CV^2) is more than 1; i.e., if the distribution of attacks is highly skewed among hosts, with a small fraction receiving most of the attacks. The equilibrium is stable when k is less than 1 because as parasitoid density increases, the distribution of attacks becomes more and more skewed. As a consequence, each individual parasitoid becomes less and less efficient, thus introducing density dependence in the parasitoid's rates of attack and per-head recruitment (Murdoch 1992; Hassell 2000). We discuss this topic further in chapter 10.

In this model there is an especially powerful trade-off between stability and suppression of the host by the parasitoid (see chapter 3). May (1978) pointed out that the host equilibrium density increases faster than exponentially as k decreases.

Refuges can also stabilize the equilibrium in the Nicholson-Bailey model. A constant density, H_r, in the refuge leads to the model

$$H_{t+1} = RH_t + R(H_t - H_r)e^{-aP_t}$$
$$P_{t+1} = (H_t - H_r)(1 - e^{aP_t}). \qquad (4.16)$$

Hassell (1978, Fig. 4.7b) shows that this type of refuge is strongly stabilizing.

A constant fraction in the refuge is also stabilizing, in contrast with the Lotka-Volterra model where a constant-fraction refuge has no effect on stability. Hassell (1978; see also 2000) analyzed a model in which a fraction, v, is outside the refuge:

$$H_{t+1} = R(1 - v)H_t + RvH_t e^{-aP_t}$$
$$P_{t+1} = vH_t(1 - e^{-aP_t}). \qquad (4.17)$$

Hassell (1978, Fig. 4.7a) showed that the region of parameter space with stable equilibria is quite small and that stability is lost if the proportion in the refuge is either too small or too large. A sufficiently large refuge leads to exponential growth of H and P and a lack of equilibrium. A small refuge, however, does stabilize the divergent Nicholson-Bailey oscillations, leading to stable limit cycles (Gurney and Nisbet 1998, Fig. 6.3) or a stable equilibrium.

An early analysis concluded that a type 3 functional response cannot stabilize the Nicholson-Bailey equilibrium (Hassell and Comins 1978), though such a response can stabilize the Lotka-Volterra equilibrium (chapter 3). This conclusion, however, arises because the functional response (the number attacked per parasitoid per unit time) in that study was assumed to depend on the *initial* number of hosts available in the generation (i.e., on H_t). In a real system, of course, the number of hosts available declines throughout the vulnerable period as parasitoid attacks accumulate. This requires us to define an instantaneous functional response and then integrate it over the vulnerable period, using the approach outlined in Box 4.1. R. M. Nisbet (unpublished) uses this approach to show that a type 3 functional response can in fact stabilize the equilibrium in a Nicholson-Bailey model.

DESTABILIZING PROCESSES

The inherent time lag in the Nicholson-Bailey model is destabilizing. The other main destabilizing process, a type 2 functional response, has an effect similar to that in Lotka-Volterra models (Hassell 1978).

EFFECTS OF ENRICHMENT

Enrichment has been taken to include increases in both the carrying capacity and the per-head rate of increase of the prey population. If we increase r in the Lotka-Volterra model, or either r or K in the continuous-time Rosenzweig-MacArthur model, we get an increase in the predator, not in the prey, population (chapter 3). The Nicholson-Bailey model is fundamentally different. The host and parasitoid equilibria in the basic model are, as noted in equations 4.13a and b, $H^* = R \ln R/(ca(R-1))$ and $P^* = \ln R/a$. If we increase the host's intrinsic rate of increase, R, both the host and the parasitoid equilibria increase.

The difference arises for the following reason. In Lotka-Volterra-like models, the predator responds continuously to a continuously changing prey population. If prey productivity increases, the prey increases temporarily, but each predator eats faster, predators increase in density, and prey density is suppressed back to the original equilibrium. By contrast, in Nicholson-Bailey models a fixed number of hosts is available at the start of the generation, and the number of parasitoids does not change through the generation. So a fixed number of parasitoids divide up the fixed number of hosts available at the start of the generation. An increase in host productivity (R in 4.13a and b) increases the number of hosts available, which can support more parasitoids.

Remarkably, this fundamental difference between continuous- and discrete-time models has not been discussed, as far as we know. Yet it is relevant to a central problem in ecology, namely how the abundance of different trophic levels should change in response to enrichment (Oksanen et al. 1981).

In at least one class of populations the response to enrichment is as predicted by Lotka-Volterra-type models. In the warm season in freshwater lakes, edible algae (the prey) and zooplankton such as *Daphnia* (the predator) have continuous births and deaths as assumed by Lotka-Volterra models. As far as can be detected, the abundances of prey and predator change as predicted: edible algal biomass increases not at all, or only weakly, with enrichment, whereas zooplankton biomass does increase (chapter 3 and Watson et al. 1992).

It does not appear to be the case in general, however, that observed changes in different trophic levels under enrichment are those predicted by Lotka-Volterra models. Yet it is also not clear that the observed changes are those predicted by competing models that assume that consumer attack rate depends on the ratio of consumers to prey (Micheli 1999). Hulot et al. (2000) present evidence that the outcome is affected by indirect effects, and this may be a general explanation for the widespread non-fit to either of the two competing classes of models. But differences between continuous and discrete breeders may also play a role where both types of species are in the same community.

The idea that, over some long time period, predators or parasitoids divide up an initially available number of prey or hosts seems to be the basis for "ratio-dependent" models (Ginzburg and Akcakaya 1992). In these models, reviewed by Abrams and Ginzburg (2000), the number of prey eaten per predator (the functional response) is determined by the ratio of prey to predators rather than by the density of prey. Such ratio dependence sometimes emerges in a natural way in a discrete-time framework, and derivations analogous to that in Box 4.1 can support the hypothesis in this context. We do not believe it makes ecological sense, however, to insert ratio-dependent functional responses into continuous-time models which, by their nature, describe instantaneous rates, except in circumstances where there is an explicit mechanism supporting the assumption. The best available approximation to predators that interfere with each other, while searching for food, is the predator-dependent functional response (equation 3.13).

Multiple Equilibria and Single-Species Dynamics in a Parasitoid-Host Model

Here we illustrate how a wide range of dynamics can emerge in a consumer-resource model when the interaction between, in this case, parasitoid and host is partially decoupled. We do this by incorporating Ricker host density dependence in the Nicholson-Bailey model. May et al. (1981) noted that equation 4.14, although commonly used, makes a particular, and in many cases unlikely, assumption (see also Hassell 2000). It assumes H_t hosts are subject to attack but that density dependence in the survivors of parasitism operates on the initial H_t hosts. If we instead assume that H_t adult hosts reproduce at rate R, that survival to the larval stage is density dependent, and that the surviving larvae are attacked by the parasitoid, we get the more realistic model

$$
\begin{aligned}
H_{t+1} &= R H_t e^{-(r H_t/K)-a P_t} \\
P_{t+1} &= R H_t e^{-r H_t/K}(1 - e^{-a P_t}),
\end{aligned}
\tag{4.18}
$$

which is a slight variant of Model 2 from May et al. (1981).

The dynamics of this system are remarkably complex but are much like those of a similar model first investigated by Neubert and Kot (1992); we refer the reader to that paper, and to Kot (2001) for details of the mathematics. Here we illustrate a few features from the large repertoire of dynamical behavior with the aid of a set of figures (Fig. 4.7), focusing on the effects of variation in K, the resource carrying capacity.

- For large values of K, the divergent oscillations of the Nicholson-Bailey model are replaced with consumer-resource limit cycles (Fig. 4.7a), consumer extinction (Fig. 4.7b), or complex, possibly chaotic, dynamics (Fig. 4.7c). These patterns are generally independent of initial conditions.
- For intermediate values of K, there is the possibility of a locally stable equilibrium (Fig. 4.7d). The equilibrium is seldom (and perhaps never) globally stable, however, and consumer extinction is a possibility with some initial conditions (Fig. 4.7e). Global stability means that population densities return to the same equilibrium from all initial values. In Figs. 4.7d and e, we see that with the same parameter values, one set of initial densities leads to persistence of the consumer whereas another set (fewer initial parasitoids) leads to consumer extinction.
- With sufficiently small K, extinction of the consumer is inevitable, and the host exhibits dynamics similar to those observed for a single species (Fig. 4.7f).

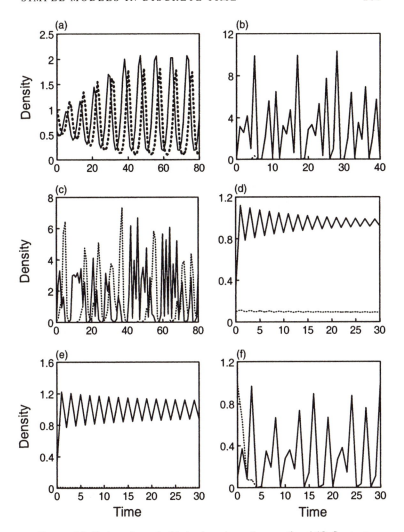

FIGURE 4.7. Various dynamical behaviors shown by equation 4.18. See text for details. Dashed line is consumer, solid line is resource. (a) Limit cycle with $r = 1, K = 3$. (b) Consumer extinction with $r = 3.5, K = 3$. (c) Complex dynamics with $r = 3$, $K = 3$. (d) Locally stable equilibrium with $r = 2, K = 1$. (e) Consumer extinction with $r = 2, K = 1$, but different initial conditions. (f) Consumer extinction and single-species-like resource dynamics with $r = 3.5, K = 0.3$.

The effect of variation in resource carrying capacity differs in several interesting ways from the effects of enrichment in continuous-time models discussed in chapter 3. The principal insights from that chapter survive: decreasing K has a stabilizing effect, and with sufficiently small values of K, the system cannot support a persisting consumer population. However, two new features emerge—the possibility of complex consumer-resource dynamics and the possibility of consumer extinction with even quite large values of K. The complex dynamics arise because the model incorporates two mechanisms capable of generating cycles: the resource density dependence that can cause cycles or chaos in single-species models via delayed feedback, and the consumer-resource interaction. It is harder to come up with a simple intuitive argument for the wide range of parameter values and initial conditions that allow for consumer extinction even with quite large values of K, but our findings with this model anticipate a property of many of the stage-structured population models discussed in later chapters, namely that a species may be unable to invade and establish itself in an environment where it can persist if previously established.

HYBRID DISCRETE-TIME/CONTINUOUS-TIME MODELS

Discrete-time models are appropriate for interactions in which the host and parasitoids reproduce once per year. The processes of searching for hosts, parasitism, and mortality, however, occur continuously during at least some portion of the year and are best represented by continuous-time models. In Box 4.1 we show that in some circumstances it is possible to integrate the equations that describe the continuous-time dynamics, over the course of a year, and we obtain an expression that gives the densities of hosts and parasitoids at the end of the year as a function of only the densities at the start of the year. In many other cases it is not possible to perform this integration analytically. In those cases it is possible to do numerical simulations of a hybrid discrete-time, continuous-time model.

As an example, consider the situation in which the host has a single generation per year but the parasitoid can potentially go through several generations within a single host generation (Briggs and Godfray 1996).

How does this affect the stability of the interaction? To model this we divide the model into a within-season and a between-season component. The life-cycle of the host is similar to that in Fig. 4.5. We define $H(\tau, t)$ and $P(\tau, t)$ as the densities of hosts and adult parasitoids on day τ during year t. $I(\tau, t)$ is the density of immature parasitoids (or parasitized hosts) on day τ within year t. We assume that hosts are in the vulnerable larval stage during the summer, so we start the year, $\tau = 0$, at the beginning of the summer, when the hosts just become vulnerable to parasitism. The within-season portion of the model includes the processes of parasitism, mortality, and production of new parasitoids:

$$\frac{dH(\tau, t)}{d\tau} = -\phi[P(\tau, t)]H(\tau, t) - \delta_L H(\tau, t)$$

$$\frac{dI(\tau, t)}{d\tau} = e\phi[P(\tau, t)]H(\tau, t) - M_I(\tau, t) - \delta_I I(\tau, t) \qquad (4.19)$$

$$\frac{dP(\tau, t)}{d\tau} = M_I(\tau, t) - \delta_P P(\tau, t)$$

where $\phi[P(\tau, t)]$ is the instantaneous per-head death rate of hosts caused by parasitoids. δ_L, δ_I, and δ_P are the constant instantaneous per-head background death rates of larval hosts, immature parasitoids, and adult parasitoids, respectively. e is the number of parasitoid eggs laid per host.

$M_I(\tau, t)$ is the rate of maturation of immature parasitoids into the adult parasitoid class on day τ within year t. We assume that immature parasitoids take a fixed T_P days to develop into adults, so the maturation rate out of the immature parasitoid stage is equal to the rate of production of immature parasitoids T_P days ago, multiplied by the probability of surviving through the stage:

$$M_I(\tau, t) = e\phi[P(\tau - T_P, t)]H(\tau - T_P, t) \exp\{-\delta_I T_P\}. \qquad (4.20)$$

An implication of this equation is that the maturation rate at time t is determined by the host density at an earlier time. Mathematically, the system of equations 4.19 thus involves time delays. Such equations are discussed in detail in chapter 5.

Each year the within-season portion of the model is integrated numerically from the start of the summer season, $\tau = 0$, to the end of the summer season, $\tau = T_S$. $T_S = T_L + T_P$, where T_L is the length of the vulnerable portion of the larval host stage during the summer, and the additional T_P days allow all of the immature parasitoids to mature into the adult parasitoid stage.

We assume that $\phi[P(\tau, t)]$ is an increasing, but saturating, function of parasitoid density:

$$\phi[P(\tau, t)] = k \ln \left[1 + \frac{aP(\tau, t)}{k} \right] \quad \text{if } 0 \le \tau \le T_L;$$

$$\phi[P(\tau, t)] = 0 \qquad\qquad\qquad \text{otherwise.}$$

(4.21)

This is equivalent to assuming that the per-parasitoid efficiency of parasitoids decreases with increasing parasitoid density, that is, there is density dependence in the parasitoid attack rate which increases as k decreases, as in equation 4.15. At k equals infinity, the function becomes a linearly increasing function of parasitoid density, $\phi[P(\tau, t)] = aP(\tau, t)$, and we recapture the assumption of random encounters between hosts and parasitoids.

The between-season component of the model consists of three discrete-time equations, mapping the densities at the end of one season onto the densities at the start of the next season:

$$H(0, t + 1) = FH(T_S, t)\sigma_H$$

$$I(0, t + 1) = 0$$

(4.22)

$$P(0, t + 1) = P(T_S, t)\sigma_P,$$

where F is the per-head density-independent fecundity of hosts and σ_H and σ_P are the survival probabilities of hosts and parasitoids, respectively, during all other seasons of the year.

In the limit, when the development time of parasitoids is longer than the time the hosts are vulnerable to parasitism ($T_P > T_L$), the parasitoids cannot go through more than one generation per host generation. The parasitoid offspring from attacks on hosts at the start of the summer will not enter the adult stage until after the larval hosts are no longer vulnerable. In this limiting case, the model can be expressed as a discrete-time system of difference equations, similar to equation 4.15. With only one parasitoid generation per year, the equilibrium is stable only if there is

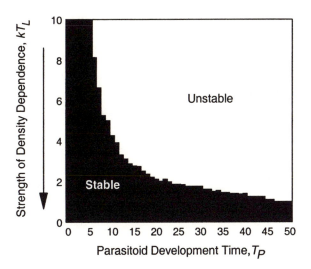

FIGURE 4.8. Region of local stability (shaded region) for the hybrid discrete-time/continuous-time model. Divergent oscillations occur outside this region. $T_L = 50$ days, $F = 2$, $\delta_P = 10^{-3}$, $\sigma_H = 1$, $\sigma_P = 0.265$, $e = 1$, $a = 0.01$/day, $\delta_L = \delta_I = 0$. © Elsevier/Academic Press, reproduced with permission from Briggs and Godfray 1996.

relatively strong density dependence in the parasitoid attack rate, such that $kT_L < 1$.

When T_P is reduced below T_L, the parasitoids can go through more than one generation within a single host generation. Attacks on hosts at the start of the summer can result in new adult parasitoids that can attack hosts later in the summer. This has a stabilizing effect on the equilibrium (Fig. 4.8). With a low T_P, less density dependence in the parasitoid attack rate is required to stabilize the equilibrium. This is presumably because the parasitoid population can respond faster to changes in the host density, suppressing potential host outbreaks more rapidly. (The stabilizing effect of a shorter parasitoid generation time is discussed again in chapter 9 in relation to biological control.)

This model illustrates an additional point that relates to the coherence of consumer-resource theory (chapter 11). It was first developed for the interaction between a virus disease and a host insect (Briggs and Godfray 1996). In the original presentation of the model, P_t was the density of free-living virus particles in the environment. They die off exponentially over time, and host larvae are infected by encountering virus particles in the environment. I_t was the density of infected host larvae and, as with parasitism, it was assumed that infection leads to host death, as is typically the case. Viruses increase in the host body and are released into the environment when the host dies, and this production rate per host is captured in the parameter e. Fig. 4.8 defines stability for a parameter set suitable for a virus-host interaction.

Local Stability Conditions and Cycle Periods for Systems of Difference Equations

This appendix outlines the methodology for stability analyses of equilibria in systems of difference equations with one and two variables. We include explicit calculations for the two major models presented in chapter 4—the Ricker and Nicholson-Bailey models. The appendix concludes with a brief analysis of the period of consumer-resource cycles near local stability boundaries, including derivation and discussion of some circumstances that lead to the general conclusion that the period will commonly be greater than 6 time steps. This result is required for chapter 11.

The texts cited in the chapter 3 appendix remain useful for the material presented here (e.g., Edelstein-Keshet 1988; Murray 1993; Hastings 1997; Adler 1998; Gurney and Nisbet 1998; Case 2000; Kot 2001).

SINGLE-SPECIES MODELS

The general form of difference equation discussed in this chapter is

$$N_{t+1} = F(N_t). \tag{A4.1}$$

Assume there is a non-zero equilibrium state N^* at which

$$N^* = F(N^*). \tag{A4.2}$$

Defining small perturbations from equilibrium by $n_t = N_t - N^*$, and linearizing about the equilibrium in a manner exactly analogous to that used in the appendixes to chapters 2 and 3, yields the equation

$$\underbrace{N^*} + n_{t+1} \approx \underbrace{F(N^*)} + n_t \frac{dF}{dN_t}\bigg|_* . \tag{A4.3}$$

The terms indicated by the braces cancel because of the equilibrium condition (A4.2). Consequently, the perturbations obey the linear dynamic equation

$$n_{t+1} = \mu n_t, \tag{A4.4}$$

with

$$\mu = \frac{dF}{dN_t}\bigg|_* . \tag{A4.5}$$

This quantity, like the related quantity λ in the analysis of continuous-time models, is sometimes called an eigenvalue.

Equation A4.4 is identical in form to equation 4.3 in the text and describes geometric growth, except that the multiplier R is replaced by the quantity μ from equation A4.5. However, μ can be positive or negative, depending on the slope of the graph of $F(N_t)$ at the equilibrium. This difference is important, as there are now four possible general forms for the dynamics:

1. $\mu > 1$:   geometric divergence from the equilibrium
2. $0 < \mu < 1$:   geometric approach to equilibrium
3. $-1 < \mu < 0$:   approach to equilibrium via damped oscillations in which the population overshoots and undershoots the equilibrium in alternate years
4. $\mu < -1$:   divergent oscillations away from equilibrium.

Options 2 and 3 correspond to a stable equilibrium.

Application to the Ricker Model

For the Ricker model in the form of equation 4.2,

$$F(N_t) = RN_t e^{-cN_t} \tag{A4.6}$$

and the equilibrium is

$$N^* = \frac{\ln(R)}{c}. \tag{A4.7}$$

The eigenvalue μ is computed by differentiating equation A4.6 to obtain

$$\frac{dF}{dN_t} = R e^{-cN_t}(1 - cN_t) = 1 - \ln(R) \tag{A4.8}$$

at equilibrium, the second equality involving use of equation A4.7 and remembering that $R e^{-cN_t}$ is equal to 1 at equilibrium.

TWO-VARIABLE MODELS

Discrete-time models of two interacting (unstructured) populations can be written in the form

$$H_{t+1} = F(H_t, P_t), \qquad P_{t+1} = G(H_t, P_t), \tag{A4.9}$$

and their equilibria are obtained by solving simultaneously the algebraic equations

$$H^* = F(H^*, P^*), \qquad P^* = G(H^*, P^*). \tag{A4.10}$$

Algebra closely analogous to that used for differential equations in chapter 3 shows that small deviations, h and p, from equilibrium satisfy the difference equations

$$h_{t+1} = A_{11}h_t + A_{12}p_t$$
$$p_{t+1} = A_{21}h_t + A_{22}p_t \tag{A4.11}$$

where, as before, the recipe for calculating the four elements of the Jacobian matrix is

$$A_{11} = \left.\frac{\partial F}{\partial H_t}\right|_*, \quad A_{12} = \left.\frac{\partial F}{\partial P_t}\right|_*, \quad A_{21} = \left.\frac{\partial G}{\partial H_t}\right|_*, \quad A_{22} = \left.\frac{\partial G}{\partial P_t}\right|_*. \tag{A4.12}$$

Like their continuous-time counterparts, equations A4.11 are linear, and it is known that their solution can be written in the form

$$h_t = C_1\mu_1^t + C_2\mu_2^t$$
$$p_t = D_1\mu_1^t + D_2\mu_2^t \tag{A4.13}$$

where C_1, C_2, D_1, and D_2 are constants whose values are determined by the initial conditions (i.e., the values of h and p at time $t = 0$), and μ_1 and μ_2 are the two eigenvalues of the Jacobian matrix. Note that we follow the convention used by Gurney and Nisbet (1998) that the eigenvalues of Jacobian matrixes are represented by λ in continuous-time models and by μ for discrete time. We make the distinction because the ecological interpretation and the relation to stability are different in the two contexts—loosely, one is a geometric multiplier, the other a coefficient in an exponential. Nevertheless, the mathematics involved in calculating the eigenvalues are the same as in chapter 3; thus the eigenvalues are the roots of the characteristic equation

$$\mu^2 + B_1\mu + B_2 = 0, \tag{A4.14}$$

where

$$B_1 = -(A_{11} + A_{22}) \quad \text{and} \quad B_2 = A_{11}A_{22} - A_{12}A_{21}. \tag{A4.15}$$

The quadratic equation A4.14 has either two real roots or a pair of complex conjugate roots. If the roots are real, then the perturbations decay to zero if and only if both roots have values between -1 and $+1$. If the roots are complex, they can be written in the form

$$\mu_1 = \rho\,e^{i\gamma}; \quad \mu_2 = \rho\,e^{-i\gamma} \tag{A4.16}$$

where ρ is the magnitude of the roots and their arguments are $\pm\gamma$. The equilibrium is stable if ρ is less than 1.

The argument, γ, of complex roots is of special interest as it contains information about the period of oscillations converging toward, or diverging from, the equilibrium. This period is $2\pi/\gamma$.

As with systems of differential equations, there are conditions for determining the stability without resorting to explicit solution of the quadratic equation. These are derived in several of the texts cited above (e.g., Kot 2001, pp. 184–86), the end result being that for stability we require

$$1 - B_1 + B_2 > 0; \quad 1 + B_1 + B_2 > 0; \quad B_2 < 1. \tag{A4.17}$$

Application to Nicholson-Bailey Model

For the Nicholson-Bailey model with $c = 1$, the Jacobian matrix elements have the form

$$A_{11} = 1$$

$$A_{12} = \frac{-R\ln(R)}{R - 1}$$

$$A_{21} = \frac{R - 1}{R} \tag{A4.18}$$

$$A_{22} = \frac{\ln(R)}{R - 1}.$$

Using equations A4.15 and performing a little algebra leads to

$$B_1 = -\left[1 + \frac{\ln(R)}{R - 1}\right]; \qquad B_2 = \frac{R\ln(R)}{R - 1}. \tag{A4.19}$$

Using these equations, it is straightforward to verify that the first two inequalities of A4.17 hold provided R is greater than 1, but the third is tricky. The algebraically proficient can prove with a little effort that in fact B_2 is greater than 1 for all R greater than 1, implying instability of the equilibrium (for a proof, see Kot 2001, pp. 194–95). The reader uninterested in the algebraic niceties can be convinced of the result by simply using A4.19 to construct a plot of B_2 against R for the ecologically relevant range of R values (0–10).

CALCULATING STABILITY BOUNDARIES

Two-dimensional difference equations can exhibit a bewildering variety of mathematical bifurcations, discussion of which goes beyond the scope of this appendix. Chapter 11 of Kot 2001 provides a good introduction. When calculating stability boundaries associated with the onset of consumer-resource cycles, however, it is useful to note that in such models, the onset of instability is usually associated with a transition from damped to divergent cycles (i.e., $\rho = 1$ at the boundary, implying $\mu = e^{\pm i\gamma}$). One useful property of the quadratic equation A4.14 is that the sum of the roots is $-B_1$ and the product of the roots is B_2. Thus on the stability boundary

$$B_2 = e^{i\gamma} e^{-i\gamma} = 1 \quad \text{and} \quad -B_1 = e^{i\gamma} + e^{-i\gamma} = 2\cos\gamma. \quad (A4.20)$$

The period on the stability boundary is calculated using the second of these equations:

$$Period = \frac{2\pi}{\gamma} \quad \text{where } \gamma = \cos^{-1}\left(\frac{-B_1}{2}\right). \quad (A4.21)$$

In summary, to compute stability boundaries and cycle periods on these boundaries, we proceed in a manner analogous to that discussed in the appendix to chapter 3 for differential equations.

1. If the equilibrium can be calculated analytically, do so. Then evaluate the Jacobian matrix. Calculate the coefficients in the characteristic equation and compute the stability boundary and periods using equations A4.20 and equation A4.21.
2. If the equilibrium cannot be calculated analytically, evaluate the Jacobian matrix and the coefficients of the characteristic equation in terms of H^* and P^*. Then work with three simultaneous algebraic equations (the two equilibrium conditions [A4.10] and the left-hand equations of A4.20). If the stability boundary is to be drawn in a plane whose axes represent two model parameters, we now have three equations in four unknowns (H^*, P^*, and the two chosen parameters). These define a curve in the relevant plane. Then compute γ and hence the cycle periods using equation A4.21.

PERIODS OF CONSUMER-RESOURCE CYCLES

Here we justify the claim that under a wide range of circumstances, the period of consumer-resource cycles will exceed 6 time units. Our starting point is equation 4.11 with c equal to 1, a simplification that leads to no loss of generality:

$$H_{t+1} = RH_t f(H_t, P_t)$$
$$P_{t+1} = H_t(1 - f(H_t, P_t)). \quad (A4.22)$$

Equilibria are solutions of

$$Rf(H^*, P^*) = 1; \qquad P^* = H^* \frac{R-1}{R}. \qquad \text{(A4.23)}$$

Stability matrix elements are

$$A_{11} = Rf(H^*, P^*) + RH^* f_H(H^*, P^*)$$

$$= 1 + RH^* f_H(H^*, P^*) \qquad \text{(A4.24a)}$$

$$A_{12} = RH^* f_P(H^*, P^*) \qquad \text{(A4.24b)}$$

$$A_{21} = (1 - f(H^*, P^*)) - H^* f_H(H^*, P^*)$$

$$= \left(\frac{R-1}{R}\right) - H^* f_H(H^*, P^*) \qquad \text{(A4.24c)}$$

$$A_{22} = -H^* f_P(H^*, P^*) \qquad \text{(A4.24d)}$$

where subscripts denote partial derivatives. For example, $f_H(H, P) = \partial f(H, P)/\partial H$.
   Now define

$$\alpha = -H^* f_P(H^*, P^*); \qquad \beta = H^* f_H(H^*, P^*). \qquad \text{(A4.25)}$$

These definitions are chosen so that "typically" we expect both α and β to be positive, as host survival should decrease as parasitoids increase, and increase as hosts increase, the principal exception being with a type 3 functional response (see Fig. 3.3).
   Equations A4.24 can then be written in the less intimidating form:

$$A_{11} = 1 + R\beta \qquad \text{(A4.26a)}$$

$$A_{12} = -R\alpha \qquad \text{(A4.26b)}$$

$$A_{21} = \frac{R-1}{R} - \beta. \qquad \text{(A4.26c)}$$

$$A_{22} = \alpha \qquad \text{(A4.26d)}$$

The characteristic equation (see equation A4.14) is

$$\mu^2 + B_1\mu + B_2 = 0 \tag{A4.27}$$

with

$$B_1 = -1 - \alpha - R\beta \quad \text{and} \quad B_2 = R\alpha. \tag{A4.28}$$

The transition to oscillations occurs when B_2 is equal to 1 (equation A4.20), so on the stability boundary the characteristic equation is

$$\mu^2 - (1 + \Omega)\mu + 1 = 0 \tag{A4.29}$$

where

$$\Omega = \alpha + R\beta > 0. \tag{A4.30}$$

The period of cycles on the stability boundary is given by

$$Period = \frac{2\pi}{\cos^{-1}\left(\dfrac{1 + \Omega}{2}\right)} \tag{A4.31}$$

with $0 < \Omega < 1$. When Ω is equal to zero the period is 6 (since $\cos^{-1}(1/2) = \pi/3$). The period is always greater than 6 if Ω is greater than zero, which is guaranteed if α and β are positive.

CHAPTER FIVE

An Introduction to Models with Stage Structure

One of our professors used to say that every biologist should visit the tropics if only to be appalled by the multiplicity of species. But one needn't go to the tropics. Insects almost everywhere are appallingly diverse. And they exhibit a huge range of life histories and lifestyles. This variety increases enormously when we come to consider interactions between insect species, and in particular between insect hosts and their parasitoids. We look at this variation systematically in the next chapter. Here we content ourselves with stating the goals that motivate the models presented in the next four chapters, which together form a unit.

We have two main questions. First, when and how do differences among individual life stages matter to population dynamics? In particular, parasitoids respond differently to different life stages of their hosts, and we want to know whether this has dynamic consequences. Second, even though there is a huge variety of life histories and lifestyles, when can we collapse these differences to mere variants of a fundamental underlying structure, thereby establishing broad generality?

We begin by exploring a model that recognizes perhaps the major difference among life stages of hosts and parasitoids—that between immature and adult stages—and we look at the consequences of the fact that typically one of these host stages is invulnerable to attack. Once we have this model, which we will refer to throughout the rest of the book as the Basic model, we will return to some of the processes from chapter 3 that affect the stability of the Lotka-Volterra model and ask if the insights gained there carry over to this stage-structured model.

The unstructured models in chapter 3 assume all host individuals have identical vital rates. This is obviously false. The most obvious differences are between the immature stages, which cannot reproduce, and adult stages, which can. Immature stages are also usually much more prone to attack by parasitoids than are adults; although there are parasitoids that attack adult hosts, the great majority attack only eggs or immatures. The population dynamics produced by this most common pattern are the main focus of this chapter. The model we develop treats the existence of distinct developmental stages more realistically than the previous chapters.

The best-known approach to modeling stage-structured populations is a generalization of the matrix models used in demography (Caswell 2001). The "projection matrix" used to update the population's age structure at each successive time takes a form that describes not only fecundity and mortality but also maturation from one stage to its successor. Although powerful and simple to use, matrix models employ discrete time steps, so the approach does not work well in truly continuous-time systems with overlapping generations and stages of unequal duration. The approach becomes intractable in situations where stage durations vary over time. Finally, there have been only a few attempts at describing the interaction between consumer and resource populations using matrix models (e.g., Kittlein 1997; Jensen and Miller 2001).

In this chapter, our primary tool for modeling stage-structured inter-actions is a continuous-time, lumped age-class technique introduced by R. M. Nisbet, W.S.C. Gurney, and co-workers (Gurney et al. 1983; Nisbet and Gurney 1983); for a review see Nisbet 1997. In this approach it is assumed that the organism's life history falls into a series of discrete developmental stages, and that all individuals within a stage have the same properties (e.g., feeding rate, fecundity, risk of death) at any given time. Real individuals within a stage of course differ, but we assume these differences are less important than dif-ferences among stages. A more general approach involving partial differential equations (de Roos 1997) can handle within-stage vari-ability, and in chapter 6 we show that our major results from this chapter are robust in the sense that relaxation of some assumptions about similarity within the stage has only a small effect on the model predictions.

The properties of insects can of course change drastically within the immature period; for example, eggs and pupae are ecologically very different from larvae. We initially ignore such differences but investigate their effects in the later part of the chapter. Adult insects do not grow, so adults tend to be relatively uniform. In addition, we deal here with populations that at least have the potential to develop overlapping generations. This is not a good framework, for example, for species that breed only once per year. Hybrid models that combine a continuous stage structure within the season, with discrete bouts of reproduction between seasons, have been developed to deal with that situation (chapter 4).

Finally, although the consumer-resource model discussed here was motivated by insect parasitoids and their hosts, the general structure has been used for true predators and prey (Hastings 1984; also see chapter 11). Indeed, the Basic model, by introducing stage structure, can be thought of as a step closer to real predator-prey interactions than is the Lotka-Volterra model.

PREAMBLE: SINGLE-SPECIES POPULATIONS WITH STAGE STRUCTURE

Stage structure in combination with *intraspecific* density dependence has effects independent of interaction with another species. We therefore explore these effects in single-species models before asking how stage structure affects parasitoid-host interactions. Single-species models also let us demonstrate without too much algebra that stage-structured models sometimes can predict real dynamics quite accurately.

Thinking about stage structure forces us to think about the species' ecology. Suppose we want to distinguish an immature and an adult stage, and we want to ask how the dynamics are affected by density dependence. As in the last chapter, we use density dependence as a proxy for the dynamics of the host's resources. So we need to ask: what are the resources of the two stages, and will an increase in a stage's density reduce its resource and thereby feed back on the stage's vital rates?

We investigate two major patterns seen in insects. (1) There are some species, in which both stages feed and have different resources, where the adult population may be more likely than the larval population to be limited by its food supply. For example, the larvae of some weevils eat leaves and the adults eat seeds and fruit, which might be in shorter supply. (2) In many species the larval food supply is more likely to be limiting. Caterpillars, for example, mainly eat leaves whereas adult moths and butterflies may be non-feeding or feed on nectar. It seems in many of these species that the larval stage will more likely suppress its food supply than will the adult, so that larval density will affect larval survival whereas adult mortality and fecundity will be independent of adult density. More complex scenarios can occur; for example, larval density can affect eventual adult size and hence adult fecundity.

We do not investigate in detail here a third possible case, which may occur in non-metamorphosing species such as locusts, in which adults and larvae share the same food supply. An increase in either stage could reduce the uptake of food by the entire population. This case is discussed in our model of *Daphnia* and algae in chapter 11.

Delayed (Between-Generation) Density Dependence in Fecundity

We illustrate the approach with a model of a famous experiment on blowflies by Nicholson (1957). These are among the most thoroughly analyzed population time series in ecology, and their dynamics are well understood. The number of adults in the population shows large-amplitude cycles with a period of about 38 days (Fig. 5.1a). Details and references are in Kendall et al. 1999.

Nicholson was interested in the dynamical effects of larval competition versus adult competition for food. In the experiment, he raised populations of blowflies by placing adults emerging from the immature population cage on to a fixed amount of meat. The eggs laid by these adults between each 2-day census were transferred to another population cage in which food for the larvae was greatly in excess. Larvae were then reared in even-aged cohorts with no competition. The adults competed for food, and fecundity declined exponentially with adult density (Fig. 5.1b), as in the Ricker model (chapter 4). Adults died off at a roughly constant rate of $d_A = 0.27$ per day—so the average life expectancy at emergence was only about $T_A = 1/d_A = 3.7$ days. The egg and pupal stages do not feed, and the larvae had excess food, so survival

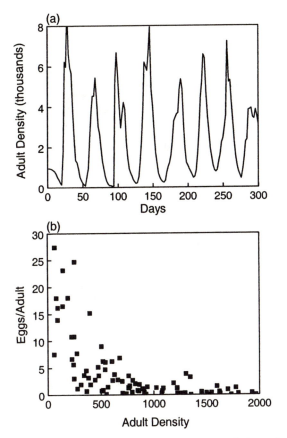

FIGURE 5.1. (a) Data from Nicholson's laboratory populations of sheep blowfly, *Lucilia cuprina* (Nicholson 1957), in which the food supply to adults was limiting. Shown are counts of adults every 2 days from Nicholson's population I (as tabulated in Brillinger et al. 1980). (b) Plot of eggs/adult versus adult blowfly density from Nicholson's data.

during the immature period (which lasted $T_J = 12$–16 days) was high ($S_J = $ about 0.9) and independent of density.

The design of this experiment induced density-dependent fecundity that acted with a delay: the density of adults in the current generation affects fecundity immediately, but given the density-independent immature survival, this effect carries through to the number of adults recruited in the next generation.

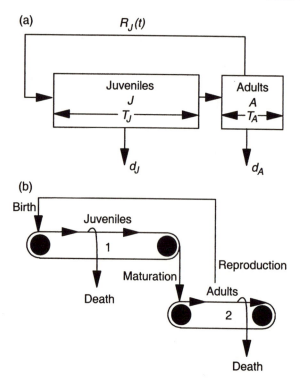

FIGURE 5.2. (a) Diagram of blowfly model showing stage structure. (b) Conveyor-belt diagram of blowfly model, described in text.

A diagram of the stage-structured model describing this interaction is shown in Fig. 5.2a. We use this type of box diagram throughout to describe the structure of our models. To visualize the model it may help to think of the stages as conveyer belts moving at a constant speed with the length of a stage represented by the length of the belt (Fig. 5.2b). Newborn individuals can be envisioned as grains of sand falling (from the reproductive adult population) on to the beginning of belt #1. The rate at which these newborns are added in this case is a density-dependent function of the number of adults present (Fig. 5.1b). Death, which occurs continuously in the immature population, is equivalent to falling off somewhere along belt #1, and maturation from immature to adult is like falling off the end of belt #1 on to the beginning of belt #2. Death during the adult stage is equivalent to falling off somewhere along belt #2.

BALANCE EQUATIONS

These facts lead to the following model, which is a continuous-time stage-structured analog of the Ricker model of chapter 4. Our two state variables are the number of juveniles, $J(t)$, and adults, $A(t)$, present at time t, and their rates of change obey the balance equations

$$\frac{dJ(t)}{dt} = R_J(t) - M_J(t) - D_J(t)$$

$$\frac{dA(t)}{dt} = R_A(t) - D_A(t). \tag{5.1}$$

Here, $R_J(t)$ represents the recruitment rate into the juvenile stage and is a density-dependent function of the current adult density; guided by Fig. 5.1b we assume

$$R_J(t) = q A(t) \exp\left(-\frac{A(t)}{A_0}\right), \tag{5.2}$$

where q is the maximum fecundity at low adult density and A_0 is a constant describing the rate at which fecundity decreases with density ($1/A_0$ is the rate of exponential decline in Fig. 5.1b).

- $D_J(t)$ is the number of juveniles dying per day at time t. There is a constant, density-independent per-head death rate, d_J, during the juvenile stage, so

$$D_J(t) = d_J J(t).$$

- This means that the fraction of individuals surviving through the juvenile stage is

$$S_J = \exp(-d_J T_J).$$

- $M_J(t)$ is the number of juveniles maturing to the adult stage per day at time t. This is equal to the recruitment rate into the juvenile stage T_J days ago multiplied by the fraction surviving through the stage:

$$M_J(t) = R_J(t - T_J) S_J.$$

(Throughout the book the notation $x(t - Z)$ means "the value of the variable x evaluated at time $t - Z$.")

- All individuals that mature out of the juvenile stage enter the adult stage, so

$$R_A(t) = M_J(t).$$

- The per-head adult death rate, d_A, is constant and density independent, so the number of adults dying per day at time t, $D_A(t)$, is

$$D_A(t) = d_A A(t).$$

Since mortality of juveniles does not depend on density, the juvenile stage simply acts as a time lag. By replacing $R_A(t)$ by $R_J(t - T_J)S_J$, the model can be fully specified by the equation for the adult population alone:

$$\frac{dA(t)}{dt} = qS_J A(t - T_J) \exp\left(-\frac{A(t - T_J)}{A_0}\right) - d_A A(t). \qquad (5.3)$$

EQUILIBRIUM

The equilibrium is obtained by setting dA/dt equal to zero at all times. Thus we set $A(t) = A(t - T_J) = A^*$ and find that

$$A^* = A_0 \ln\left(\frac{qS_J}{d_A}\right). \qquad (5.4)$$

MODEL DYNAMICS

Stability analysis of this model follows exactly the same principle as in the simpler models of the previous chapters but is somewhat more complicated (appendix). One complication is that whereas the Lotka-Volterra model had only two solutions (the eigenvalues) to its characteristic equation, the characteristic equation of a delay-differential model may have an infinite number of eigenvalues. The stability of the equilibrium is again determined by only the largest of these eigenvalues, but the presence of others can affect both the transient dynamics and the type of fluctuations seen.

Derivation of the stability criteria is explained in the appendix. The "parameter space" in Fig. 5.3a is separated into a region in which the equilibrium is unstable and one in which it is stable. The curve bisecting the plot is the stability boundary; i.e., it is the set of values of T_A/T_J and qT_A at which the real part of the dominant eigenvalue is zero and equilibria are neutrally stable (appendix). For any given adult stage

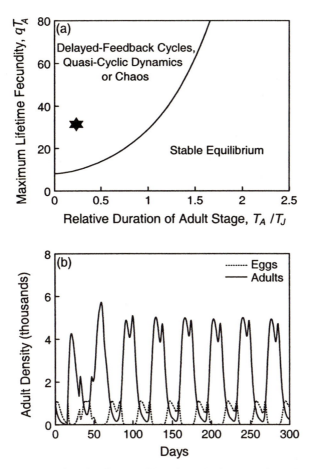

FIGURE 5.3. (a) Stability diagram of blowfly model in terms of duration of the adult stage relative to that of the juvenile stage, T_A/T_J, and maximum lifetime fecundity: daily fecundity multiplied by average duration the adult female lives, q^*T_A. ★ indicates parameters for real system. (b) Simulation using parameters for real system, showing cycles with a period of approximately 37 days. $T_J = 15.6$ days, $S_J = 0.9$, $q = 8.5$, and $A_0 = 600$; all other parameter values are given in text.

duration, the equilibrium is more likely to be stable if the maximum lifetime fecundity, qT_A, is low (Fig. 5.3a). The value of the parameter defining the decline of fecundity with adult density, A_0, has no effect on stability. A_0 acts only as a scaling parameter, determining the equilibrium density.

It is useful to think of each point in the stability diagram in Fig. 5.3a as representing a different hypothetical population. Each population has a uniquely defined fecundity and adult duration, but all populations have the same values for parameters that are fixed (e.g., immature development time). The different populations might be different species. For example, some insects have relatively long-lived juveniles but adults that live for a single day, and their populations would be clustered near the y-axis. But the different points could also be the same species in different environments. For example, if fecundity increases with temperature (but juvenile and adult durations do not), then a species' position in Fig. 5.3a would shift upward with temperature.

Simulations of the model at the point defined by the blowfly system show large-amplitude limit cycles with a period of 37 days (Fig. 5.3b), which is very similar to that observed in the real data. This is slightly more than twice the blowfly's juvenile developmental period. Cycles with a period of 2–4 times the developmental period turn out to be characteristic of models such as this one in which there is delayed density dependence: the effect of density dependence on the stage in which it occurs is delayed until the following generation. We therefore call these *delayed-feedback cycles*.

The process driving these cycles is as follows. When larval density is high, subsequent adult density is also high. This results in a reduction not only in per-head fecundity (eggs/female) but also in total number of eggs produced. The descendants of this dense group of adults therefore form a small group of adults. These adults therefore have large amounts of food per head, high fecundity, and the group of adults arising from them is abundant. Successive "generations" of adults therefore tend to alternate between high and low densities, and the period between peaks is at least 2 development-times ($2T_J$). The period will be longer than $2T_J$ if there is a relatively long-lived adult stage that can suppress reproduction for an extended period of time.

This model is the analog of the discrete-time Ricker model in which the basic period is exactly two single-generations when adults die off after reproduction, but the period can increase when adults survive to breed more than once (chapter 4). In models such as the discrete-time Ricker the instability is often described as resulting from overcompensation. This is because high adult densities actually result in fewer total progeny, not only in fewer progeny per head. In the absence of such overcompensation, the populations will not cycle.

Delayed-feedback cycles provide an example of how models can help us define *classes of dynamics* that transcend the details of any particular biological mechanism. Nicholson was able to change the experimental protocol for blowflies so that the larvae, rather than the adults, competed for food (Nicholson 1957). When this was done, larvae at high density relative to the food supply grew more slowly, and smaller larvae were less likely to pupate successfully. Furthermore, the larval stage in blowflies is relatively short ($\sim$ 5 days) compared to the time ($\sim$ 15 days) required for an egg to develop to a mature adult. Thus larval competition in blowflies occurs among individuals hatching over a relatively narrow time window, and its consequences are delayed (they are observed at the pupal stage). However, its effect on population dynamics is similar to that of density-dependent fecundity described above. This is because the dynamical cause—delayed density dependence—is the same, even though the biological details differ greatly. We discuss this and related issues in chapter 11.

Immediate (Within-Generation) Density Dependence in Immature Survival

The Indian meal moth (*Plodia interpunctella*), a pest of stored grain, provides an example of a system in which the resources are limiting for only the juvenile stage. Populations of the moth have been raised in the laboratory with a fixed amount of food supplied biweekly. The populations show regular cycles with a period of about 39 days, which is slightly longer than the generation time, that is the time between the average newly laid egg in two successive generations (Fig. 5.4a) (Gurney et al. 1983; Sait et al. 1994).

In the *Plodia* life cycle there are eggs, larvae, pupae, and adults, but only the larvae feed. Rearing experiments showed that the combined immature stage lasts $T_J = 28$ days, the adults live about 5 days on average, and females lay about $b = 9.4$ viable female eggs per moth per day (Gurney et al. 1983). The larvae compete for food, and in our model of the system we make the simplest assumption, that the daily per-head death rate of immatures increases linearly with the current density of immatures, so the survival of a cohort decreases exponentially with increasing density. Of course, in the real population the eggs and pupae do not feed, and a large larva depresses the food supply more than does a small larva, but as a first approximation this model ignores these facts.

BALANCE EQUATIONS

Once again our two state variables are the number of juveniles, $J(t)$, present at any time t, and the number of adults, $A(t)$. The model is in exactly the same form as equation 5.1:

$$\frac{dJ(t)}{dt} = R_J(t) - M_J(t) - D_J(t)$$

$$\frac{dA(t)}{dt} = R_A(t) - D_A(t).$$

(5.5)

- $R_J(t)$ is the recruitment rate of juveniles and is just $bA(t)$ and, unlike the first example, is density independent.
- $D_J(t)$ is the number of juveniles dying per day at time t, which as decided above is $[cJ(t) + d_J]J(t)$, where d_J is the density-independent background death rate and c is the intraspecific competition coefficient. It is difficult to measure the parameter c, but it turns out not to be important to the form of the dynamics.
- $R_A(t)$ is the number recruiting per day to the adult stage at time t and is equal to $M_J(t)$.
- $D_A(t)$ is the number of adults dying per day at time t. In a departure from earlier models, we assume the adults all live 5 days and then die. Therefore, $D_A(t)$ is simply the number that recruited $T_A = 5$ days previously, $R_A(t - 5)$. This gives us a variant of the standard assumption that life expectancy at emergence is exponentially distributed (i.e., that adults die at a constant per-head rate).

The only new technicality in the model is calculating $M_J(t)$. The number of juveniles maturing into adults per day, at time t, is the number of eggs produced T_J days previously, times the fraction surviving to time t:

$$M_J(t) = R_J(t - T_J)S_J(t),$$

(5.6)

where $S_J(t)$ is the fraction surviving the period $t - T_J$ to t. If the death rate of juveniles were simply a constant, d_J, then the fraction surviving of those born T_J days previously would be $\exp(-T_J d_J)$. But in fact the death rate at a particular time depends on the density of juveniles at that time. Therefore the fraction that survives through the juvenile stage depends on the instantaneous mortalities summed over the entire juvenile period:

$$S_J(t) = \exp\left[-\int_{t-T_J}^{t} (cJ(x) + d_J)\, dx\right]. \qquad (5.7)$$

We encounter this same idea when determining the through-stage survival in parasitoid-host models. Its main message is that in general the fraction surviving the time period $t - T_J$ to t is

$$\exp\left[-\int_{t-T_J}^{t} (\text{INSTANTANEOUS DEATH RATE FROM ALL CAUSES})\, dx\right].$$

EQUILIBRIA

As before, the equilibria are obtained by setting the right-hand side of all the differential equations equal to zero, $dJ/dt = 0, dA/dt = 0$, with $J(t) = J^*$ and $A(t) = A(t - T_J) = A^*$. This means that at equilibrium, $M_J^* = bA^* \exp[-(cJ^* + d_J)T_J]$. The equilibrium juvenile population is

$$J^* = \frac{\ln(bT_A) - d_J T_J}{cT_J}, \qquad (5.8)$$

and the equilibrium adult population is

$$A^* = \frac{(cJ^* + d_J)\, J^* T_A}{(bT_A - 1)}. \qquad (5.9)$$

MODEL DYNAMICS

We have plotted the stability boundary (Fig. 5.4b) in terms of two parameter groups: the duration of the adult stage relative to that of the juvenile stage (T_A/T_J), and the lifetime fecundity of adult females (bT_A). This allows us to look at the effect of changing the duration of the adult stage

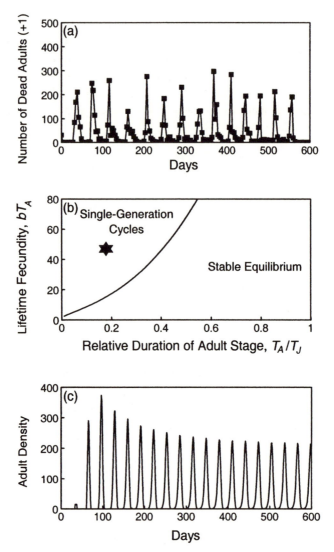

FIGURE 5.4. (a) Data from real *Plodia* population, in which juveniles compete for food (data from Sait et al. 1994). (b) Stability diagram for *Plodia* model, with ⋆ denoting parameters for real system, with $T_J = 28$ days, $d_J = 0$/day, $c = 5 \times 10^{-5}$/juvenile/day. (c) Simulation of *Plodia* model, with $T_A = 5$ days, $b = 9.4$/day, all other parameters as in (b). Simulation started at time $t = 0$ with 15 newborn larvae entering the juvenile class.

without also changing the number of eggs an individual produces in her lifetime. This stability boundary is found through the methods discussed in the appendix.

We find that the equilibrium is stable when the lifetime fecundity is relatively low and when the adults are long-lived relative to the juveniles. The fact that a high lifetime fecundity, which translates into a high rate of increase, is destabilizing fits our previous insight from the blowfly model, which also combines a developmental lag and density dependence. We will see this insight reinforced in parasitoid-host models. As we will see below, the long-lived adult stage is stabilizing because it smears out the discrete-single-generation structure that occurs in the single-generation cycles in the unstable region.

We can gain some general insight into the causes of fluctuations in populations by looking at dynamics in this model when the equilibrium is unstable. Near the stability boundary in Fig. 5.4b there are cycles with a period of 28 to 42 days. Thus the period of the cycles is about 1 to 1.5 times T_J. A period equal to 1 to less than 2 times T_J turns out to be a general feature of models of this sort in which density dependence has its effects immediately on the abundance of the stage in which it operates (Gurney and Nisbet 1985; Kendall et al. 1999). We therefore call this direct feedback and the resulting cycles single-generation cycles. We saw in the previous example that distinctly different cycles appear when the feedback is delayed.

The cause of these cycles in the model is as follows. A large single-aged cohort of young juveniles entering the juvenile class causes a high mortality rate on all juveniles, driving down the abundance of juveniles of all ages, including the dominant age cohort and all older juveniles. Only individuals that enter the juvenile stage slightly after the peak in density, i.e., after it has collapsed, are able to survive in high numbers to adulthood. The production of new juveniles is thus suppressed until the maturation of the individuals that entered the juvenile stage slightly after the dominant age cohort. These adults then give rise to a new large cohort of juveniles. Thus, the result is cycles that are slightly longer than the development times of the organisms involved.

Simulations of the model at the spot that defines the *Plodia* population show large-amplitude stable limit cycles with a period of approximately 32 days (Fig. 5.4c). This is close to the period seen in the data (Fig. 5.4a), and we conclude that the model does a reasonable first-order job of explaining the dynamics, even though it omits some detail.

In this simple model we have assumed that all juvenile *Plodia* are identical. The real system does differ in three important respects. First, competition between young and old larvae is asymmetrical, with the per-head effect of an old (large) larva on a young (small) larva being much greater than the effects of young on old. Second, larvae of all stages cannibalize eggs. Third, after the larval stage there is a pupal stage that neither enters into, nor is affected by, competition for food. A model including all of these processes, although somewhat more complicated, predicts cycles with a period slightly longer than the development time of the species, and closer to that observed in the real populations. The details of the mechanism by which cycles are maintained, however, are slightly different (see Briggs et al. 2000).

THE BASIC STAGE-STRUCTURED HOST-PARASITOID MODEL

In this section we investigate the dynamical effect of the following general feature of host-parasitoid systems: it is usually the juvenile host stage that is attacked by the parasitoid. The adult host in most systems is invulnerable. This model and its variants were initially motivated by the interaction between California red scale and its parasitoid *Aphytis melinus* (chapter 2), which is described in the next two paragraphs.

The life history of (female) red scale and the approximate use of its various developmental stages by *A. melinus* are illustrated in Fig. 5.5. Female adult red scale, which are invulnerable to parasitism by *Aphytis*, produce crawlers that move a few inches, settle on the tree, insert their mouthparts, and never move again unless they are adult males seeking females. Female scale grow through two molts and three instars, the last of which becomes an invulnerable mature female after it is inseminated. After a lag, these females begin to produce crawlers.

Aphytis attacks most of the immature scale stages (Fig. 5.5). However, the probability of attack increases and the nature of the attack changes as the scale grows. Smaller (younger) stages are eaten by *Aphytis* ("host-fed"), the nutrients being used partly for maintenance but also to mature more eggs. Intermediate-sized scales (large instar 2)

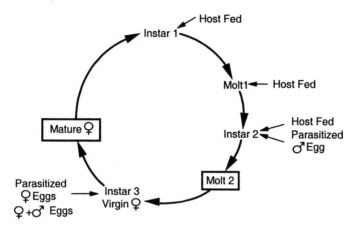

FIGURE 5.5. Diagram of female red scale life history and pattern of attacks by *Aphytis melinus*. Shown are types of attacks on different immature scale stages. Boxes indicate invulnerable stages.

typically receive a male egg, whereas larger, older hosts (instar 3 females) mainly receive a female egg or a male and a female egg. Parasitoids can determine the sex of the egg as they lay it, fertilizing female but not male eggs. Finally, *Aphytis*'s response to an encountered scale also depends on the parasitoid's own state, and in particular the number of mature eggs a female carries. Models in chapters 5–7 ultimately incorporate all these aspects of natural history.

In this model, we have in mind a host population that is suppressed to such a low density by its parasitoid population that changes in host density do not affect the host's own resources. So as in the previous chapters, we do not model the host's resources, nor initially, do we make the host's per-head growth rate depend on host density. In the absence of the parasitoid the host grows exponentially, at a rate determined by food quality, temperature, and so on.

The model (Murdoch et al. 1987) is a stage-structured analog of the Lotka-Volterra model (Fig. 5.6a). The four stages are: unparasitized juvenile hosts $J(t)$, adult hosts $A(t)$, immature parasitoids $I(t)$, and adult parasitoids $P(t)$. Parasitized hosts become immature parasitoids immediately, as occurs in most parasitoid species; parasitized hosts cease to feed and develop and have no effect on host population dynamics. (Parasitoids that immediately kill or paralyze their host in this way are termed idiobionts).

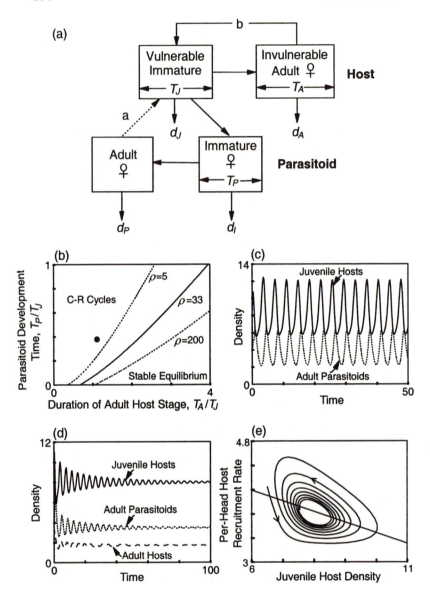

(a)

b

Vulnerable Immature T_J

Invulnerable Adult ♀ T_A

Host

a

d_J

d_A

Adult ♀

Immature ♀ T_P

Parasitoid

d_P

d_I

(b)

Parasitoid Development Time, T_P/T_J

C-R Cycles

$\rho=5$

$\rho=33$

$\rho=200$

Stable Equilibrium

Duration of Adult Host Stage, T_A/T_J

(c)

Density

Juvenile Hosts

Adult Parasitoids

Time

(d)

Density

Juvenile Hosts

Adult Parasitoids

Adult Hosts

Time

(e)

Per-Head Host Recruitment Rate

Juvenile Host Density

We continue the assumption that all individuals in a stage are the same. In particular, we do not distinguish the sexes. We could assume that the sex ratio is 50:50 and follow both males and females, but it is simpler to have the convention that we are following only females of both the host and the parasitoid. This is particularly appropriate for parasitoids, in which only the females kill hosts. We include sex in later models.

We assume development to adult takes the same time for all individuals in each species: T_J for the juvenile host and T_P for the immature parasitoid. In chapter 3 we investigated a simple model (equation 3.21) in which hosts matured out of the juvenile stage at a constant rate, m. This meant that the time spent in the vulnerable juvenile stage was exponentially distributed with a mean duration of m^{-1}. Clearly, an exponential distribution is an unrealistic assumption (because, for example, the modal development time is zero). In this chapter we look at whether a fixed development time gives different predictions from the exponential development time, and also examine the effects of a parasitoid developmental delay.

FIGURE 5.6. (a) Diagram of Basic host-parasitoid model. (b) Stability diagram for Basic host-parasitoid model. Lines show stability boundaries for three different values of lifetime host fecundity, ρ. Equilibrium is stable to the right of the line for each value of ρ and unstable to the left. Dynamics to the left of the boundary are dominated by consumer-resource (C-R) cycles (either stable limit cycles or diverging oscillations). Value of $\rho = 33$, shown in solid line (from Murdoch et al. 1987), is the default for all further graphs in the figure. A reasonable estimate of real parameter values places the red scale–*Aphytis* interaction (indicated by the dot) firmly in the unstable region. (c) Simulation of Basic host-parasitoid model in unstable region, with time shown in units of juvenile host stage durations, T_J, for $\rho = 33$, $T_A/T_J = 1.5$, $T_P/T_J = 0.3$. (d) Simulation of Basic host-parasitoid model in stable region, with time shown in units of juvenile host stage durations, T_J, for $\rho = 33$, $T_A/T_J = 2$, $T_P/T_J = 0.3$. (e) For the simulation in (d), per-head recruitment rate into the juvenile host stage, $R_J(t)/J(t)$, versus juvenile host density, $J(t)$. In all cases, $d_P = 8/T_J$, $d_J = 0/T_J$, $S_I = 1$.

Balance Equations

The above ecology leads to the following model. Apart from the addition of stage structure, we retain all of the assumptions of the standard Lotka-Volterra model.

$$\frac{dJ(t)}{dt} = R_J(t) - M_J(t) - X(t) - D_J(t)$$

$$\frac{dA(t)}{dt} = R_A(t) - D_A(t)$$

$$\frac{dI(t)}{dt} = R_I(t) - M_I(t) - D_I(t)$$

$$\frac{dP(t)}{dt} = R_P(t) - D_P(t)$$

(5.10)

where $R_i(t)$ is the recruitment rate into each stage, i, $M_i(t)$ is the maturation rate out of that stage, and $D_i(t)$ is the death rate during that stage that is due to all causes unrelated to parasitism. $X(t)$ is the rate of loss of unparasitized juvenile hosts due to parasitism. Next we give a specific form to each of the terms.

Host birth rate. Insects such as moths, midges, and mayflies live as adults for one or a few days, but in that time they may lay hundreds of (typically small) eggs. By contrast, adult beetles, grasshoppers, and scale insects, for example, live for weeks or months, and each day they lay few (typically large) eggs. Over their lifetimes, however, these different types of insects may not lay hugely different numbers of eggs. We take this into account in the following way.

We assume the average adult lifetime fecundity is ρ. Then the number of eggs laid per unit time is $b = \rho/T_A$, where T_A is the average adult lifespan. Thus ρ is bT_A. Now we can fix lifetime fecundity, ρ, and then vary the duration of the adult stage. As in the *Plodia* model above, this allows us to vary adult duration without confounding its effect with changes in fecundity.

Juvenile host recruitment. We assume the host's intrinsic growth rate is constant with fecundity, b, so each unit time the number of juvenile hosts recruited is simply $R_J(t) = bA(t)$.

Parasitism rate. As in the Lotka-Volterra model, we assume parasitoids encounter and parasitize hosts at random at rate a and have no handling time, so there are $X(t) = aJ(t)P(t)$ hosts parasitized per unit time.

Background juvenile host death rate. We assume a constant rate of mortality of juvenile hosts, d_J, that is due to causes unrelated to parasitism, so $D_J(t) = d_J J(t)$.

Maturation rate of juvenile hosts to the adult stage. The number of juvenile hosts maturing into adults each unit time is the fraction surviving, $S_J(t)$, of those that were born T_J days previously:

$$M_J(t) = bA(t - T_J)S_J(t), \tag{5.11}$$

where $S_J(t)$ is the fraction surviving the period $t - T_J$ to t. As discussed in the *Plodia* example above, this is

$$S_J(t) = \exp\left(-\int_{t-T_J}^{t} [aP(x) + d_J]\,dx\right). \tag{5.12}$$

Adult host recruitment rate. This is simply the number of immature hosts becoming adults each unit time, $R_A(t) = M_J(t)$.

Adult host death rate. We assume adults suffer a constant per-head background death rate, d_A, caused by factors unrelated to parasitism, so that $D_A(t) = d_A A(t)$. Adults are therefore dying off exponentially, so the mean life expectancy of an adult is $T_A = 1/d_A$.

Juvenile parasitoid recruitment rate. We assume that each parasitized host gives rise to one immature parasitoid. Thus the number of immature parasitoids produced per unit time is simply the number of juvenile hosts that have been parasitized, $R_I(t) = X(t)$. Making the conversion rate something other than 1 (e.g., if more than 1 parasitoid egg is oviposited per host) alters only the equilibrium densities, not the form of the dynamics.

Adult parasitoid recruitment rate. Notice that the probability of survival through the immature parasitoid stage, S_I, is fixed since the background per-head death rate is density independent and constant and the immature parasitoid stage duration is constant. Because of this, the explicit equation for the rate of change of immature parasitoids is unnecessary; the stage operates purely as a time lag. This stage thus acts in the same manner as the immature stage in the blowfly example. Because there is no density dependence within the stage, the stage can be subsumed into the equation for the adult stage as a time lag. The number maturing out of the immature parasitoid stage at time t is simply a constant fraction of the number created through parasitism T_P days previously. The recruitment rate into the adult parasitoid stage then becomes $R_P(t) = X(t - T_P)S_I$, where $S_I = \exp(-d_I T_P)$.

Adult parasitoid death rate. Finally, we assume that adult parasitoids also die off at a constant rate, d_P, so the number dying per unit time is $D_P(t) = d_P P(t)$.

This model, which we call the Basic model, serves as our baseline stage-structured host-parasitoid model. For convenience, the equations are summarized in Box 5.1.

EQUILIBRIA

The equilibria are calculated as in the *Plodia* model above:

$$\text{Juvenile hosts: } J^* = \frac{d_P}{a S_I}$$

$$\text{Adult hosts: } A^* = \frac{J^*(a P^* + d_J)}{d_A(\rho - 1)} \qquad (5.13)$$

$$\text{Adult parasitoids: } P^* = \frac{\ln(\rho) - d_J T_J}{a T_J}.$$

The equilibria are simpler than they appear, and two of them are precise analogs of the Lotka-Volterra equilibria. First, it helps to notice that the host's per-head rate of increase in the absence of the parasitoid (the analog of r in the Lotka-Volterra model) is $\ln(\rho)/T_J - d_J$. Now we see (1) that the parasitoid equilibrium is analogous to the Lotka-Volterra parasitoid equilibrium of r/a; and (2) if the parasitoid juveniles all survive, S_I is equal to 1 and the equilibrium density of unparasitized

BOX 5.1.
BASIC HOST-PARASITOID MODEL (MURDOCH ET AL. 1987)

Juvenile hosts are attacked by the parasitoid, and adults are invulnerable.

Parameters

ρ	per-individual lifetime host fecundity
$T_A = 1/d_A$	average duration of adult host stage
T_J	duration of juvenile host stage
T_P	duration of juvenile parasitoid stage
a	parasitoid attack rate
S_I	juvenile parasitoid through-stage survival
d_J	background death rate of juvenile hosts
d_A	death rate of adult hosts
d_P	death rate of adult parasitoids

Equations

$dJ(t)/dt = R_J(t) - M_J(t) - X(t) - D_J(t)$ juvenile hosts

$dA(t)/dt = R_A(t) - D_A(t)$ adult hosts

$dP(t)/dt = R_P(t) - D_P(t)$ adult parasitoids

$R_J(t) = (\rho/T_A)A(t)$ recruitment rate into juvenile host stage

$M_J(t) = (\rho/T_A)A(t - T_J)$
$\exp\left\{-\int_{t-T_J}^{t}[aP(x) + d_J]dx\right\}$ maturation rate out of juvenile host stage at time t

$X(t) = aJ(t)P(t)$ parasitism rate

$D_J(t) = d_J J(t)$ background juvenile host death rate

$R_A(t) = M_J(t)$ recruitment rate into adult host stage

$D_A(t) = d_A A(t)$ adult host death rate

$R_P(t) = aJ(t - T_P)P(t - T_P)S_I$ recruitment rate into adult parasitoid stage

$D_P(t) = d_P P(t)$ adult parasitoid death rate

hosts is just d_P/a, which is the Lotka-Volterra host equilibrium. Thus all of the properties of Lotka-Volterra equilibria in Box 3.1 carry over to this model. For example, the immature host equilibrium depends only on parasitoid parameters.

STABILITY

Dynamic equations can be rescaled, thereby reducing the number of parameters and both simplifying analysis and making presentation easier (Edelstein-Keshet 1988; Gurney and Nisbet 1998). In this case, we can redefine time in terms of the duration of the juvenile host stage, T_J. When we do so, the basic time unit becomes T_J, and the durations of the adult host and parasitoid stages, and of the immature parasitoid stage, are in units of juvenile host development times.

For most of this book, we constrain the duration of the immature parasitoid stage to ≤ 1. This is because parasitoids almost always develop faster than their hosts. Only in chapters 6 and 11 do we spend a little time considering conditions where the duration is more than 1.

Whether or not the equilibrium is stable depends on the actual parameter values. Because there are many parameters in the model, and we have only two dimensions in which to represent stability on paper, we need to choose how to portray the effect of parameter values on stability. We choose to concentrate on the duration of the adult stage, since its potential stabilizing effect is a main reason for developing the model, and on the length of the parasitoid development time, which we expect to be destabilizing.

Fig. 5.6b shows a standard stability plot for this model. In several figures in this chapter and the next we present stability boundaries in terms of two sets of relative stage durations: the duration of the adult host stage relative to the duration of the juvenile host stage, T_A/T_J, and the duration of the immature parasitoid stage relative to the duration of the juvenile host stage, T_P/T_J. The parameter space is again separated into a region in which the equilibria are unstable and one in which they are stable. The stability boundary is the set of values of T_A/T_J and T_P/T_J at which the real part of the dominant eigenvalue is zero and equilibria are neutrally stable (appendix). Boundaries are shown for three different values of the lifetime host fecundity, ρ. Notice that because we can vary only T_A/T_J and T_P/T_J in this two-dimensional graph, we need to fix all the other parameter values.

We can think of each point on the graph as defining a particular parasitoid-host interaction. Consider, for example, a set of host insects whose juvenile stages all last 100 days. Interactions involving mayflies and midges (which typically live 1 day as adults) would be points clustered extremely close the y-axis. Those parasitoids whose immature stages took 100 days to develop would be at $Y = 1$, those taking 50 days would be at $Y = 0.5$, etc. Some species of insects take a year to develop but the adults can then live for 2 or even 3 years; so, for example some beetles and their parasitoids would occur as a column of points around the value of 2 or 3 on the x-axis.

We first look at the origin in Fig. 5.6b, $T_A = 0$, $T_P = 0$, to show that the dynamics are related to those of the Lotka-Volterra model. To do this, we assume the adult host lives for only an instant, but in that instant produces its lifetime complement of eggs (appendix). We also let T_P equal zero, so parasitized hosts become new adult parasitoids immediately. Under these circumstances, the equilibrium is neutrally stable just as in the Lotka-Volterra model. Furthermore, the period of the cycles at this point is the exact analog of the Lotka-Volterra period (chapter 3). The dominant eigenvalue here is

$$\lambda = \pm i \sqrt{[\ln(\rho)/T_J - d_J]d_P}$$

which is simply $\sqrt{rd_P}$. Hence the period of the cycle is $2\pi/\sqrt{rd_P}$, which of course is the value in the basic Lotka-Volterra model.

We now investigate what happens as we move away from the origin. First, moving up the y-axis reintroduces the parasitoid's developmental time lag. As we would expect, this immediately renders the equilibrium unstable. Furthermore, the fluctuations increase in amplitude with time, so there is no adequate stabilizing mechanism operating to "capture" these fluctuations to produce limit cycles. These are unstable consumer-resource cycles. A second set of eigenvalues that appear at the origin and along the y-axis is of interest. They give rise to oscillations with a period equal to the host development time, T_J, hence we term them single-generation cycles. At the origin the single-generation cycles are neutrally stable, but for values of T_P/T_J greater than zero the single-generation cycles are obscured by the divergent consumer-resource cycles. We return to single-generation cycles below.

Next we move along the x-axis, i.e., we increase the duration of the invulnerable adult host stage from zero while keeping the parasitoid development time equal to zero. Here we can compare the results of this model, in which all host individuals spend a fixed amount of time in the juvenile stage before maturing into the adult stage, with equation 3.21 in chapter 3 in which the time spent in the juvenile stage was exponentially distributed. Recall from chapter 3 that in the latter case, the addition of an invulnerable adult stage to the Lotka-Volterra model had a stabilizing effect and changed the dynamics from neutrally stable to stable. Here we find that with a vulnerable juvenile stage of constant duration, the addition of an invulnerable adult stage can have either a stabilizing or a destabilizing effect, depending on its relative duration. The initial effect of increasing the adult duration is to destabilize the equilibrium: the neutrally stable equilibrium at the origin gives way to diverging cycles and chaos. As the duration of the invulnerable adult stage is increased further, the invulnerable adult stage has a stabilizing effect, leading to damped oscillations and a stable equilibrium. Thus our intuition about the effect of an invulnerable adult stage from the Lotka-Volterra model in chapter 3 does not completely carry over in this case.

Away from the two axes in Fig. 5.6b there are both a parasitoid developmental lag and a finite adult host stage. To the left of the stability boundary the equilibrium is unstable. Close to the y-axis, with a short invulnerable adult stage, the dynamics are dominated by divergent consumer-resource oscillations. As we move toward the stability boundary, the consumer-resource cycles become stable limit cycles (Fig. 5.6c). Once across the boundary we find damped cycles leading to a stable equilibrium (Fig. 5.6d). Thus, over most of the parameter space increasing the duration of the invulnerable adult stage has a stabilizing effect.

Table 5.1 lists the effects on stability of increasing the various parameters. Notice that two, a and S_I, have no effect on stability— they are simply scaling parameters that influence only the value of the equilibrium. Table 5.1 again shows that insights gained from modifications of the Lotka-Volterra models discussed in chapter 3 carry through. Increases in parameters determining the host's rate of increase in the absence of the parasitoid are destabilizing, i.e., they reduce the region of parameter space in which a stable equilibrium occurs (Fig. 5.6b). Increases in adult parasitoid longevity are also

TABLE 5.1. Effect of increases in parameters on stability
of the host-parasitoid equilibrium

Parameter	Effect
ρ	destabilizing
$T_A = 1/d_A$	stabilizing
T_P	destabilizing
a	no effect
S_I	no effect
d_J	stabilizing
d_P	stabilizing

destabilizing. And of course, increases in the parasitoid time lag are destabilizing.

Finally, we reinforce a point about density dependence and stability made in chapter 3, where a Lotka-Volterra-type model with a long-lived invulnerable adult stage can have a stable equilibrium even though no vital rate is apparently density dependent. Stability in the Basic model has the same source. The recruitment rate to the immature host stage has a density-dependent component (Fig. 5.6e), which we argue is induced by the asynchronous trajectories of the vulnerable immature and invulnerable adult host stages (Fig. 5.6d). The relatively long-lived adult stage includes individuals that were born over a range of time periods and that survived parasitism by a range of parasitoid densities. Therefore the adult density effectively integrates recruitment over a range of immature dynamics. The density of this stage is necessarily more constant than that of the juvenile host stage, and its dynamics become, to a greater or lesser extent, uncoupled from the dynamics of the juvenile stage, and hence of the parasitoid. As a consequence, the per-juvenile recruitment rate to the immature stage is somewhat independent of immature density at any particular time.

A range of models has shown that a recruitment rate that is constant, or at least uncorrelated with the dynamics of the receiving population, can translate into a density-dependent per-head recruitment rate and have a stabilizing effect (chapter 3, "Simple Models of Stage and Spatial Structure"). The mechanism through which a long-lived invulnerable adult stage stabilizes a consumer-resource interaction is thus related to the way movement among asynchronous subpopulations can stabilize spatially subdivided consumer-resource interactions, reinforcing this point from chapter 3.

Our early analyses of this model (Murdoch et al. 1987) suggested that an invulnerable stage alone could not explain the observed stability of the real red scale–*Aphytis* interaction. Preliminary parameter estimates placed the real system (the dot in Fig. 5.6b) well into the unstable region where large-amplitude consumer-resource cycles are expected. Although these parameter estimates are not accurate, it would take probably unrealistically large changes in parameter values to alter this conclusion.

Cycle Periods

We noted above that the period of the neutrally stable cycles at the origin (with $d_J = 0$ and $T_J = 1$) in Fig. 5.6b is exactly that of the basic Lotka-Volterra model, namely $2\pi/\sqrt{rd_P}$. With the "default" set of parameter values for the interaction between red scale and *Aphytis* ($\rho = 33$, $d_P = 8/T_J$), shown in Fig. 5.6b, this gives a period of 1.2 (where the unit of time is the host's development time, T_J). The stable consumer-resource limit cycles away from the origin always have a period longer than this. As we move up the stability boundary for $\rho = 33$ in Fig. 5.6b, the period increases, from about 1.8 to 7, almost linearly with the duration of both the adult host stage and the parasitoid development delay. Since these two variables increase concurrently along the stability boundary, we cannot separate their effects quantitatively. However, investigations of the relationship between cycle period and adult host (resource) longevity in other models (Murdoch et al. 2002) suggest that the increase in cycle period here is caused by the increase in parasitoid development delay.

Fig. 5.6c illustrates the stable consumer-resource limit cycles seen in most of the unstable space in Fig. 5.6b. Notice that even in this stage-structured model, these cycles retain the basic form we saw in chapter 3. The adult parasitoids still lag $\frac{1}{4}$ cycle period behind the "host," which in this case is the immature host.

ECOLOGICAL PROCESSES INDUCING INSTABILITY

Time Lags

In chapter 3 we found that a developmental delay in a consumer population had a destabilizing effect. In the Basic model of Box 5.1, which includes time lags in both the host and parasitoid populations, we once again find that the lag in the parasitoid development has the expected destabilizing effect. This can be seen by comparing the dynamics of this model along the y-axis with those of the Lotka-Volterra model in chapter 3, where such a time lag is missing: our model with a finite parasitoid development lag is always unstable, whereas the Lotka-Volterra model is always neutrally stable. In the Basic model, we cannot distinguish the effect of a time delay in the host population from the confounding effect of an invulnerable adult stage. Here we look briefly at the effect of a delay between the time a host is born and the time at which it begins to produce offspring, without also changing the age-specific susceptibility of the host to parasitism.

The following model includes all of the assumptions of the standard Lotka-Volterra model, except that, as above, each host individual cannot produce offspring until after it matures at the age of T_J days. As above, $J(t)$ and $A(t)$ are the densities of juvenile and adult hosts at time t, and $P(t)$ is the density of parasitoids. To investigate only the effect of a host developmental delay we assume, as in the Lotka-Volterra model, that all hosts are vulnerable to parasitism. Parasitoids have the same attack rate on juvenile and adult hosts, and all parasitized hosts immediately become new searching parasitoids. In addition, for simplicity, we assume that the only source of mortality to the host is the parasitoid.

$$\frac{dJ(t)}{dt} = rA(t) - M_J(t) - aJ(t)P(t)$$

$$\frac{dA(t)}{dt} = M_J(t) - aA(t)P(t) \qquad (5.14)$$

$$\frac{dP(t)}{dt} = aP(t)[J(t) + A(t)] - d_P P(t)$$

where

$$M_J(t) = r A(t - T_J) \exp\left(- \int_{t-T_J}^{t} a P(x)\, dx \right).$$

We recover the neutrally stable Lotka-Volterra model if we set T_J equal to zero, and we can look at the effect of a host developmental delay by increasing T_J from zero. When we do this, we find that the equilibrium remains neutrally stable. Thus we find that when no intraspecific density dependence is included in any of the host demographic rates (and there is no invulnerable host stage), a host developmental time lag on its own has no effect on stability of the host-parasitoid interaction. In the stage-structured model above, we found our first clue that this might be the case, where at the origin in Fig. 5.6b neutrally stable cycles were observed, even though the host produced all of their offspring in one instant when they reached the age of T_J days.

Thus, the conclusion from this section is that a time lag is destabilizing only when it produces a delay in a density-dependent process. The delay in parasitoid development produces a delay in the mortality of the host caused by the parasitoid. The process is (indirectly) density dependent because the host mortality induced when the parasitoids mature now depends on previous host density: current host death rate becomes dependent on its past density. When all stages of the host are vulnerable to parasitism, however, the delayed host reproduction does not affect any density-dependent process, and therefore has no effect on stability.

Type 2 Functional Response

In real predators, as we discussed in chapter 3, the notion of a type 2 functional response is straightforward. It is derived by assuming that each captured prey requires a "handling time," whether this involves real handling of the prey or whether it arises from satiation (gut fullness), or both. Here we introduce such a type 2 functional response to the Basic model. But the effects of satiation and egg depletion are potentially more complicated than a simple handling-time effect, particularly in parasitoids that host-feed, and we return to these concerns in chapter 7.

To include a type 2 functional response, the parasitism rate in equation 5.10 above becomes

$$X(t) = \frac{aJ(t)P(t)}{1 + (aJ(t)/J_{MAX})} \tag{5.15}$$

where J_{MAX} is the maximum rate of parasitism by the parasitoids (analogous to 1/handling time). Likewise this instantaneous mortality rate is incorporated into the through-stage juvenile host survival:

$$S_J(t) = \exp\left\{ -\int_{t-T_J}^{t} \left(\frac{aP(x)}{1 + aJ(x)/J_{MAX}} + d_J \right) dx \right\}. \tag{5.16}$$

A type 2 functional response has the expected result of reducing the region of parameter space in which stable equilibria occur. When J_{MAX} is equal to infinity, we recapture the stage-structured model above. Fig. 5.7 shows that as J_{MAX} is reduced, the stability boundary is shifted to the right, so that a longer invulnerable adult stage is required for the equilibrium to be stable.

ECOLOGICAL PROCESSES INDUCING STABILITY

We next explore the dynamical effects of adding to the Basic model some of the processes we examined in the simpler models of chapter 3. Our purpose in part is to show that they have similar dynamical effects in the two types of model. But in addition, the stage-structured model forces us to recognize that processes such as host density dependence take qualitatively different forms in different types of insects.

Host Density Dependence

Here we add to the host population the same two forms of density dependence that we looked at in the single-species models.

DENSITY-DEPENDENT HOST FECUNDITY

First, we make the fecundity of adult hosts dependent on the current adult density as seen in the blowfly model above. The only change to

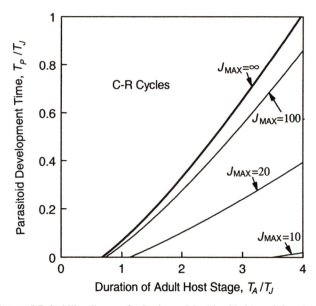

FIGURE 5.7. Stability diagram for Basic model with added type 2 functional response, as in Fig. 5.6b, in terms of relative durations of the adult host stage and parasitoid development time (both relative to duration of juvenile host development time). For each value of J_{MAX}, the equilibrium is locally stable to the right of the boundary and unstable to the left. To the left of the boundary, dynamics are dominated by consumer-resource (C-R) cycles. $\rho = 33$, all other parameters as in Fig. 5.6b. Heavier line for $J_{MAX} = \infty$ is the boundary for the Basic model with the default parameter values.

the Basic model (Box 5.1) that this requires is replacing $R_J(t)$ with

$$R_J(t) = \frac{\rho}{T_A} A(t) \exp\left[-\frac{A(t)}{A_0}\right]. \tag{5.17}$$

In the single-species model with density-dependent fecundity above, we found that changing the value of A_0 affected only the equilibrium density of the population and did not affect stability of the equilibrium. Now, when both intraspecific density dependence and parasitism act on the host population, increasing or decreasing A_0 relative to the parasitoid attack rate (a) changes the relative magnitude of the effects of intraspecific competition and parasitism on the host population. The host moves from the extreme of being self-limited when a is equal to zero to being limited wholly by the parasitoid, as in the Basic model, when $(1/A_0)$ is equal to zero. In the absence of the parasitoid, in the blowfly model (equations 5.1–5.4), we found that the equilibrium was stable for species with long-lived adult stages (high T_A) and low values of lifetime fecundity (low ρ). In the unstable region, delayed-feedback cycles with a period of 2–4 host development times occurred. The Basic model without host self-limitation ($(1/A_0) = 0$) is also stable for species with long-lived adult stages and low lifetime fecundity. But in this case, the unstable region is dominated by consumer-resource, rather than delayed-feedback, cycles.

For a given parasitoid attack rate, decreasing A_0 leads to an increased effect of intraspecific density dependence relative to parasitism. Adding a small amount of intraspecific density dependence (high A_0) has the effect of suppressing the consumer-resource cycles, leading to a stable equilibrium for parameter values where the model without density dependence would produce large-amplitude, long-period cycles. But for high amounts of intraspecific density dependence (low A_0), the parasitoid is driven extinct, and the host on its own displays shorter period delayed-feedback cycles.

Thus, unlike the non-delayed processes in chapter 3, delayed density dependence in the host can be destabilizing as well as stabilizing in parasitoid-host models. However, we would expect this from our consideration of simpler models with delayed density dependence (chapter 4 and the first section of this chapter).

DENSITY-DEPENDENT JUVENILE HOST SURVIVAL

We now look at the case where there is intraspecific density dependence in juvenile host survival, rather than adult host fecundity. In this case we modify the Basic model by adding an intraspecific density-dependent mortality term to the balance equation for the juvenile host population. We assume, as in the single-species *Plodia* model (equations 5.5–5.7), that the daily per-head death rate of juvenile hosts increases linearly with their current density. We further assume that hosts discontinue feeding when they become parasitized, so that parasitized hosts do not enter into the density-dependent term. The death rate on juvenile hosts from causes unrelated to parasitism then becomes

$$D_J(t) = [cJ(t) + d_J]J(t). \tag{5.18}$$

As before, this affects the probability of surviving through the juvenile stage:

$$S_J(t) = \exp\left(-\int_{t-T_J}^{t} [aP(x) + cJ(x) + d_J]\, dx\right). \qquad (5.19)$$

The equilibrium parasitoid density is

$$P^* = \frac{\ln(\rho) - d_J T_J - cJ^* T_J}{aT_J}, \qquad (5.20)$$

where

$$J^* = \frac{d_P}{aS_I}. \qquad (5.21)$$

Thus, the parasitoid can persist and have a positive equilibrium density on this host if

$$\frac{1}{c}\left(\frac{\ln(\rho)}{T_J} - d_J\right) > \frac{d_P}{aS_I}. \qquad (5.22)$$

This implies that persistence of the parasitoid is possible if the equilibrium density of juvenile hosts set by intraspecific density dependence in the absence of the parasitoid is greater than the equilibrium density of juvenile hosts set by the parasitoid. This condition is the analog of the general requirement for persistence of consumers (chapter 3). When the parasitoid does persist, the immature host equilibrium density is unchanged from that in the Basic model.

Increasing the strength of intraspecific host density dependence (increasing c) causes a linear increase in the adult host equilibrium density:

$$A^* = \frac{J^*(aP^* + d_J + cJ^*)}{d_A(\rho - 1)}. \qquad (5.23)$$

As in the case of density-dependent host fecundity, above, changing the value of c relative to the parasitoid attack rate, a, varies the relative effects of intraspecific host density dependence versus parasitism on the host population. When c is equal to zero we recapture the Basic parasitoid model. When a is equal to zero, we approximate the single-species model with density-dependent immature survival discussed in the *Plodia* example above. (The only difference is that here we are assuming that the adult host has a constant mortality rate, rather than a constant stage duration. The effect of this is that the model with the constant adult mortality rate produces single-generation cycles over a more restricted range of parameters than the model with the constant adult stage duration.)

To appreciate the effect of adding density dependence in the juvenile host, first recall that both the single-species model with density-dependent immature survival (equations 5.5–5.7) and the Basic parasitoid host model without host density dependence (Box 5.1) predict a stable equilibrium when the lifetime host fecundity (ρ) is low and adult hosts are long-lived (high T_A). In the unstable region, the single-species model predicts single-generation cycles, whereas the Basic parasitoid model predicts consumer-resource cycles or divergent oscillations.

We can observe the effect of increasing intraspecific host density dependence relative to parasitism as follows. Choose a point in parameter space where the adult host is short-lived and has a high fecundity, and where the Basic host-parasitoid model ($c = 0$) predicts divergent consumer-resource cycles (Fig. 5.6b), then increase the value of c (Fig. 5.8).

Adding a small amount of density dependence (low c) has the effect of keeping the host densities in the parasitoid-host cycles from exploding to extremely high densities. This stabilizes the divergent consumer-resource cycles into stable host-parasitoid limit cycles. In some cases where the host-parasitoid interaction is very unstable, when the host is at very high densities it can take several host generations for the parasitoid population to recover from a trough in density. In these cases, the host population can undergo single-generation cycles at high densities for a few generations, before the density is brought back down by the parasitoid (Fig. 5.8a). Notice that these cycles are intrinsic to the host, and become manifest when the parasitoid has only negligible effect. They are like the single-generation cycles in the single-species models above, rather than those discussed in the next two sections.

Further increases in c increase the stabilizing effect, resulting first in stable consumer-resource limit cycles (Fig. 5.8b) and then in a stable host-parasitoid equilibrium (Fig. 5.8c). As c increases further, the threshold is reached where the parasitoid can no longer persist on

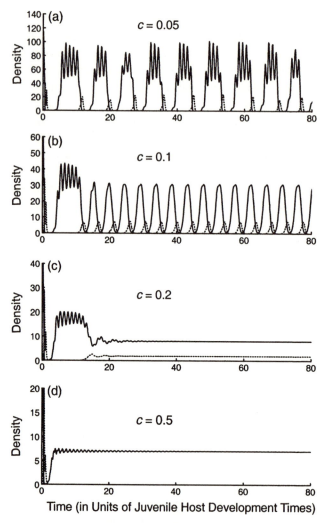

FIGURE 5.8. For the host-parasitoid model with density-dependent juvenile host survival, simulations showing the effect of increasing the strength of intraspecific competition, c, relative to parasitism (the attack rate, a, is kept constant). Solid lines represent juvenile hosts, dotted lines represent adult female parasitoids throughout. $T_A/T_J = 0.1$, $T_P/T_J = 0.5$, $\rho = 33$, $d_P = 8/T_J$, $a = 1/T_J$, with $T_J = 1$; all other parameters as in Fig. 5.6b.

the host population, and the parasitoid goes extinct (Fig. 5.8d). For values of c above this threshold, the single-species dynamics of either a stable equilibrium or single-generation cycles occur.

Once again, parasitoid-host models with this type of within-generation host density dependence behave as we would expect from the single-species model in the *Plodia* example above. Within-generation density dependence is less likely to destabilize, and indeed does so only by suppressing the hosts to a level that cannot support parasitoids. However, we do see a dynamic that is new to host-parasitoid interactions, namely the intermittent single-generation fluctuations in the host at high host, but low parasitoid, densities.

Density Dependence in the Parasitoid

DENSITY DEPENDENCE IN THE PARASITOID DEATH RATE

The Basic model assumes that the parasitoid is limited only by the availability of hosts. We can look at the effect of an adult parasitoid death rate that increases with parasitoid density by adding a density-dependent term to the function describing the number of parasitoids dying per unit time (see chapter 3):

$$D_P(t) = [cP(t) + d_P]P(t). \qquad (5.24)$$

The equilibrium densities for this modification of the model are:

$$J^* = \frac{d_P + c\,P^*}{aS_I}, \quad A^* = \frac{J^*(aP^* + d_J)}{d_A(\rho - 1)}, \quad P^* = \frac{\ln(\rho) - d_J T_J}{aT_J}. \qquad (5.25)$$

Thus the equilibrium parasitoid density is unchanged from the Basic model, but the equilibrium densities of both juvenile and adult hosts increase linearly with increasing c, the strength of density dependence in the parasitoid death rate. As expected from chapter 3, adding density dependence to the consumer population causes the resource equilibrium density to increase with consumer equilibrium density, and thus with resource rate of increase. Also as expected from chapter 3, density dependence in the parasitoid death rate has a stabilizing effect, resulting in an increase in the region of parameter space in which a stable equilibrium is observed (Fig. 5.9a).

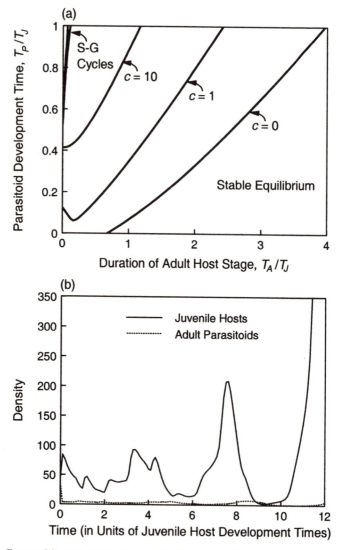

FIGURE 5.9. (a) Stability diagram for host-parasitoid model with density-dependent parasitoid death rate. Stability boundaries are shown for three different values of the strength of density dependence, c. For each value of c, the equilibrium is locally stable to the right of the boundary and dynamics are dominated by consumer-resource cycles to the left. Single-generation cycles occur in small regions near the y-axis. Single-generation cycle boundaries are shown for $c = 1$ and $c = 10$ (these are the lines delineating the combinations

The density-dependent parasitoid death rate also lets us see more of the potential single-generation cycles that lurk near the y-axis in Fig. 5.6b, i.e., when the adult host is short-lived (Fig. 5.9a). For some values of ρ and d_P, single-generation cycles are observed as transient dynamics on top of the consumer-resource cycles (Fig. 5.9b). These cycles are discussed further in the next section.

PARASITOID-DEPENDENT ATTACK RATE

We noted in chapter 3 that the effective attack rate of the individual parasitoid may decrease as parasitoid density increases. We can incorporate this into the model in the same way as in the simple Lotka-Volterra model. A number of forms for this function have been used in the literature. Here we investigate the effects of using the following function describing the number of hosts dying per unit time because of parasitism (Godfray and Hassell 1989):

$$X(t) = k \ln \left(1 + \frac{aP(t)}{k} \right) J(t). \tag{5.26}$$

The rest of the model is identical to the Basic model in Box 5.1. This formulation for the parasitism function assumes that attacks are distributed among hosts according to a negative binomial distribution, as discussed in chapter 4. The smaller the value of the parameter k, the more attacks are concentrated into a small fraction of the population, the higher is the fraction that are thus "wasted," and the more strongly does the attack rate per parasitoid decline as parasitoid density increases. As k approaches infinity we recapture the Basic model. As k decreases, the severity of parasitoid-dependence in the functional response increases. The equilibrium densities for this modification of the model are:

$$J^* = \frac{d_P P^*}{r S_I}, \quad A^* = \frac{J^*(r + d_J)}{d_A(\rho - 1)}, \quad P^* = \left(\frac{k}{a} \right) \left\{ \exp \left(\frac{r}{k} \right) - 1 \right\}, \tag{5.27}$$

of parameters at which the eigenvalues with imaginary parts corresponding to a period of approximately one host development time have real parts equal to zero). $\rho = 33$, $d_P = 8/T_J$, $a = 1/T_J$, $d_J = 0/T_J$, $S_I = 1$, with $T_J = 1$. (b) Simulation showing single-generation cycles only as transient dynamics for parameters that otherwise display divergent consumer-resource cycles; $T_A/T_J = 0.05$, $T_P/T_J = 0.7$, $c = 10$.

where

$$r = \frac{\ln(\rho)}{T_J} - d_J.$$

As k is decreased, the equilibrium density of all stages of hosts and parasitoids increases rapidly, as we would expect (chapter 4).

Decreasing k has the expected effect of increasing stability. As the parasitoid becomes less efficient, it is less able to overexploit the host population, and therefore does not generate consumer-resource cycles. Specifically, for any particular duration of the invulnerable adult stage, T_A, the consumer-resource cycles are suppressed over a region of parameter space whose extent increases as k decreases (Fig. 5.10a).

The new phenomenon we noted in the density-dependent parasitoid death rate model above now appears if parasitoid dependence in the attack rate is sufficiently strong; when the adult host lives for only a short time, and the consumer-resource cycles are suppressed, stable single-generation cycles can occur (Fig. 5.10b). Stable single-generation cycles were first demonstrated in a consumer-resource model by Auslander et al. (1974) and were discovered in a parasitoid-host model by Godfray and Hassell (1987; 1989).

Recall that near the y-axis of Fig. 5.6b there are eigenvalues associated with cycles with a period of T_J, but that these cycles are swamped by the unstable consumer-resource cycles. The parasitoid density dependence added to the functional response lets us see these cycles by

---

FIGURE 5.10. (a) Stability diagram for host-parasitoid model with parasitoid-dependent attack rate. For each value of k, roots of the characteristic equation corresponding to consumer-resource cycles have negative real parts to the right and below the solid boundary, i.e., consumer-resource cycles are suppressed in these regions. To the right and below the solid boundary the equilibrium is locally stable for each value of k, except for the small shaded region near the y-axis when $k = 1$. In that shaded region, persistent single-generation (S-G) cycles occur. Above and to the left of the solid boundary in each case, dynamics are dominated by persistent consumer-resource cycles. $\rho = 33, d_P = 8/T_J, a = 1/T_J, d_J = 0/T_J, S_I = 1$, with $T_J = 1$. (b) Simulation showing single-generation cycles, with $T_A/T_J = 0.01, T_P/T_J = 0.75, k = 1$. (c) Example of single-generation cycles in *Opisina arenosella*, a web-forming caterpillar on coconut in Sri Lanka, redrawn from Godfray and Hassell 1989 with permission from the British Ecological Society.

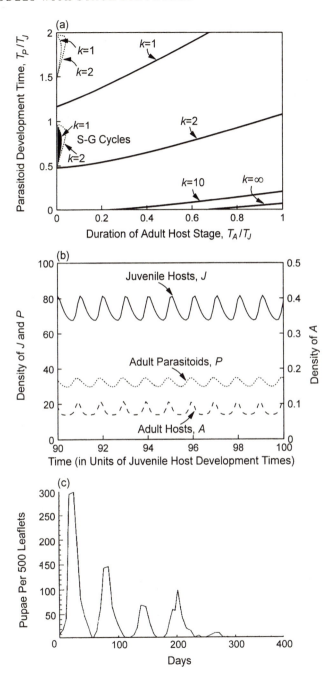

suppressing the unstable consumer-resource cycles in this region of parameter space. The cycles are most prevalent in the region where the parasitoid time delay is a fraction, or 1+ a fraction, of the host development time. Finally, the single-generation cycles are themselves suppressed if the parasitoid density dependence is sufficiently strong. Thus these stable limit cycles occur in a fairly narrow region of parameter space. The shaded region in Fig. 5.10a (for $k = 1$) is the only case in which the single-generation cycles are not swamped out by the consumer-resource cycles, and are therefore persistent. Godfray and Hassell (1989) have suggested that these cycles correspond with roughly single-generation cycles seen in several tropical insect pests whose adult stage is brief (e.g., Fig. 5.10c).

Overall, this section tells us that insights from simpler models, both non-delayed and delayed, generally carry through to our stage-structured parasitoid-host models. We also see several new types of dynamics, however, which result from the stage-structured interaction itself. Of these the most interesting are the single-generation cycles that are exposed when consumer-resource cycles are suppressed.

SINGLE-GENERATION CYCLES IN PARASITOID-HOST MODELS

Causes of Single-Generation Cycles

Godfray and Hassell (1989) provide a clear account of how such cycles might arise in their model. The same reasoning applies here.

First, notice that the period of the cycle is one host generation; i.e., T_J (duration of the juvenile host stage) plus a portion of T_A (duration of the adult host stage). Now ignore for a moment the parasitoid. If we perturb the system by increasing the number of juvenile hosts at one point in time, this abundant cohort will produce another abundant cohort one host-generation time later, and this pattern will repeat itself (though the peaks could be expected to spread out through time). Now add the effect of the parasitoid. In the model, the peak in host density will immediately produce a peak of immature parasitoids. These will develop into a peak of adult parasitoids T_P days later. These parasitoids will cause a higher-than-equilibrium fraction of host deaths. So a crucial requirement for the cycles to occur is that this peak in adult parasitoid density occur at a time when both the hosts that gave rise to it, and the offspring of these hosts (the next host peak), are not present as juveniles. That is, the hosts in the abundant cohorts need to be in the adult stage. If they are not, the peak of hosts will produce a peak in searching parasitoids T_P days later, and these parasitoids will annihilate either the peak of juveniles or their offspring. If T_P has the correct value, the peak of searching adult parasitoids will create a trough of host density that alternates with the peak of juvenile hosts, thus reinforcing the cycles.

The duration of the parasitoid juvenile stage is therefore crucial. In our model, if the adult hosts live effectively for an instantaneously short time, producing their lifetime births in a moment, the cycles can only occur if T_P is about 0.5 to 0.9. If T_P is shorter, a peak in parasitoid density will kill the high-density cohort of hosts just before it matures. If T_P approaches 1, the second-generation peak of hosts will appear just as the adult parasitoid peak emerges and will again be annihilated. Fig. 5.10b illustrates how a peak in juvenile host density leads to a peak in adult parasitoid density T_P days later, which coincides with the next peak in (safe) adult hosts, rather than juvenile hosts.

These cycles thus contrast with the consumer-resource limit cycles, in which adult parasitoid density lags $\frac{1}{4}$ of a cycle period behind the immature hosts (Fig. 5.6c; see also Figs. 3.1 and 3.5).

Window of Vulnerability

The Basic model assumes there are vulnerable juvenile and invulnerable adult hosts. But in real insects the vulnerable stage is typically restricted to only part of the immature host period. Most obviously, in metamorphosing species with eggs, larvae, and pupae, usually only one of these stages is attacked by a given parasitoid species (Fig. 5.11a). Even if this is the larval stage, there is typically only a segment of the larval period that is vulnerable to any particular parasitoid species. In insects that do not metamorphose, and so have no or fewer qualitatively distinct stages, the vulnerable stage is typically also restricted to only a segment of the immature period. The interaction between California red scale and *Aphytis melinus* provides an example (Fig. 5.5).

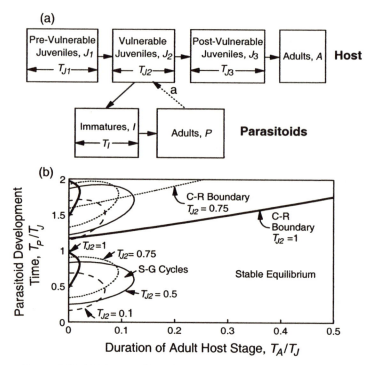

FIGURE 5.11. (a) Diagram of the window of vulnerability model. Only stage J_2 is vulnerable to attack by the parasitoid. (b) Stability boundaries for the window of vulnerability model, showing regions with single-generation (S-G) cycles, consumer-resource (C-R) limit cycles, and a stable equilibrium (sets of curves shown with the same line type—thin solid, thick solid, dashed, and dotted—are for a single value of T_{J2}). For each value of T_{J2}, roots of the characteristic equation corresponding to consumer-resource of cycles have negative real parts to the right and below the line labeled, i.e., consumer-resource cycles are suppressed in these regions. For depicted values of T_{J2} less than 0.75, consumer-resource cycles are suppressed throughout all of the parameter space shown. Single-generation cycles occur for parameters to the left of the curved regions for each value of T_{J2}. $\rho = 33, d_P = 8/T_J,$ $k = 1, T_{J1} + T_{J2} + T_{J3} = T_J = 1, d_J = 0, S_I = 1.$

The difference between metamorphosing species and those with direct development is thus quantitative rather than qualitative from the point of view of vulnerability to parasitism. The window of vulnerability can be characterized by two features: its duration and its position in the life cycle. In this subsection we explore the effects of variation in these two aspects.

To investigate this we look at a model in which the juvenile host stage is divided into a pre-vulnerable stage (J_1), a vulnerable stage (J_2), and a post-vulnerable stage (J_3), of durations T_{J1}, T_{J2}, and T_{J3}, respectively (Fig. 5.11a); these could be egg, larval, and pupal stages. Adult hosts remain invulnerable. As before we scale all parameters relative to the total juvenile host development time, $T_J (= T_{J1} + T_{J2} + T_{J3})$, and explore the effects of altering the fraction of the host immature period that is vulnerable to parasitism: T_{J2}/T_J. We also compare the effects of an invulnerable stage that comes before the vulnerable stage in the life history with one that comes after the vulnerable stage.

Here we need to modify the Basic model only slightly. The two additional invulnerable stages can be incorporated simply as two additional time lags: in the recruitment rate into the vulnerable juvenile host stage and the recruitment rate into the adult stage, as no density-dependent processes occur during these pre- and post-vulnerable immature stages. For simplicity, we assume that all juvenile host stages have the same per-head background mortality rate, d_J, due to causes unrelated to parasitism.

$$\frac{dJ_2(t)}{dt} = R_{J2}(t) - M_{J2}(t) - f[P(t)]J_2(t) - d_J J_2(t)$$

$$\frac{dA(t)}{dt} = R_A(t) - d_A A(t) \tag{5.28}$$

$$\frac{dP(t)}{dt} = f[P(t - T_P)]J_2(t - T_P)S_I - d_P P(t)$$

where $f[P(t)]$ is a function to be described below. The recruitment rate into the vulnerable stage is simply b times the number of adults present T_{J1} days ago multiplied by the probability of survival through the pre-vulnerable juvenile stage:

$$R_{J2}(t) = bA(t - T_{J1})\exp\{-d_J T_{J1}\}. \tag{5.29}$$

The maturation rate out of the vulnerable juvenile stage is

$$M_{J2}(t) = R_{J2}(t - T_{J2})\exp\left(-\int_{t-T_{J2}}^{t} [f[P(x)] + d_J]dx\right), \tag{5.30}$$

and the recruitment rate into the adult stage is

$$R_A(t) = M_{J2}(t - T_{J3}) \exp\{-d_J T_{J3}\}. \qquad (5.31)$$

When the per-head mortality of hosts due to parasitoids is simply $f[P(t)] = aP(t)$, as in the Basic model, the single-generation cycles are still swamped by unstable consumer-resource cycles regardless of the duration of the window of vulnerability. However, Godfray and Hassell (1989) investigated a version of this model in which the attack rate depended on parasitoid density, using the function in equation 5.26:

$$f[P(t)] = k \ln \left(1 + \frac{aP(t)}{k} \right).$$

They showed, as noted above, that such density dependence tends to suppress the consumer-resource limit cycles and expose single-generation cycles. We next explore the effect of the duration of the window of vulnerability in this case.

Shortening the duration of the vulnerable window has two effects. First it tends to increase the likelihood of single-generation cycles. Fig. 5.11b shows that decreasing the fraction of the juvenile stage that is vulnerable to parasitism has the effect of increasing the region of parameter space in which the roots corresponding to single-generation cycles have positive real parts. In contrast to the Basic model with density-dependent parasitoid attack rate, discussed above, a peak of immature hosts can avoid being parasitized by the adult parasitoids they give rise to by being in any of the three invulnerable stages (pre-vulnerable juveniles, post-vulnerable juveniles, or adults). Thus the brevity of the window of vulnerability, combined with the short adult stage, increases the likelihood of single-generation cycles. (See also Briggs and Godfray 1995b for a discussion of an analogous insect-pathogen model in which single-generation cycles as well as cycles that have a period of $\frac{1}{2}$, $\frac{1}{4}$, and other fractions of the host development time can occur.) However, the second effect of a short window of vulnerability is a stabilizing one. Reducing the fraction of the juvenile host period that is vulnerable to parasitism tends to reduce the region of parameter space in which consumer-resource cycles occur, and for very short windows of vulnerability it can also decrease the region in which single-generation cycles can occur.

When no density-dependent processes occur during the invulnerable juvenile stages, the position of the vulnerable stage in the juvenile period has no effect on either the equilibrium densities or the stability boundaries. We recover the same picture as in Fig. 5.11b regardless of whether the invulnerable portion of the development time occurs before or after the vulnerable stage.

Adult Host Stage Is Vulnerable

Thus far we have considered only parasitoids that attack during the host's juvenile stage. There are some parasitoid species that attack only the adult host stage, and we comment briefly on the dynamic outcome. This case was first explored by Hastings (1984). The model is simpler because, as in the case of the juvenile parasitoids above, the juvenile hosts die off at a constant background rate and therefore act as a simple time lag. So now we need specify explicitly only the dynamic equations for adult hosts and parasitoids.

BALANCE EQUATIONS

$$\frac{dA(t)}{dt} = bA(t - T_J)S_J - aA(t)P(t) - d_A A(t)$$

$$\frac{dP(t)}{dt} = aA(t - T_P)P(t - T_P)S_I - d_P P(t) \tag{5.32}$$

where $S_J = \exp\{-d_J T_J\}$ is the probability of surviving through the juvenile host stage. All other parameters are as in the Basic model.

EQUILIBRIA

The equilibria are precise analogs of the Lotka-Volterra equilibria:

$$A^* = \frac{d_P}{aS_I}, \quad P^* = \frac{bS_J - d_A}{a}. \tag{5.33}$$

STABILITY

In the absence of the parasitoid, the average duration of the adult stage is $T_A = 1/d_A$. The actual duration of the adult host stage will be shorter than T_A, however, because of mortality by the parasitoid. We again set total host fecundity in the absence of the parasitoid to $\rho = bT_A = b/d_A$

and investigate the effect of varying T_A while keeping ρ constant, but realizing in this case that actual lifetime fecundity will be reduced by the action of the parasitoid.

Fig. 5.12a shows that an invulnerable juvenile host stage suppresses the consumer-resource cycles, so that the parameter space is now dominated by single-generation cycles, with a period of approximately one host development time. In addition, smaller regions of parameter space, in which cycles with a period of $\frac{1}{2}$, $\frac{1}{3}$, and $\frac{1}{4}$ of a host single-generation, are now prominent. So single-generation cycles can arise without any density dependence in the parasitoid population: the invulnerable juvenile stage suppresses the consumer-resource cycles, and the shortened duration of the adult stage (caused by parasitoid attacks) makes single-generation cycles more likely.

The mechanism leading to single-generation cycles (and fractional-generation cycles) is the same as in the model specified in equation 5.28, in which only part of the juvenile stage is attacked by the parasitoid. If the development time of the parasitoid is a fraction of that of the host, then the adult parasitoids produced from attacks on a peak in adult hosts will not occur at the same time as the offspring of the peak in adult hosts. This will lead to high parasitism at times when the majority of hosts are not in a vulnerable stage, reinforcing the troughs in host density. The fractional-generation cycles can arise because for certain values of the parasitoid development time, it is possible for more than one "dominant cohort" of hosts to be established within the development time of the host. See Briggs and Godfray 1995b for a full discussion.

Single-Generation Cycles in Real Populations

Godfray and Hassell (1989) collected examples of cycles in tropical insects where the cycle period is roughly equal to the development time of the host. They suggest that these exemplify the single-generation cycles predicted above, when parasitoid search rate is assumed to decrease with parasitoid density. Several of these examples, in moth species, fit the requirements: the juvenile host stage is parasitized and the adult is invulnerable and short-lived (of course we do not know if parasitoid search rate was density dependent).

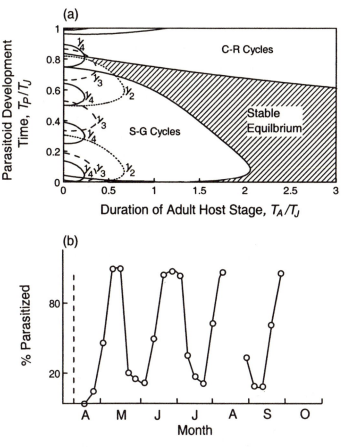

FIGURE 5.12. (a) Stability diagram for the model in which adult hosts are attacked by the parasitoid and juvenile hosts are invulnerable. Shown are regions in which there is a stable equilibrium, consumer-resource (CR) cycles, single-generation (S-G) cycles, and cycles with a period of $\frac{1}{2}$, $\frac{1}{3}$, and $\frac{1}{4}$ of a host development time. $\rho = 10$, $d_P = 5$, $T_J = 1$, $S_J = 1$, $S_I = 1$, $a = 1$. Pattern of generation and fractional-generation cycles begins to repeat when T_P / T_J approaches 1 (solid line at top left), but these cycles are not expressed because they are swamped by consumer-resource cycles. (b) Single-generation cycles in West Indian cane-fly, as indicated by the fraction of adult cane-flies parasitized by *Stenocranophilus quadratus*. Reproduced from Metcalfe 1971 with the permission of CAB International.

A few of the examples, however, do not correspond well with such a model. One of these we discuss here, and another in chapter 6. Metcalfe (1971) found clear evidence for single-generation cycles in the West Indian cane-fly (a delphacid leafhopper, *Saccharosydne saccharivora*), a pest of cane sugar. The cycles, as reflected in the fraction of adults parasitized, are illustrated in Fig. 5.12b. The nymphs and adults are attacked by a stylopid parasitoid in the genus *Stenocranophilus*, but the main effect is to sterilize the adult. The egg stage, which lasts about as long as the adult, is not attacked. The parasitoid takes somewhat less time to develop than does the host.

The life histories in this interaction thus correspond better with the model just presented, in equations 5.32, where the juvenile hosts are invulnerable and the adults are attacked. This model, furthermore, generates single-generation cycles over a large region of parameter space (Fig. 5.12a) and is thus a feasible explanation of these real cycles.

Koinobiotic Parasitoids

In the Basic model and all of its modifications discussed so far, we have been assuming an idiobiontic lifestyle by the parasitoids. Idiobionts are parasitoids that immediately kill or paralyze the hosts they parasitize, so the hosts do not continue to grow or develop. For idiobionts it is often reasonable to accept our assumption that the development time of the immature parasitoid, T_P, does not depend on the age of the juvenile host attacked. Koinobionts, by contrast, do not immediately kill their host. They instead allow the host to continue to develop for some time. In some cases, the juvenile parasitoid does not start to develop within the host until the juvenile host has reached a particular stage of development. This means that for many koinobiontic species, the development time of the parasitoid may depend on the age of the host when it was parasitized, with parasitoids taking longer to develop on young juvenile hosts than on old hosts.

Gordon et al. (1991) investigated the effect of this aspect of the parasitoid life history in a variation of the window of vulnerability model described in the preceding section. The host life cycle is divided into a pre-vulnerable (J_1), vulnerable (J_2), and post-vulnerable (J_3), juvenile stage and an invulnerable adult stage (A), as in the window of vulnerability model. Adult parasitoids attack the J_2 stage, but the juvenile parasitoid does not start to develop until the host reaches the age where it would mature out of the J_2 stage. Thus, unparasitized and parasitized hosts in the J_2 stage continue to develop at the same rate. When they reach the age of $T_{J1} + T_{J2}$, unparasitized hosts enter the J_3 stage but parasitized hosts enter the immature parasitoid stage. The immature parasitoid stage then lasts a fixed duration of T_P days, before the parasitoids mature into the adult parasitoid stage.

The only modification to the window of vulnerability model described above is as follows. The recruitment rate into the immature parasitoid stage is equal to the rate that hosts entered the J_2 stage T_{J2} days ago, multiplied by the probability of surviving background mortality but then being parasitized (the term in the { } below). Thus,

$$R_I(t) = R_{J2}(t - T_{J2}) \exp(-d_J T_{J2})$$

$$\times \left\{ 1 - \exp\left(-\int_{t-T_{J2}}^{t} [f[P(x)]]dx \right) \right\}. \qquad (5.34)$$

Gordon et al. (1991) found that this had little effect on the host-parasitoid dynamics. If parasitoid density dependence is included in the attack rate, the stability boundaries for the model are very similar to those in Fig. 5.11b. The only difference is that the region of parameter space in which single-generation cycles occur is shifted slightly to the lower values of T_P. This can be explained by the fact that the actual time between attack by a parasitoid and emergence of a new adult parasitoid is slightly longer than T_P, because of the variable delay before the parasitoid enters the immature parasitoid stage.

Technicalities Involving
Delay Differential Equations

Delay differential equations (DDEs) share many properties with ordinary differential equations (ODEs), but there are several important differences, which must be appreciated before trying to compute explicit numerical solutions. By the "numerical solution" of a DDE, we mean a table of values of the state variables over some time interval—for example, the data used to construct many of the figures in chapter 5. In this appendix, we first discuss some features of the solutions of DDEs and then comment on subtleties relating to the formulation of stage-structure models that use DDEs. We outline methodology for local stability analysis and the determination of cycle periods.

For further discussion of the formulation and analysis of stage-structured models using DDEs, see Nisbet 1997. For a serious mathematical treatment of DDEs, see Gyori and Ladas 1991, Kuang 1993, or Diekmann et al. 1995. MacDonald 1989 is almost completely devoted to local stability analysis, with particular emphasis on models with distributed delays. Gurney and Nisbet (1985) and Jones et al. (1988) analyze cycle periods in single-species models.

"SOLVING" DDEs

Arguably the most important difference between DDEs and ODEs is that a solution to a DDE is not specified uniquely by the initial value of the dependent variables. For example, in the blowfly model described by equation 5.3, it is not sufficient to give the value $A(0)$ of the adult population at the start of a simulation; further information is required.

This normally will come in one of two forms, the first possibility being specification of the adult population over a time interval equal in length to the time delay, T_J, e.g., the interval $-T_J \leq t < 0$. We call this extra information the *initial history*. Armed with knowledge of the initial history, we can visualize the mechanics of constructing a numerical solution to a first-order DDE by considering successive intervals of duration T_J as follows. The initial history specifies the values of $A(t - T_J)$ for $0 \leq t < T_J$; for example, if T_J is equal to 15 days, t is equal to 5 days, then $t - T_J$ is equal to -10 days, which lies in the range -15 to 0 over which the history was specified. We can now integrate the DDE and obtain values of $A(t)$ for all times up to $t = T_J$ using any one from the multiplicity of numerical algorithms on offer for solving ODEs. But $A(t)$ for $0 \leq t < T_J$ is equal to $A(t - T_J)$ for $T_J \leq t < 2T_J$, so we can now integrate the DDE up to t equal to $2T_J$; and so on for as long as we want.

A second form for the additional information required to solve a DDE is applicable in many ecologically important situations where an initial history is unavailable, or even meaningless. For example, a laboratory population may be set up at some particular time (which we call $t = 0$), and we cannot assign a value to population size at times prior to this starting time. In such circumstances, it is natural to regard the population dynamics in a two-stage model as being determined by the DDE together with the *initial juvenile age distribution* (i.e., the number of individuals per unit age interval of age a at time 0), which we denote by $J_i(a)$, where $0 \leq a < T_J$. In the blowfly model, the adult recruitment rate at times in the interval $0 \leq t < T_J$ is then given by

$$R_A(t) = J_i(T_J - t) \exp[-(T_J - t)d_J]. \tag{A5.1}$$

To make sense of this expression, notice that $T_J - t$ is the initial age of a juvenile maturing at time $t \leq T_J$. To obtain the adult population $A(t)$ during this initial time interval $0 \leq t < T_J$, we solve the ODE

$$\frac{dA}{dt} = R_A(t) - d_A A(t) \tag{A5.2}$$

with $R_A(t)$ evaluated using equation A5.1. For $t \geq T_J$, the integration of the DDE proceeds as in the previous paragraph, the solution from 0 to T_J constituting the initial history. This procedure is particularly simple if the initial juvenile population is 0. Then $J_i(a)$ is equal to zero, for all ages $a \leq T_J$, and there is no adult recruitment prior to t equal to T_J.

There are several packages available for numerical solution of DDEs, given an initial history. For example, the Solver package, described by Gurney and Nisbet 1998, contains data structures that are particularly convenient for handling stage-structured models; it was used for all DDE solutions presented in this book and can be downloaded from http://www.stams.strath.ac.uk/ecodyn/solvman.html. A Windows-based DDE–solving program written by Simon Wood of the University of St. Andrews is available from http://dolphin.mcs.st-and.ac.uk/simon/dde.html; this code has superior numerics to Solver but lacks some of the special features for stage-structured models.

MATHEMATICAL TRICKS WITH
STAGE-STRUCTURED MODELS

There are some unpleasant technicalities associated with evaluation of the integral that defines the through-stage survival in models with time-dependent juvenile mortality (as is present in almost all host-parasitoid models and in some single-species models). Similar problems arise in models with dynamically varying time delays such as the *Daphnia* model discussed in chapter 11. For example, calculating the maturation rate from the juvenile to the adult stage in the *Plodia* model presented in this chapter involves evaluating the survival term defined in equation 5.7, namely

$$S_J(t) = \exp\left[-\int_{t-T_J}^{t} (cJ(x) + d_J)dx\right]. \qquad (A5.3)$$

This can be evaluated, at each time step, by numerical integration, but this is computationally very inefficient. A preferable approach uses one of several mathematical tricks to circumvent the need to evaluate the integral. All involve differentiating an integral whose limits depend on time. The key is a formula that mathematically adept readers can readily derive for themselves but is only occasionally written out explicitly in calculus texts, namely that if

$$y(t) = \int_{a(t)}^{b(t)} f(x, t)dx \qquad (A5.4)$$

then

$$\frac{dy}{dt} = \int_{a(t)}^{b(t)} \frac{\partial f(x, t)}{\partial t} dx + f(b(t))\frac{db(t)}{dt} - f(a(t))\frac{da(t)}{dt}. \quad \text{(A5.5)}$$

The trick for handling equation A5.3 and similar equations is to differentiate with respect to time and obtain a DDE for $S_J(t)$ (e.g., Gurney et al. 1983; Nisbet and Gurney 1983; Gurney and Nisbet 1985; Murdoch et al. 1987). We rewrite equation A5.3 as

$$\ln[S_J(t)] = -\int_{t-T_J}^{t} (cJ(x) + d_J)dx, \quad \text{(A5.6)}$$

and then make use of equation A5.5 with $f(x, t) = cJ(x) + d_J, a(t) = t - T_J, b(t) = t$, implying $\partial f(x, t)/\partial t = 0, da(t)/dt = 1, db(t)/dt = 1$ to obtain:

$$\frac{d}{dt}\ln[S_J(t)] = \frac{1}{S_J(t)}\frac{dS_J(t)}{dt}$$

$$= -([cJ(t) + d_J] - [cJ(t - T_J) + d_J]),$$

i.e.,
$$\frac{dS_J(t)}{dt} = S_J(t)(cJ(t - T_J) - cJ(t)). \quad \text{(A5.7)}$$

Equation A5.7 is solved simultaneously with the other model equations, the initial condition being determined from equation A5.6 using the initial history. The calculation is particularly simple for the (common) situation where the system is assumed to be empty prior to $t = 0$. Then $S_J(0) = e^{-d_J T_J}$.

An alternative way of dealing with equation A5.3 is especially useful in simulations that start with an initial age distribution. We define a new variable:

$$\Phi(t) \equiv \int_{0}^{t} (cJ(x) + d_J)dx. \quad \text{(A5.8)}$$

For times $t < T_J$, the new variable represents the proportion of individuals aged $T_J - t$ in the initial age distribution that survive and mature at time t (Nisbet 1997). Application of equation A5.5 to differentiate equation A5.8 is straightforward, and we find that the new variable obeys the ODE

$$\frac{d\Phi(t)}{dt} = cJ(t) + d_J \quad \text{with initial condition } \Phi(0) = 0. \quad (A5.9)$$

The juvenile through-stage survival with $t > T_J$ is given by

$$S_J(t) = \exp[\Phi(t - T_J) - \Phi(t)]. \quad (A5.10)$$

The trick now is to solve equation A5.9 simultaneously with the other DDEs that make up the model, and to store the history of Φ as we go along. The through-stage survival can then be evaluated when needed using equation A5.10.

LOCAL LINEARIZATION AND THE CHARACTERISTIC EQUATION

Local stability analysis for DDEs rests on the same principles we outlined for ODEs in chapter 3. The dynamic equations are linearized about the equilibrium, and solutions proportional to $e^{\lambda t}$ are assumed. This enables calculation of a characteristic equation all of whose roots must have negative real parts for the equilibrium to be locally stable. The practicalities are very different, however, and much more difficult, than with ODEs for several reasons, the most important being that the characteristic equations for DDEs are typically transcendental equations whose form includes a mix of polynomials and exponentials. These equations normally have an infinite number of roots.

For a small number of simple models, it is possible to rigorously prove stability (see references above), but for most of the more complex stage-structured models we are limited to calculating stability boundaries. Here we detail the process in a simple context, the model of Nicholson's blowflies described in equation 5.3. For this model, it is possible to calculate stability boundaries explicitly (Gurney et al. 1980; Nisbet and Gurney 1982; Nisbet 1997), but here we present the calculation in a form that admits generalization to the tougher models.

We rewrite the model in the following form:

$$\frac{dA(t)}{dt} = S_J R_J(t - T_J) - d_A A(t) \quad (A5.11a)$$

$$R_J(t) = qA(t)\exp(-A(t)/A_0). \quad (A5.11b)$$

The equilibrium is given by equation 5.4:

$$A^* = A_0 \ln\left(\frac{q S_J}{d_A}\right); \qquad R_J^* = q A^* \exp(-A^*/A_0). \qquad \text{(A5.12)}$$

We linearize equations A5.11 by considering small perturbations from equilibrium:

$$a(t) = A(t) - A^*; \qquad r_J(t) = R_J(t) - R_J^*. \qquad \text{(A5.13)}$$

Then after a little algebra (which makes use of the approximation $e^{-x} \approx 1 - x$ when x is small), we find:

$$\frac{da(t)}{dt} = S_J r_J(t - T_J) - d_A a(t) \qquad \text{(A5.14a)}$$

and
$$r_J(t) = d_A\left(1 - \frac{A^*}{A_0}\right) a(t). \qquad \text{(A5.14b)}$$

Now assume the familiar form for solutions

$$a(t) = \tilde{a} e^{\lambda t}; \qquad r_J(t) = \tilde{r}_J e^{\lambda t} \qquad \text{(A5.15)}$$

where $\tilde{a}$ and $\tilde{r}_J$ are constants and λ is the eigenvalue. These forms imply

$$\frac{da(t)}{dt} = \lambda \tilde{a} e^{\lambda t}; \qquad r_J(t - T_J) = e^{-\lambda T_J} \tilde{r}_J e^{\lambda t}. \qquad \text{(A5.16)}$$

Substituting in equations A5.14 yields:

$$(\lambda + d_A)\tilde{a} - S_J e^{-\lambda T_J} \tilde{r} = 0 \qquad \text{(A5.17a)}$$

$$-d_A\left(1 - \frac{A^*}{A_0}\right) \tilde{a} + \tilde{r} = 0 \qquad \text{(A5.17b)}$$

which can be rewritten in matrix notation as:

$$\mathbf{m}(\lambda)\begin{pmatrix} \tilde{a} \\ \tilde{r} \end{pmatrix} = \mathbf{0} \quad \text{with} \quad \mathbf{m}(\lambda) = \begin{bmatrix} \lambda + d_A & -S_J e^{-\lambda T_J} \\ -d_A\left(1 - \frac{A^*}{A_0}\right) & 1 \end{bmatrix}.$$
$$\text{(A5.18)}$$

The equations only have non-zero solutions if

$$\text{Det}(\mathbf{m}(\lambda)) = 0. \qquad \text{(A5.19)}$$

Equation A5.19 is the characteristic equation for this system.

With larger systems the 2×2 matrix in equation A5.18 is replaced with a matrix of higher dimension.

COMPUTING STABILITY BOUNDARIES AND
CYCLE PERIODS

Our approach to the calculation of stability boundaries is closely analogous to that taken in chapter 3. On a stability boundary, either a root of the characteristic equation is zero, or there is a complex conjugate pair of pure imaginary roots, which we write as $\lambda = \pm i\omega$. Thus from equation A5.19,

$$\mathrm{Det}(\mathbf{m}(i\omega)) = 0. \tag{A5.20}$$

This relationship in fact defines two equations in real variables, obtained by setting the real and imaginary parts of the determinant separately to zero.

So if, for example, we want to calculate the stability boundary for the blowfly model in the $q - d_A$ plane, we choose values for the other model parameters, and are left with two equations relating the three variables q, d_A, and ω. These define the stability boundary, and the period of cycles on the boundary is calculated from the usual equation $P = 2\pi/\omega$. Mathematica code implementing this process is given in Box 5A.1. Even readers unfamiliar with Mathematica are encouraged to read this example, as interpreting the code is probably the easiest way to understand the steps involved in the calculations.

Unfortunately, there are several problems that must be addressed in order to have confidence in the results of such a calculation. There are an infinite number of roots to the characteristic equation, but our recipe only finds the parameter combinations at which one conjugate pair are pure imaginary. There is no guarantee that these are the dominant roots. It is possible to devote considerable mathematical effort to resolving these problems, and in some circumstances this effort is necessary. However, we can commonly use a judicious mixture of numerical solutions of the DDEs and runs of a program such as that shown in Box A5.1 to obtain boundaries that are valid "beyond reasonable doubt." Numerical experiments will help achieve good initial guesses for the roots (especially for the periods). Then once a candidate stability boundary has been computed, it can be checked by obtaining numerical solutions with parameter values chosen above and below the boundary. If all checks out, we trust the boundary.

BOX A5.1.

MATHEMATICA CODE FOR EVALUATING THE LOCAL STABILITY
BOUNDARY AND CYCLE PERIODS FOR THE BLOWFLY MODEL

Clear[M, m, f, lambda, fr, fi, Period, w, Tj, q, da, A0, Astar, LSB,
LSBPLOT, BOUNDPERIOD, PERIODPLOT]
Array[M, {2, 2}];

(* **Define equilibrium population and matrix elements** *)
Astar := A0 Log[q Sj/da]; (* **Adult equilibrium** *)
M[1, 1] := lambda + da; M[1, 2] := -Sj Exp[-lambda Tj];
M[2, 1] := -da (1 - Astar/A0); M[2, 2] := 1
m = Array[M, {2, 2}];

f[lambda_] := Det[m]; (* **Characteristic equation** *)

(* **Specify pure imaginary roots and break out real and
imaginary parts of CE** *)
lambda := I w; w := 2 Pi/Period; (* **relationship between omega and
period** *)
g := ComplexExpand[f[lambda]]
fr := ComplexExpand[Re[g]]
fi := ComplexExpand[Im[g]]

(* **Values of fixed parameters** *)
A0 = 1; Sj = 0.9; Tj = 15.6;

(* **Switch off warning messages -
USE THIS OPTION WITH CARE ** AFTER ** DEBUGGING!!** *)
Off[FindRoot::frnum]
Off[ReplaceAll::reps]
Off[FindRoot::regex]
Off[FindRoot::frns]
Off[FindRoot::frsec]

(* **Solve CE and plot stability boundary** *)
LSB[da_] := q /. FindRoot[{fr == 0, fi == 0}, {Period, 32, 50}, {q, 2, 3}];
LSBPLOT =
    Plot[LSB[da], {da, 0.01, 0.3}, PlotRange -> {0, 10}, AxesLabel ->
    {da, q}]

(Box A5.1. continued)

(* Calculate and plot period on the stability boundary *)
BOUNDPERIOD[da_]:=
  Period/. FindRoot[{fr == 0, fi == 0}, {Period, 32, 50}, {q, 2, 3}];
PERIODPLOT= Plot[BOUNDPERIOD[da], {da, 0.01, 0.3},
  PlotRange -> {0, 100}, AxesLabel -> {da, Period}]

This code produces the output plots shown in Fig. A5.1.

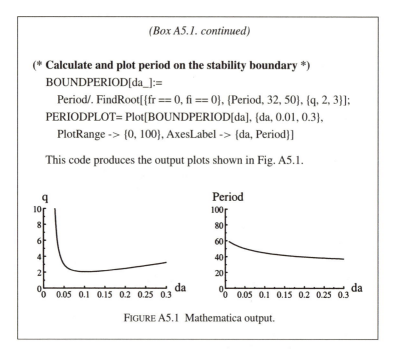

FIGURE A5.1 Mathematica output.

We end this introduction to stability analysis for stage-structured models with a warning. The repertoire of complex dynamic behavior that can occur in these models is much greater than with ODEs, and there is much less mathematical literature to guide our intuition than exists for ODEs. The limitations of intuition based on local stability analyses must be kept in mind; for example, several models in chapter 6 have multiple attractors (a stable point and a limit cycle), a feature not revealed by local stability analysis.

CHAPTER SIX

Dynamical Effects of Parasitoid Lifestyles

As noted in chapter 5, host life histories show enormous variation. Since parasitoid natural history is also highly variable, the diversity of forms of parasitoid-host interactions is great. Our task in chapters 6 and 7 is to bring order to this diversity and establish that we can encompass most of it in a general and relatively simple theory.

The life histories of parasitoids, the ways they respond to the variety of host attributes, the "decisions" they make in allocating reproductive effort, etc., all show variation from one group of parasitoids to another. There is also variation in how any particular parasitoid responds to the different host individuals it meets, both because the hosts can vary, for example in size, and because the state of the parasitoid can vary. We can think of the particular way that each parasitoid species deals with the variation it meets as its lifestyle. Although each species is unique, lifestyles can be grouped into classes with distinctive features.

Our aim in the next few chapters is to present a general dynamical theory that takes as its building blocks the major lifestyles seen in parasitoids. We will show that the great mass of variation can be collapsed to a single framework with a small range of resulting population dynamics. Although parasitoids are our focus in this chapter, it is likely that the dynamic consequences of the behaviors discussed are paralleled in predators, which we comment on at the end of the chapter.

In this section we survey the major lifestyles for the reader unfamiliar with parasitoid natural history. We review the main types of differences among individual hosts, the main types of parasitoids, and the major kinds of interactions that result. We cannot of course include every variation in lifestyle, but we include a large number of those commonly observed.

First, we remind the reader that the great majority of host insect species go through a distinct metamorphosis between larval and adult stage, usually in a non-feeding pupal stage. Examples of such *holometabolous* insects are moths, beetles, and flies. The *hemimetabolous* remainder grow gradually more adultlike as they develop, with the final stage being the adult. Examples are scale insects, grasshoppers, and plant-sucking bugs. Both types typically have an egg stage, and both types typically have flying adults and relatively immobile juveniles, though sometimes a juvenile stage can disperse, for example by being windblown, and sometimes the adults are immobile (e.g., adult female scale insects). As noted in chapter 5, the adult stage of some holometabolous insects is very short-lived relative to the juvenile stages, whereas in others, and in most hemimetabolous species, it is long-lived.

PARASITOID LIFESTYLES

Parasitoids have a fascinating range of, sometimes deplorable, life cycles. The immature stages live in or on their host, and the adults are usually free flying. They may be entirely parthenogenic. Those that have sexes can choose the sex of the egg as it is being laid by fertilizing it (producing a female) or not (producing a male), because of their *haplodiploid* nature. Parasitoids may produce only one offspring from a host, or many, and some (*polyembryonic*) species even produce many from one egg. In one group, the *autoparasitoids*, the male is a parasitoid of the female. Among the more bizarre variants, which we do not model, are sons that can develop precociously inside their mother inside the host, eat her, and then mate with their sisters.

A major distinction between parasitoids concerns maturation of eggs; i.e., the way oocytes are transformed to eggs containing nutrients that can support the initial development of the larva. From the point of view of population dynamics, there are two main classes.

- *Non-feeders* need no external source to mature all of their eggs. The nutrients needed all derive from reserves developed in the larval stage.

- *Host-feeders* need to feed on individuals of the host species to obtain nutrients (presumably mainly protein) to mature eggs in addition to those they have soon after emergence (which depend on larval resources). Hosts are often killed in the process of host-feeding.

(Entomologists distinguish between pro-ovigenic species, which emerge with their full complement of mature eggs, and synovigenic species, which emerge with some mature eggs but develop more as time goes on. For a variety of reasons, however, this distinction is less useful for our purposes than is that between host-feeders and non-feeders; Jervis et al. 2001).

Most of what we say in this chapter applies to the two main types, non-feeders and host-feeders, so we do not distinguish between them. It will turn out, for example, that the dynamical effect of host-feeding in some cases is manifest simply as a limiting case of the other parasitoid behaviors we investigate. However, one particular aspect of host-feeding, the possibility of satiation, does have a distinct dynamical effect, which we will discuss in chapter 7. There is one dynamical effect that is special to feeding on resources other than the host, which we will also discuss in chapter 7.

A second distinction, discussed briefly in chapter 5, relates to the response of the host after it is parasitized. Most hosts stop growing when they are parasitized, so the size of the resource package for the parasitoid is fixed at that moment. Parasitoids inducing this response are termed idiobionts. Parasitism by members of the second group, the koinobionts, does not prevent further growth in the host, which continues to feed for at least some of the remaining time. In both cases, as we shall see, larger (older) hosts typically have a larger "yield" for the parasitoid. Once again, we will show that this difference between parasitoid types turns out to matter rather little dynamically.

In addition to these broad differences among parasitoid species, an individual parasitoid typically responds to differences in the individual hosts it encounters. The response is determined largely by the host individual's size or stage. For example, most species attack juvenile and not adult hosts, and within any stage there is usually a restricted window of vulnerability. In addition, within the attacked range, parasitoids also typically respond differently to different-sized hosts, so we complete our brief survey by looking at such variation.

Parasitoid Responses to Host Attributes

We expect parasitoid decisions to reflect selective pressures. For example, parasitoids should lay male eggs in smaller (lower quality) hosts and female eggs in larger hosts, because the fitness of a female offspring should increase faster with host size than that of a male offspring (Charnov 1982). In general, they should commit more reproductive resources (and especially female eggs) to higher quality hosts. A host's quality as a resource tends to increase with its size. Typically, size increases with age, so host quality tends to increase with age, and we will see that response to host age is the key to dynamical consequences. Throughout this chapter we use size as the key host characteristic, but the implication is that typically the relationship also holds for host age. In some cases, host quality declines with host age, a relationship whose dynamic consequences we also consider.

Clutch size depends on host size. Parasitoids frequently lay more eggs in larger hosts, which provide a larger package of nutrients (Luck et al. 1982; Hardy et al. 1992; Vet et al. 1993; Godfray 1994; Mayhew 1998). The difference may be modest. For example, *Aphytis* lays one egg in the smallest suitable stage of California red scale but typically not more than three in the largest. In other species, such as *Colpoclypeus florus* on the tortricid moth, *Adoxophes orana,* the largest host may receive up to 40 times as many eggs as the smallest (Dijkstra 1986).

Juvenile parasitoids survive better in larger hosts. Juvenile parasitoids tend to survive better in larger and older host larvae, in part because such hosts provide more nutrients, but partly because older hosts tend to suffer less mortality from other sources (Salt 1941; Vet et al. 1993; Srivastava and Singh 1995). This is particularly true in idiobionts that kill or paralyze their host at the time of oviposition. This can also be true in koinobionts, however, for at least three reasons: (1) in larger (older) hosts the juvenile parasitoid develops faster, and so is exposed to mortality forces for a shorter time; (2) older hosts tend to have a lower instantaneous risk of mortality; and (3) there is a higher likelihood of completing development if the juvenile parasitoid receives more nutrients.

Sex allocation is size dependent. Although both sexes may gain an increase in fitness with increasing size, Charnov (1982) pointed out that the gain is likely to be greater for females than for males, as body size is likely to affect female fecundity more than male mating success. There is therefore selection for using larger hosts for female eggs, and for relegating male eggs to smaller hosts (King 1994; Boavida et al. 1995; Wen et al. 1995). In *Aphytis*, for example, female eggs are laid mainly in third instar scales, but male eggs are laid in large second instars. Sex allocation and clutch size can respond to host size in tandem. For example, as host size increases, *Aphytis* clutches may progress through a sequence such as: 1 male egg, 1 female egg, 1 male + 1 female egg, 2 female eggs.

Host-feeding is size-dependent. Host-feeding parasitoids typically feed on smaller hosts to gain nutrients and oviposit on larger hosts. In *Aphytis*, for example, host-feeding occurs mainly in first and early second instars, with oviposition occurring only in later instars. Oviposition yields an immediate gain in fitness, whereas host-feeding yields a gain in fitness only if the parasitoid lives to encounter suitable hosts after the new eggs have been matured (Collier 1995a). It takes less nutrient to produce an egg than to allow development of a juvenile parasitoid, so for evolutionary reasons similar to those in size-dependent sex allocation, the general tendency is for smaller hosts to be used for host-feeding and larger ones for parasitism (Bartlett 1964; Abdelrahman 1974; Bokonon-Ganta et al. 1995).

Juvenile parasitoids develop faster in older hosts. This pattern is common in koinobionts (Bokonon-Ganta et al. 1995; Croft and Copland 1995; Srivastava and Singh 1995; Harvey and Vet 1997; Jones and Greenberg 1999; Plarre et al. 1999; Bertschy et al. 2000) but has also been observed rarely in idiobionts (Islam 1994).

Larger hosts are attacked more frequently. There is remarkably little information on relative attack rates of hosts that differ in size. Attack rate depends on both encounter and acceptance rates. Less frequent rejection of larger hosts has been seen in both idiobionts (Yoo and Ryoo 1989; Islam 1994) and koinobionts (Karamaouna and Copland 2000; Neveu et al. 2000). The reverse may also happen (Chau and

Mackauer 2000). In laboratory and field experiments we found that *Aphytis* encounters larger hosts at a higher rate and typically (but not always) rejects them less often (W. Murdoch and S. Swarbrick, unpubl. data). We might expect rejection rates to vary with host abundance, but this does not seem to have been much studied.

Host quality declines with age. The above pattern is occasionally reversed. In parasitoids of eggs and pupae, survival of the immature parasitoid is sometimes reduced in older hosts (Honda and Luck 2000), sometimes because the older hosts have a better immune response (Bauer et al. 1998; Sagarra et al. 2000). King (1990) found a more male-biased sex ratio in offspring laid in older hosts. This was associated with older hosts (pupae) weighing less than younger pupae. Development in koinobionts is sometimes faster in smaller hosts (Sequeira and Mackauer 1992; Harvey et al. 1999; Hu and Vinson 2000; Nussbaumer and Schopf 2000), though the emerging adults are smaller. This reverse pattern is easily included in the theory to follow.

Some or all aspects of gain may change concurrently with host age. For example, in *Aphytis* older hosts are less likely to be host-fed but yield more future eggs per meal, are more likely to receive female eggs, and are more likely to receive multiple eggs. Thus, although gain may increase as a step function from one stage to the next as a result of any one mechanism, it may change more gradually when they are all combined. To illustrate one possible pattern we calculated a gain curve for *A. melinus* attacking red scale (Fig. 6.1). Our models cover both step functions and smoothly changing gain functions.

A Simplification: Host-Feeding Sometimes
Has No Effect on Stability

Some of the apparent complexity can be removed immediately: in chapter 7 we find that in some cases host-feeding per se has no effect on stability. Host-feeding does, however, interact with host stage structure to affect stability, but it does so in a simple way that fits into our comprehensive framework.

Aphytis provides a typical example of a host-feeder. A day or so after the female emerges she has a full complement of mature eggs

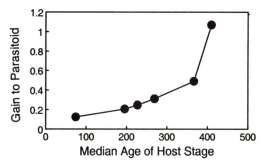

FIGURE 6.1. Estimated gain, in female-egg equivalents, to a female *Aphytis* from attacking hosts of different ages. Host age is the median age (in degree-days) of each of the 6 vulnerable stages of red scale. For each stage, a gain value was calculated from the probability, given an attack, of host-feeding versus parasitism, number of egg-units per meal, probability of laying a given number of male and/or female eggs, and hypothesized values for a male egg and for future eggs derived from current meals.

(typically 6–12, depending on her size), which have developed from nutrients gained as a larva. After laying a few eggs on encountered hosts, she may host-feed, depending on the quality of the encountered host. To host-feed she drills a hole in the scale cover, penetrates the scale body, thrusts with the ovipositor apparently to break up the host body, then turns around and feeds from the wound. The nutrients are turned into new mature eggs within a day or so.

To model host-feeding we need to break the searching female parasitoid population into classes corresponding to the number of mature eggs carried. The dynamics of such *state-dependent* behavior are the subject of the next chapter, so we content ourselves here simply by noting that host-feeding often has no effect on stability if the host population is not subdivided into classes—i.e., all host individuals are the same. Host-feeding reduces the efficiency of transferring attacked hosts to new parasitoids. Some fraction of attacked hosts do not become juvenile parasitoids, so the conversion efficiency is lower than it would be if all hosts were converted to new parasitoids. (We alert the reader here to physiological constraints on conversion rate, which we discuss in chapter 7.) In models such as the Lotka-Volterra model of chapter 3, or our Basic model of chapter 5, altering the conversion efficiency has no effect on stability. Therefore, in those types of models host-feeding tends to have no effect (discussed further in chapter 7). Host-feeding does affect host and parasitoid equilibrium densities, which we also discuss in chapter 7.

This summary glosses over the fact that adding the time delay between a host meal and the creation of a new mature egg is destabilizing (chapter 7). But the delay is small compared with others in the interaction, and we ignore its effect here.

FOUR MECHANISMS INDUCING GREATER GAIN FROM OLDER HOSTS

We now illustrate the theory with four examples (Murdoch et al. 1997). We do this as a way of developing intuition about the general theory of parasitoid lifestyles. As we proceed, it will become clear that the mechanisms are falling into a single theoretical framework, and in the next section we explain the generality of that framework.

Mechanism 1. Size-/Age-Dependent Sex Allocation: The Sex-Allocation Model

Parasitoids tend to lay male eggs in smaller (younger) hosts and female eggs in older hosts. We make two temporary assumptions. (1) An attack on a younger (smaller) host yields a juvenile male parasitoid and makes no contribution to the future female parasitoid population, so we assume there is no "gain" from a host that yields a male. This is a reasonable first-order approximation since male parasitoids do not kill hosts, and it has been argued (though the data are sparse) that virginity is evanescent in adult female parasitoids in the field since there are always more males than are needed to fertilize the female population (Godfray 1994). At the end of this chapter we show that this assumption has no qualitative effect on the result, and its quantitative effect is as expected. (2) An attack on an older juvenile host produces a single female parasitoid. We re-examine this assumption in mechanism 3.

We divide the juvenile host stage into two age classes corresponding to these assumptions. The younger stage, J_1, lives for T_1 days and is attacked and killed by the parasitoid at rate a_1 but makes no contribution to the juvenile female parasitoid population, whereas the older stage, J_2, lives T_2 days, is attacked and killed at rate a_2, and contributes a female juvenile when attacked (Fig. 6.2a). The total juvenile host development time is $T_J = T_1 + T_2$. The model is thus a quite small modification of the Basic model of chapter 5 (see Box 6.1).

The most important factor determining the dynamical effect of age-dependent sex allocation is the relative mortality imposed by the parasitoid on the two juvenile host stages, namely $(a_1 T_1)/(a_2 T_2)$. The product, $a_1 T_1$, is a measure of the intensity of parasitoid mortality on the younger age class. We can increase mortality on this stage by increasing the instantaneous rate, a_1, or the duration over which that rate is experienced, T_1. The two parameters do not have precisely the same effect on dynamics, but the effects are close enough that it is useful to think of them as effectively interchangeable. A few minutes consideration of this parameter group will give deep insight into the dynamics of all of the models examined in this chapter.

First, however, consider the dynamical implications of the structure of the interaction (Fig. 6.2a). Suppose the system is at equilibrium and we perturb it by increasing the density of searching female parasitoids at some time t. This will cause more young hosts to die than would otherwise be the case. As the survivors develop into older juveniles, some days later, the density of older juveniles will be lower than it would otherwise have been—and lower than its equilibrium density. Each female parasitoid searching at that time will therefore contact fewer older juveniles than she would do at equilibrium, and her per-head rate of producing juvenile female parasitoids will therefore be lower. Still later, the per-head rate at which adult parasitoids recruit will also be reduced. Thus, an increase in adult parasitoid density now will lead to a reduction in the per-head rate of parasitoid recruitment later. The host age structure and parasitoid development delay therefore lead to delayed density dependence in the parasitoid recruitment rate.

Now consider what happens when we change the relative mortality imposed on the two age classes. For example, if we increase the ratio, $(a_1 T_1)/(a_2 T_2)$, the strength of the above effect will be increased: i.e., the intensity of the density dependence and/or the length of the delay will be greater.

Finally, consider the effects of delayed density dependence in the parasitoid. Density dependence alone should stabilize the interaction

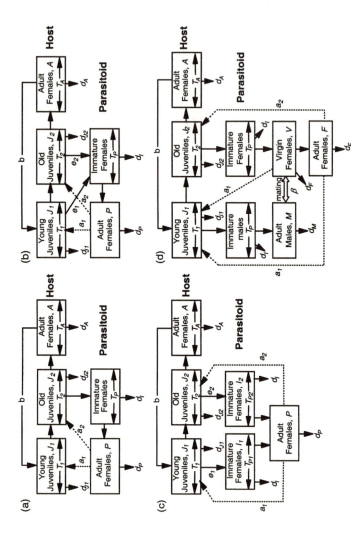

FIGURE 6.2. Diagrams of several of the host-size-/age-dependent behaviors of parasitoids discussed in text. (a) Size-/age-dependent sex allocation or size-/age-dependent host-feeding: young juvenile hosts receive male eggs or are fed upon, old juvenile hosts receive female eggs. (b) Size-/age-dependent clutch size: young juvenile hosts receive e_1 eggs, old juvenile hosts receive e_2 eggs. (c) Koinobiont model with immature parasitoids developing faster on older juvenile hosts. (d) Size-/age-dependent sex allocation in which male parasitoids are explicitly modeled.

BOX 6.1.

SEX-ALLOCATION MODEL

Male parasitoid eggs are laid on young juvenile hosts, female parasitoid eggs on old juvenile hosts. Adult hosts are invulnerable.

Populations

$J_1(t)$ number of young juvenile hosts at time t (receive male eggs)

$J_2(t)$ number of old juvenile hosts at time t (receive female eggs)

$A(t)$ number of adult hosts at time t (invulnerable)

$P(t)$ number of adult female parasitoids at time t

Properties of Individuals

T_1	duration of young juvenile host stage
T_2	duration of old juvenile host stage
$T_J = T_1 + T_2$	total juvenile host development time
T_P	duration of immature parasitoid stage
$T_A = 1/d_A$	average duration of adult host stage
ρ	per-individual lifetime adult host fecundity
a_1	parasitoid attack rate on young juvenile hosts
a_2	parasitoid attack rate on old juvenile hosts
$s_I = \exp(-d_I T_P)$	immature parasitoid through-stage survival
d_{J1}	background death rate of young juvenile hosts
d_{J2}	background death rate of old juvenile hosts
d_A	death rate of adult hosts
d_I	death rate of immature parasitoids
d_P	death rate of adult parasitoids

Vital Processes

$$R_{J1}(t) = (\rho/T_A)A(t)$$ recruitment rate of young juvenile hosts at time t

$$X_1(t) = a_1 J_1(t) P(t)$$ total parasitism rate on J_1 at time t

$$S_{J1}(t) = \exp\left\{ -\int_{t-T_1}^{t} (a_1 P(x) + d_{J1})\, dx \right\}$$ survival through J_1 stage at time t

$$R_{J2}(t) = R_{J1}(t - T_1) S_{J1}(t)$$ recruitment rate of old juvenile hosts at time t

(Box 6.1. continued)

$X_2(t) = a_2 J_2(t) P(t)$ total parasitism rate on J_2
                                              at time t

$S_{J2}(t) = \exp\left\{-\int_{t-T_2}^{t} (a_2 P(x) + d_{J2}) \, dx\right\}$ survival through J_2 stage
                                              at time t

$R_A(t) = R_{J2}(t - T_2) S_{J2}(t)$ recruitment rate into adult
                                              host stage at time t

$R_P(t) = X_2(t - T_P) S_I$ recruitment rate of adult
                                              female parasitoids at
                                              time t

Balance Equations

$dJ_1(t)/dt = R_{J1}(t) - R_{J2}(t) - X_1(t)$ young juvenile hosts
$\qquad\qquad - d_{J1} J_1(t)$

$dJ_2(t)/dt = R_{J2}(t) - R_A(t) - X_2(t)$ old juvenile hosts
$\qquad\qquad - d_{J2} J_2(t)$

$dA(t)/dt = R_A(t) - d_A A(t)$ adult hosts

$dP(t)/dt = R_P(t) - d_P P(t)$ adult female parasitoids

(chapter 3). If the density dependence is delayed, however, as in this case, it may destabilize the interaction by inducing overcompensation in the parasitoid population, in analogy with the single-species models we discussed in chapters 3 and 5. This tension between stabilizing and destabilizing effects is the key to understanding this group of models.

The persisting fundamental Lotka-Volterra-like nature of the model can be seen in that the equilibrium of the older host age class (which is the available host population from the parasitoid's point of view) is an exact analog of that of the host population in the Lotka-Volterra model; the adult parasitoid equilibrium is also an analog of that in the Lotka-Volterra model (Table 6.1). Age structure, however, affects the equilibrial densities of the young immature and adult hosts. These equilibria increase as the ratio a_1/a_2 increases because the parasitoid

TABLE 6.1. Sex-allocation model equilibria

Adult female parasitoids	$P^* = \dfrac{\ln(\rho) - d_{J1}T_1 - d_{J2}T_2}{a_1T_1 + a_2T_2}$
Old juvenile hosts	$J_2^* = \dfrac{d_P}{a_2 s_I}$
Adult hosts	$A^* = \dfrac{(a_2 P^* + d_{J1})\, J_2^*}{d_A\, \{\rho \exp[-(a_1 P^* + d_{J1})\, T_1] - 1\}}$
Young juvenile hosts	$J_1^* = \dfrac{\rho d_A A^*\, \{1 - \exp[-(a_1 P^* + d_{J1})\, T_1]\}}{a_1 P^* + d_{J1}}$

Note, for P^* and J_2^*, analogy with the Lotka-Volterra equilibria. Numerator in P^* is proportional to host rate of increase when adult stage is short, including juvenile background death rates, and denominator is proportional to total per-head attack rate on the two stages. Numerator in J_2^* is parasitoid death rate, and denominator is parasitoid attack rate times conversion rate (i.e., survival rate of immature parasitoids).

is then essentially "wasting" more attacks on younger hosts that do not produce a female offspring, and hence the density of these stages has to increase to maintain the parasitoid at the density that allows it to control the older hosts. An increase in T_1/T_2 has a similar effect. As might be expected, an increase in the parasitoid rate of attack on older hosts causes a decrease in equilibrium density of all host and parasitoid stages, in accordance with simple Lotka-Volterra models (chapter 3).

LOCAL STABILITY

Fig. 6.3 shows the model behavior in the same parameter space as in the Basic model in chapter 5; i.e., in each panel of the picture we show for a particular value of $(a_1T_1)/(a_2T_2)$ the effects of varying the duration of the invulnerable adult stage (relative to the duration of the juvenile host stage) and the duration of the parasitoid developmental delay (also relative to the duration of the juvenile host stage), with all of the other parameters held constant. We then look at the effect of the ratio $(a_1T_1)/(a_2T_2)$ by letting it take different values in the different panels. The results are what we would expect from our discussion above.

First, as expected, if we remove the young immature stage, we get back the Basic model: as the ratio $(a_1T_1)/(a_2T_2)$ goes to 0 ($T_1 = 0$), the stability boundary for this model is identical to that of the Basic model (Fig. 6.3a). That is, the Basic model is the limiting case as the duration

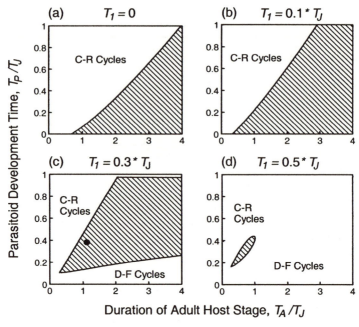

FIGURE 6.3. Local stability diagram of the sex-allocation model for four different values of $(a_1T_1)/(a_2T_2)$. In (a) $T_1 = 0$, in (b) $T_1 = 0.1*T_J$, in (c) $T_1 = 0.3*T_J$, and in (d) $T_1 = 0.5*T_J$. In each case $T_2 = T_J - T_1$, and $a_1 = a_2 = 1/T_J$. In each panel the effects of varying the duration of the invulnerable adult stage and the parasitoid developmental delay (both relative to duration of the juvenile host stage) are shown. In upper left portion of each graph, dynamics are dominated by consumer-resource (C-R) cycles. In lower right regions of (c) and (d) dynamics are dominated by delayed-feedback (D-F) cycles. At top of (c) for relative parasitoid development times approaching 1, a region with parasitoid single-generation cycle starts to appear (discussed later in text). Using parameter estimates from Fig. 5.6b, the real red scale–*Aphytis* system is now the dot within the locally stable region of parameter space in (c). In (d) there is no distinct boundary between delayed-feedback and consumer-resource cycles. Default parameters for this and all other figures in this chapter are $\rho = 33$, $d_{J1} = d_{J2} = 0$, $s_I = 1$, $d_P = 8/T_J$.

of, or attack rate on, the small host stage goes to zero. This turns out to be a useful result in later models.

Second, and also as expected, increases in the ratio $(a_1T_1)/(a_2T_2)$ increase the strength of the delayed parasitoid density dependence, and are initially stabilizing in the sense that the space in which consumer-resource cycles occur decreases. For example, with T_1 equal to 0.1^*T_J, the stability boundary moves to the left, so that a stable equilibrium is predicted for a larger region of parameter space (Fig. 6.3b). This stabilizing effect is consistent with the suppression of the consumer-resource cycles in the Basic model that occurred when we added non-delayed density dependence to the parasitoid attack rate.

Third, further increases in $(a_1 T_1)/(a_2 T_2)$ yield some additional suppression of the space occupied by consumer-resource cycles, but they also induce a second type of instability, which we term delayed-feedback cycles because they are analogous to such cycles seen in the stage-structured single-species models we discussed at the start of chapter 5. Delayed-feedback cycles are large-amplitude stable limit cycles with a period on the order of 1 to 3 times the host developmental duration. With T_1 equal to 0.3^*T_J, a locally stable equilibrium occurs in the shaded region shown in Fig. 6.3c, the consumer-resource region occurs above and to the left of the upper boundary, and delayed-feedback cycles occur below the lower boundary. There is also a second region of cycles for large values of the parasitoid development time, which we discuss in the section "The Nature and Origins of Delayed-Feedback Cycles" later in this chapter. A further increase in delayed density dependence ($T_1 = 0.5^*T_J$) causes the lower stability boundary delineating delayed-feedback cycles to sweep higher and to the left, reducing the region in which the equilibrium is locally stable to the very small shaded area in Fig. 6.3d.

In general, then, the effect of increasing the relative mortality on the young juvenile hosts is to shift the consumer-resource stability boundary to the left, reducing this type of instability, and to move the new delayed-feedback stability boundary upward and to the left, increasing the area in which this type of instability occurs.

The trade-off between these two types of instability is seen in other parameter values also. Thus increases in T_A, and decreases in T_P, also tend to suppress consumer-resource cycles and simultaneously induce delayed-feedback cycles.

Figs. 6.4a and b illustrate the consumer-resource and delayed-feedback cycles. We explore the latter more fully below, but note here that the period of the delayed-feedback cycles varies roughly between 1 and 3 times the host immature development time, whereas the consumer-resource cycle period can be more than twice this long. Notice too that the adult host fluctuates very little in the delayed-feedback cycles in Fig. 6.4b; we discuss this further below.

MULTIPLE ATTRACTORS

Unfortunately, in this model local stability analysis is not always an accurate guide to dynamical behavior when the system is perturbed more than a small distance from equilibrium. Although small perturbations in the shaded areas of Fig. 6.3 are followed by a return to the equilibrium, large perturbations in some regions can lead to delayed-feedback stable limit cycles. Indeed, these can also be seen after large perturbations in some of the area designated as consumer-resource cycles; for example, in Fig. 6.4c delayed feedback cycles are observed following a large perturbation for parameters in the "C–R cycles" region of Fig. 6.3c (Briggs et al. 1999b). Since large perturbations are perhaps the rule in real systems, the dynamical regimes expected on the basis of this model are thus less stable than is suggested by local stability analysis. This behavior occurs because for some parameter values in the shaded region there is a subcritical Hopf bifurcation leading to two attractors, one a stable point, the other a stable limit cycle.

REAL RED SCALE AND *APHYTIS*

It appears that the combination of an invulnerable adult host stage and size-selective sex allocation may explain the stability of the real red scale–*Aphytis* interaction. With the same parameter values as in Fig. 5.6b, the real system is now contained in the locally stable region of parameter space, as indicated by the dot in Fig. 6.3c. (We will see below that incorporating size-/age-dependent host-feeding does not change this conclusion.)

FIGURE 6.4. Examples of population dynamics from the sex-allocation model. In all panels solid line is old juvenile hosts, dashed line adult hosts, dotted line adult parasitoids. In (a) consumer-resource cycles occur with $T_A = 1{}^*T_J$, $T_P = 0.4{}^*T_J$, $T_1 = 0.1{}^*T_J$. In (b) delayed-feedback cycles occur with $T_A = 1{}^*T_J$, $T_P = 0.05{}^*T_J$, $T_1 = 0.3{}^*T_J$. In (c) multiple attractors occur for $T_A = 1{}^*T_J$, $T_P = 0.5{}^*T_J$, $T_1 = 0.3{}^*T_J$; the trajectory is first attracted to a consumer-resource cycle, but as the amplitude of the cycle increases, the trajectory is captured by the delayed-feedback attractor. All other parameters as in Fig. 6.3.

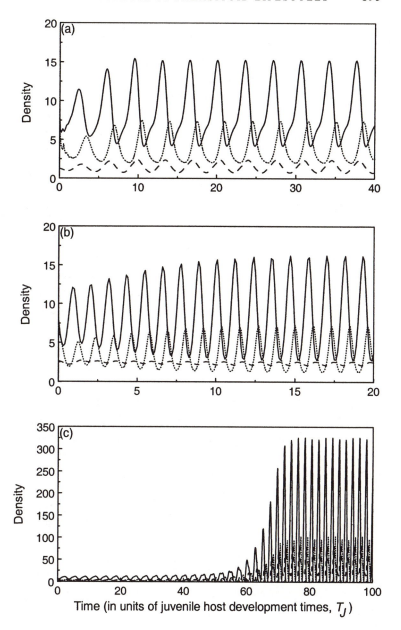

This conclusion needs to be treated with caution. The presence of a region with multiple attractors renders the conclusion fragile, since of course the real system will be subject to large perturbations. Clearly, we need more accurate parameter estimates and a model that contains more details of the real system (chapter 12).

Mechanism 2. Size-/Age-Dependent Host-Feeding

Next, let us ignore sex allocation and consider a host-feeding parasitoid. Such parasitoids tend to host-feed on younger hosts and parasitize older hosts. We make the temporary assumption that host meals make no contribution to the future female parasitoid population, and that each attack on an older host yields one (female) parasitoid.

We now have exactly the situation in mechanism 1. There is no gain from young hosts and a gain of 1 from older hosts. The same model holds for this case, and the results are exactly the same. Indeed, when this model was first introduced, it treated size-dependent sex allocation and host-feeding as a single process (Murdoch et al. 1992b).

Mechanism 3. Size-/Age-Dependent Clutch Size

In mechanism 1 we assumed that oviposition led to only one (female) parasitoid offspring. Since the young stage yielded no female offspring, the gain from attacking an old immature host was effectively infinitely greater than that from a young immature host. But as noted, many species lay more eggs in older (larger) hosts.

Suppose we now consider only the range of host ages that receive a clutch containing at least one female egg. Now the relative gain from an older host can take finite values, for example from 2 or 3 in *Aphytis* to at least 40 (in *Colpoclypeus*).

Fig. 6.2b shows a diagram for this kind of behavior. Notice that we still have stage-specific durations (T_i) and attack rates (a_i), but now there is also a parameter for each stage, e_i, that defines how many female eggs are laid in hosts of stage i. So now the number of new adult female parasitoids recruiting at time t is

$$R_P(t) = e_1 a_1 J_1(t - T_P) P(t - T_P) s_1$$
$$+ e_2 a_2 J_2(t - T_P) P(t - T_P) s_2. \tag{6.1}$$

Initially we set $a_1 = a_2 = 1$ and $T_1 = T_2 = 0.5*T_J$ and vary the ratio of the number of female eggs laid in the two stages: e_2/e_1, where e_1 is the number laid in a young immature host, e_2 is the number laid in an old immature host, and e_2 is greater than e_1. Later we will ask what happens if we change a_1/a_2.

Fig. 6.5 shows that increasing e_2/e_1 has the same qualitative effects as increasing T_1 in Fig. 6.3. If the parasitoid does not distinguish between the two stages and each receives the same number of female eggs ($e_2/e_1 = 1$), we regain the Basic model of chapter 5, regardless of the ratio $(a_1 T_1)/(a_2 T_2)$. Increasing the number of eggs placed in large hosts relative to the number placed in small hosts is initially stabilizing. Thus, with e_2/e_1 equal to 3 the stability boundary in Fig. 6.5 shifts to the left, increasing the area in which a stable equilibrium occurs and reducing the area of parameter space in which consumer-resource cycles occur. Additional increases in e_2 relative to e_1 lead to relatively small gains in stability and result in the appearance of delayed-feedback cycles. For example, when the larger host individuals receive 10 eggs each and younger hosts receive 1 ($e_2/e_1 = 10$), the equilibrium is locally stable in the shaded region in Fig. 6.5; there are consumer-resource cycles above the region and delayed-feedback cycles below. As e_2/e_1 increases further, the region with delayed-feedback cycles is increased and the stable region is reduced. For example, with e_2/e_1 equal to 100, the equilibrium is locally stable in only the very small region shown in Fig. 6.5. For a given value of T_1, as the ratio e_2/e_1 goes to infinity the stability boundary approaches that shown in Fig. 6.3d for the sex-allocation model at the equivalent value of T_1. The results depend only on the ratio e_2/e_1 and not on the absolute values of e_2 and e_1.

This mechanism introduces a general result: *the Basic model (chapter 5) and the sex-allocation model (mechanism 1) are limiting cases when the relative gain to the future female parasitoid population increases with the age of the attacked juvenile host.* In the Basic model the gains are the same; in the sex-allocation model the gain from an old juvenile is effectively infinitely greater than that from a young juvenile. Increasing the clutch size on older hosts relative to that on younger hosts (e_2/e_1) increases the strength of the delayed density dependence in the parasitoid attack rate in a way that is entirely analogous to increasing the relative parasitoid-induced mortality on younger hosts $(a_1 T_1)/(a_2 T_2)$.

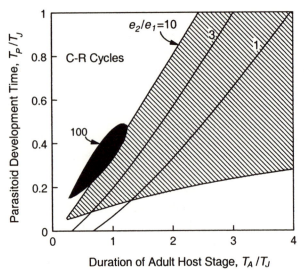

FIGURE 6.5. Local stability diagram of the size-/age-dependent clutch size model, showing the effects of varying the ratio of clutch size on old versus young juvenile hosts, e_2/e_1. Boundaries are shown for three different values of e_2/e_1 (indicated by numbers on the lines). For $e_2/e_1 = 3$ or 1, the equilibrium is locally stable to the right of the respective boundary, with consumer-resource (C-R) cycles to the left. For $e_2/e_1 = 10$, the equilibrium is locally stable in the hatched region, with delayed-feedback cycles below the lower boundary. For $e_2/e_1 = 100$, the equilibrium is locally stable only in the small black region. $T_1 = T_2 = 0.5*T_J$. All other parameters as in Fig. 6.3. Modified from Fig. 2c in Murdoch et al. 1997 with permission from the British Ecological Society.

Mechanism 4. Survival of Juvenile Parasitoids is Higher in Older Hosts

This relationship is common in koinobionts but may also occur in some idiobionts. We can deal with it easily. In mechanism 3, e_1 and e_2 are the number of eggs laid in young and old immature hosts, respectively. But these parameters could just as well stand for several other aspects of gain, or for a combination of them. For example, we could incorporate into e_i both the number of eggs laid in a host of age i and the fraction of these that survive to become an adult. Then this mechanism is exactly the same as mechanism 3.

A UNIFYING FRAMEWORK AND EXTENSIONS

The reader will by now have realized that, first, all of the mechanisms above are actually manifestations of a single broader phenomenon: the gain to the future female population (the number of adult female parasitoids produced per host) increases with the age of the attacked host. Second, they all have the same dynamic effect and fit the single quantitative framework depicted in Fig. 6.3. Attacks by parasitoids on young host stages that kill the host, but give relatively little gain to the future female parasitoid population, lead to delayed density dependence. Small amounts of delayed density dependence have a stabilizing effect, reducing the region in which consumer-resource cycles occur. But larger amounts of delayed density dependence result in the formation of a new type of delayed-feedback cycle. We investigate the nature and origins of these delayed-feedback cycles at the end of this chapter. But first, we note that other mechanisms that lead to the gain increasing with host age also fit into the same framework. These are summarized in Table 6.2, which also lists four examples of mechanisms that have the opposite dynamical effect.

Here, briefly, are explanations for the additional mechanisms in Table 6.2.

Mechanism 8. Juvenile Parasitoids Develop Faster in Older Hosts

As noted above, koinobiontic parasitoids typically develop faster when they are laid as eggs in older (and therefore larger) hosts (Fig. 6.2c). Although this has the same dynamic effect as the other mechanisms discussed here, it has the effect in a different way.

This mechanism requires a small change to the model of host-stage-dependent clutch size. Now juvenile parasitoids from the younger host stage take longer to mature into adults than do those from the older host stage (Box 6.2). To see that the dynamical effect is the same as in the other cases, we fix the development time on younger hosts at T_{P1} and let the development time on older hosts, T_{P2}, be a fraction of T_{P1}, namely f^*T_{P1}, where f is smaller than 1. (This mechanism does not require different clutch sizes from different-sized hosts, but we have illustrated it here with $e_2/e_1 = 10$.) Now, with the y-axis

TABLE 6.2. Size-dependent parasitoid mechanisms that fit into a common
dynamical framework

A. Mechanisms that increase delayed density-dependent effect
Idiobiont parasitoids
  1. Sex ratio of parasitoid offspring becomes increasingly female with
     host age.
  2. Parasitoid feeds on young hosts and parasitizes old hosts.
  3. Clutch size increases with host age.
  4. Immature parasitoids survive better in older hosts.
  5. Mechanisms 1–3 depend on parasitoid egg load (see chapter 7).
  6. Combinations of 1–4.

Koinobiont parasitoids
  7. Juvenile parasitoids survive better in older hosts.
  8. Juvenile parasitoids develop faster in older hosts.

B. Mechanisms that decrease delayed density-dependent effect
  9. Gain decreases with host age (smaller clutches or lower survival of
     immature parasitoids on older hosts).
  10. Younger (older) hosts are more (less) likely to be rejected.
  11. Male limitation (explicit gain from production of male parasitoid
      offspring).
  12. Egg limitation (explicit gain from host-feeding); see chapter 7.

---

of our standard stability diagram (Fig. 6.6) changed to T_{P1}, we see
that decreasing f moves the consumer-resource boundary to the left;
i.e., the consumer-resource limit cycles are confined to a smaller area
of parameter space, and there is a very small increase in the region
with delayed feedback cycles. This result is expected since decreasing f effectively decreases the average parasitoid development delay.
A constant mortality rate during the immature parasitoid development
time further strengthens the effect: faster development of parasitoids on
older juvenile host stages increases the parasitoids' through-stage survival. This has the same effect as increasing the clutch size from older
hosts, as seen above.

Mechanism 9. Gain Decreases with Host Age

Decreasing gain with host age is no doubt the exception, but it does
occur. When gain decreases with host age, we simply get the opposite

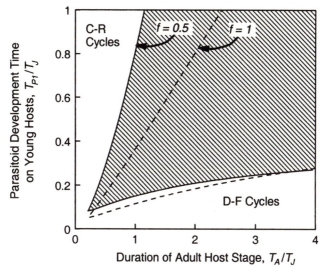

FIGURE 6.6. Local stability diagram of the koinobiont model with immature parasitoids developing faster on older juvenile hosts (Box 6.2). Parasitoid development time on old juvenile hosts, T_{P2}, is f^*T_{P1}. Shaded area shows the region of parameter space with a locally stable equilibrium for $f = 0.5$. Also shown in dashed line is the boundary for $f = 1$. $e_2/e_1 = 10, T_1 = T_2 = 0.5^*T_J$; all other parameters as in Fig. 6.3.

result to those above. If gain is lower in older immatures, the consumer-resource stability boundary (e.g., Fig. 6.3) is pushed to the right, thus increasing the region of parameter space with consumer-resource cycles.

Mechanism 10. Younger Hosts Are More Likely to Be Rejected

A higher rejection rate on stage i is reflected in a lower attack rate, a_i. So in the above models, greater rejection of younger hosts reduces $(a_1T_1)/(a_2T_2)$ and hence reduces the strength of delayed density dependence in the parasitoid attack rate. It therefore has a dynamical effect similar to that of gain declining with host age. Preferential rejection of older hosts has the opposite effect; i.e., it reinforces delayed density dependence.

BOX 6.2.
KOINOBIONT MODEL WITH IMMATURE PARASITOIDS DEVELOPING
FASTER ON OLDER HOSTS

Same as Box 6.1 except:

Populations

$J_1(t)$ number of young juvenile hosts at time t (receive e_1 female
         eggs)

$J_2(t)$ number of old juvenile hosts at time t (receive e_2 female eggs)

Properties of Individuals

T_{Pi} development time of juvenile parasitoids laid on J_i
         (for $i = 1, 2$)

e_i clutch size of female parasitoid eggs on host stage J_i
         (for $i = 1, 2$)

s_i through-stage survival of juvenile parasitoids laid on J_i
         (for $i = 1, 2$)

Vital Processes

$s_1 = \exp(-d_I T_{P1})$ through-stage survival of immature
                                    parasitoids on J_1 hosts

$s_2 = \exp(-d_I T_{P2})$ through-stage survival of immature
                                    parasitoids on J_2 hosts

$R_P(t) = e_1 s_1 X_1(t - T_{P1})$ adult parasitoid recruitment rate
       $+ e_2 s_2 X_2(t - T_{P2})$ at time t

Mechanism 11. Effect of Producing Males:
An Assumption Revisited

As a first approximation we assumed in the sex-allocation model above
that an encounter with a young juvenile host that leads to production of
a male parasitoid makes no contribution to the future female population.
We show here that taking into account the true effect of male production

does not change our qualitative conclusions. It does, however, reduce the relative gain to the future parasitoid population from encountering an older juvenile host. It therefore reduces the strength of the delayed density dependence in parasitoid per-head recruitment, and hence reduces the likelihood that delayed-feedback cycles occur. We see in the next chapter that explicitly taking into account nutrients gained from host meals has a similar effect (mechanism 12 in Table 6.2).

We now modify the sex-allocation model of Box 6.1 to explicitly follow the dynamics of both male and female parasitoids (see Box 6.3 and Fig. 6.2d). Attacks on young juvenile hosts result in male parasitoid offspring. Attacks on old juvenile hosts result in unmated virgin female parasitoids. Because of the haplodiploid genetic system of hymenoptera, virgin females can produce only male offspring until they become mated. Mating occurs as random encounters between adult male and virgin female parasitoids, with an encounter rate β.

We assume that virgin females are able to oviposit at the same rate as inseminated females but that they confine their attacks to the younger, male-producing, juvenile hosts. There is in fact some evidence that virgin females sometimes have a slightly lower attack rate (N. Mills pers. comm.).

The mating rate, β, determines the rate of random encounters between virgin females and adult male parasitoids, and hence determines the degree of male limitation in the model. If β is large, there is a high contact rate between virgin females and males, and on average, individuals spend little time in the virgin female class. In this case, in which the production of new females is not limited by mating, producing additional male parasitoid offspring provides little gain to the future female parasitoid population. In the limit as β gets large, we recapture the sex-allocation model of Box 6.1, in which there was zero gain from production of a male parasitoid offspring.

By contrast, if β is small, individuals on average spend longer in the virgin female class and the female parasitoid population is limited by mating. In this case the production of male parasitoids from small hosts contributes significantly to the future female parasitoid population. This has the effect of reducing the delayed density-dependent effect of laying male eggs on young hosts, as there is now a substantial gain from attacking either young or old juvenile hosts.

BOX 6.3.
EFFECT OF PRODUCING MALES

Same as Box 6.1 except:

Populations

$J_1(t)$ number of young juvenile hosts at time t (receive male eggs)
$J_2(t)$ number of old juvenile hosts at time t (receive female eggs)
$M(t)$ number of adult male parasitoids at time t
$V(t)$ number of virgin adult female parasitoids at time t
$F(t)$ number of mated adult female parasitoids at time t

Properties of Individuals

β mating rate (encounter coefficient between males and
         virgin females)
d_M death rate of adult male parasitoids
d_F death rate of adult female parasitoids

Vital Processes

$$X_1(t) = a_1 J_1(t)[V(t) + F(t)]$$ total parasitism rate on
                                       J_1 at time t

$$X_2(t) = a_2 J_2(t) F(t)$$ total parasitism rate on
                               J_2 at time t

$$S_{J1}(t)$$ survival through J_1

$$= \exp\left\{ -\int_{t-T_1}^{t} (a_1 V(x) + a_1 F(x) + d_{J1})\, dx \right\}$$ stage at time t

$$S_{J2}(t) = \exp\left\{ -\int_{t-T_2}^{t} (a_2 F(x) + d_{J2})\, dx \right\}$$ survival through J_2
                                                                              stage at time t

$$R_M(t) = X_1(t - T_P)s_I$$ recruitment rate of
                               male adult
                               parasitoids at time t

$$R_V(t) = X_2(t - T_P)s_I$$ recruitment rate of
                               virgin female
                               parasitoids at time t

$$\phi(t) = \beta M(t) V(t)$$ mating rate

(Box 6.3. continued)

Balance Equations

$dV(t)/dt = R_V(t) - \phi(t) - d_F V(t)$ virgin adult female parasitoids

$dF(t)/dt = \phi(t) - d_F F(t)$ mated adult female parasitoids

$dM(t)/dt = R_M(t) - d_M M(t)$ adult male parasitoids

A MORE GENERAL MODEL: THE GENERIC GAIN MODEL

Here we outline a model in which gain is assumed to change with the age of the encountered host, but we need not specify a mechanism. Gain can change continuously with host age, and the gain function can have any of a wide range of shapes. In all of the preceding analyses we have assumed that the gain to the future female parasitoid population increases with host size in discrete jumps. But in many parasitoids the gain increases more or less gradually, for the reasons outlined next. We show that our results hold up in the more general case of a gain that changes continuously with host age.

We can expect that gain will often increase gradually because various processes are operating concurrently, and because there really is a gain from host-feeding, and perhaps from male production. For example, as a host ages, the probability it will be host-fed may decrease, the number of eggs produced from a host meal will increase, the probability of receiving a female egg will increase, the number of female eggs in the clutch will increase, and the probability of an oviposited egg surviving to maturity will likely increase. These processes operate together, so the number of female parasitoid offspring arising from an encounter with a host of age *a* will be the net expected value of all of these processes, and often will be non-integer; the gain will therefore increase smoothly with host age. We used *Aphytis* responses to red scale to provide an example (Fig. 6.1). (We are able to plot the gain for *Aphytis* against scale age, rather than scale size, because scale size is directly related to scale physiological age; in the field populations are far below the limits set by their food supply and individual growth is temperature dependent.)

Gain in *Aphytis* accelerates to a maximum. Clearly, the gain function might take different shapes in other interactions, and some hypothetical

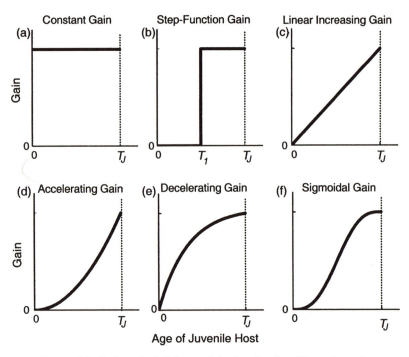

FIGURE 6.7. Six hypothetical forms of the function describing gain to the future female parasitoid population as a function of the age of juvenile host attacked. Redrawn with permission from Fig. 1 in Briggs et al. 1999b.

forms are shown in Fig. 6.7. If the gain is constant with juvenile host age (Fig. 6.7a), we recapture the Basic model (chapter 5). The sex-allocation model (Box 6.1) includes a gain that increases as a discrete jump (the step function of Fig. 6.7b). Gain functions that increase continuously may take on a linear (Fig. 6.7c), accelerating (Fig. 6.7d), decelerating (Fig. 6.7e), or sigmoidal (Fig. 6.7f) form with host age. They may even be humped, which we do not show.

In order to include a gain function that increases continuously with host age, we must abandon the delay differential equation formulation and instead express the interaction as a system of partial differential equations with respect to time and age. We do not go into detail about this formulation here but refer interested readers to Briggs et al. 1999b, and only briefly summarize the results obtained from this generic gain model.

The main conclusion is that our previous results generalize to this new formulation. A gain that is constant and independent of host age of course regains us the results of the Basic model. Gain increasing with host age, regardless of the shape of the function, induces delayed density dependence in the parasitoid recruitment function. This reduces the region of parameter space in which consumer-resource cycles occur, and can lead to delayed-feedback cycles if the delayed density dependence is strong enough. A decelerating gain function, or one that increases only linearly with host age, induces only weak delayed density dependence. The delayed density dependence is strongest when the gain function accelerates with host age (as in the *Aphytis* example). A sigmoidal gain function, which is closest to our previous step function, produces an intermediate level of delayed density dependence, as it includes both accelerating and decelerating regions. A decreasing gain function would increase the range of parameter values in which consumer-resource cycles are found.

THE NATURE AND ORIGINS OF DELAYED-FEEDBACK CYCLES AND SINGLE-GENERATION CYCLES: INSIGHTS FROM A SIMPLIFIED MODEL

Delayed-Feedback Cycles

We can get insight into the mechanism producing delayed-feedback cycles by looking at both a simplified version of the above models and a single-species model (Briggs et al. 1999b). Delayed-feedback cycles can occur when the adult host stage is very long-lived relative to the juvenile development time (large T_A; Figs. 6.3, 6.5, 6.6). Inspecting the form of the delayed-feedback cycle (e.g., Fig. 6.4b) reveals that the cycles in the parasitoid and juvenile host stages occur when the long-lived adult host stage fluctuates very little. This suggests that a simplification used first in single-species models might be useful. Hastings (1987) and Hastings and Costantino (1987) in models of cannibalism in the flour beetle *Tribolium* approximated the long-lived adult host stage by assuming a constant inflow of newborns into the immature stage. A simplified parasitoid-host model is given in Box 6.4. It is equivalent to the sex-allocation model of Box 6.1, except the recruitment rate into

BOX 6.4.

SIMPLIFIED SEX-ALLOCATION MODEL

Populations

$J_1(t)$ number of young juvenile hosts at time t (receive male eggs)

$J_2(t)$ number of old juvenile hosts at time t (receive female eggs)

$P(t)$ number of adult female parasitoids at time t

Properties of Individuals

T_i duration of juvenile host stage J_i (for $i = 1, 2$)

$T_J = T_1 + T_2$ total juvenile host development time

T_P duration of immature parasitoid stage

R constant recruitment rate of young juvenile hosts

a parasitoid attack rate on all juvenile hosts

s_I juvenile parasitoid through-stage survival

d_{Ji} background death rate of juvenile host stage J_i

                                   (for $i = 1, 2$)

d_P death rate of adult parasitoids

Vital Processes

$X_1(t) = a J_1(t) P(t)$ total parasitism rate on J_1 at time t

$S_{J1}(t)$ survival through J_1 stage at time t
$= \exp \left\{ - \int_{t-T_1}^{t} (a P(x) + d_{J1}) \, dx \right\}$

$R_{J2}(t) = R S_{J1}(t)$ recruitment rate of old juvenile
                                                               hosts at time t

$X_2(t) = a J_2(t) P(t)$ total parasitism rate on J_2 at time t

$S_{J2}(t)$ survival through J_2 stage at time t
$= \exp \left\{ - \int_{t-T_2}^{t} (a P(x) + d_{J2}) \, dx \right\}$

$M_{J2}(t) = R_{J2}(t - T_2) S_{J2}(t)$ maturation rate out of old juvenile
                                                               stage at time t

$R_P(t) = X_2(t - T_P) s_I$ recruitment rate of adult parasitoids
                                                               at time t

---

(Box 6.4. continued)

Balance Equations

$dJ_1(t)/dt = R - R_{J2}(t)$ young juvenile hosts
$\qquad\qquad - X_1(t) - d_{J1}J_1(t)$

$dJ_2(t)/dt = R_{J2}(t) - M_{J2}(t)$ old juvenile hosts
$\qquad\qquad - X_2(t) - d_{J2}J_2(t)$

$dP(t)/dt = R_P(t) - d_P P(t)$ adult female parasitoids

---

the young juvenile host stage is replaced by a constant, R, and individuals mature out of the older juvenile stage into oblivion. Also, for simplicity, we assume that the parasitoid attacks all juvenile hosts with the same attack rate, a.

In this simplified model, true consumer-resource cycles cannot occur, because the recruitment rate into the host population is no longer influenced by mortality from the parasitoid. However, as shown in Fig. 6.8, delayed-feedback cycles can occur. Periodic variation in the recruitment of individuals into the juvenile class (the mechanism for single-generation cycles in chapter 5) is not required for the maintenance of delayed-feedback cycles; they result from only cyclic variation in the survival through the juvenile stage.

To see what is happening in these cycles, we look in detail at the age structure of the host population during the cycle. In Fig. 6.9b, the juvenile host stage is divided into 10 substages, each 0.1^*T_J days broad. In this example, T_1 is equal to 0.6^*T_J, so attacks on the first six substages (which comprise J_1) do not produce female parasitoid offspring, but attacks on the last four substages (which make up J_2) do lead to a female parasitoid offspring. Also shown in Fig. 6.9 are the per-head host death rate (which is simply $aP(t)$), the per-parasitoid birth rate of immature parasitoids ($aJ_2(t)$, the total birth rate of immature parasitoids ($aJ_2(t)P(t)$), and the density of parasitoids.

During the cycle, high densities of adult parasitoids mean a high per-head host death rate (the point marked A in Fig. 6.9a), suppressing the densities of juvenile hosts of all ages. The densities of all juvenile hosts will remain suppressed until the density of adult parasitoids declines below its equilibrium value (point B in Fig. 6.9a) through mortality

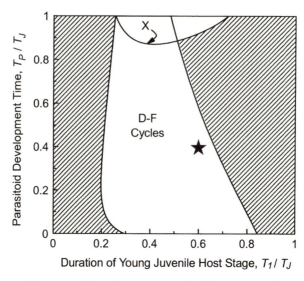

FIGURE 6.8. Local stability diagram for the simplified sex-allocation model with a constant recruitment rate of juveniles into the youngest age class (Box 6.4). The equilibrium is locally stable in the shaded region, delayed-feedback cycles occur in unshaded region, parasitoid single-generation cycles occur above the boundary X. The star marks the parameters used in the simulation of Fig. 6.9. $R = 100/T_J, d_P = 8/T_J, a = 1/T_J, s_I = 1, d_{J1} = d_{J2} = 0$, with $T_J = 1$. Redrawn with permission from Fig. 7b in Briggs et al. 1999b.

(at rate d_P). Since there is a constant input of newborn hosts entering the youngest age class, after the adult parasitoid density drops below equilibrium, the hosts entering the juvenile class will have a higher than average survival rate. Thus a larger than average cohort of juvenile hosts will begin to progress through the age classes. The parasitoid population cannot begin to benefit from the pulse of hosts until the hosts reach the age at which they can be used for the production of female parasitoid offspring, $T_1 = 0.6*T_J$. When the density of hosts in the J_2 stage builds up above equilibrium, the per-head birth rate of immature parasitoids, $aJ_2(t)$, rises above equilibrium (point C in Fig. 6.9c). The total birth rate of immature parasitoids, however, is $aJ_2(t)P(t)$, and at point C the density of adult parasitoids, $P(t)$, is still very low. Thus there is a delay between when the per-head birth rate of parasitoids rises above equilibrium (point C) and when the total birth rate of parasitoids rises above equilibrium (point D in Fig. 6.9d), which depends on the parasitoid development time. The total birth rate of immature parasitoids, $aJ_2(t)P(t)$, reaches its maximum (point E in Fig. 6.9d) as J_2 is decreasing and P is increasing, and the maximum adult parasitoid density occurs T_P days later (point F in Fig. 6.9e). This results in a high per-head host death rate, and the cycle repeats.

Looking at the form of the cycles, we can see that four processes go into determining the period of the delayed feedback cycles, and a rough approximation of the cycle period is: $\frac{1}{d_P} + T_1 + \Phi T_P + T_P$. The first term represents the time that it takes a high parasitoid density to decay below equilibrium when the host density is low. The second term represents the time between when the parasitoid density decays below equilibrium and when high densities of hosts start to enter the J_2 stage. T_P then enters the final two terms. The first of these is the time it takes after hosts enter the J_2 stage before the adult parasitoid population starts to increase (this also depends on several other factors, including the host recruitment rate and parasitoid attack rate, which have been included in the scaling factor Φ). The final T_P is the time between the peak in the total birth rate of immature parasitoids and the peak in the density of adult parasitoids. Fig. 6.10a shows the period increasing roughly linearly with T_P, at a slope between 2 and 3 for T_P/T_J less than 1.

Returning to the original full sex-allocation model of Box 6.1, Fig. 6.10b shows that the above results about the period of the delayed-feedback cycles gained from the simplified models carry over. This figure shows the period of the cycles measured from simulations at various points in Fig. 6.3c, for different values of T_A. For low values of the parasitoid development time ($T_P/T_J < 1$), the period of the cycles increases with T_P/T_J at a slope of between 2 and 3. We comment in the next section on the cycle period when T_P/T_J is greater than 1.

Single-Species-Like Dynamics: Delayed-Feedback Cycles

This model, with constant host recruitment rate, is similar to *chemostat* and *semi-chemostat* models. A chemostat model defines the dynamics of a laboratory population, such as a bacterium (consumer), being supplied at a constant rate with nutrients (resource) from a reservoir. Both bacteria

Hosts

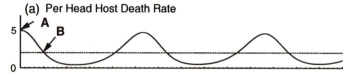

(a) Per Head Host Death Rate

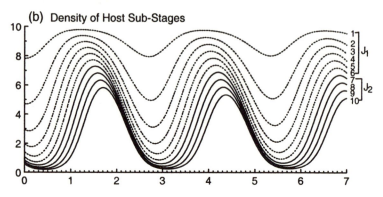

(b) Density of Host Sub-Stages

Parasitoids

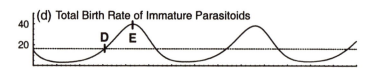

(c) Per Parasitoid Birth Rate of Immature Parasitoids

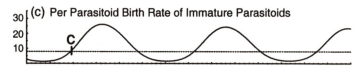

(d) Total Birth Rate of Immature Parasitoids

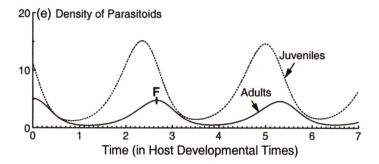

(e) Density of Parasitoids

Time (in Host Developmental Times)

and nutrient are assumed to wash out at the same per-capita rate. A semi-chemostat does not allow wash-out of the consumer, or the wash-out rates are different for consumer and resource. In our models, consumer and resource background death rates play the role of wash-out rates in a semi-chemostat analog.

The level of nutrient in chemostats fluctuates in response to consumption by the bacteria, but the nutrient's renewal rate does not depend on either nutrient or consumer density. Bacteria are thus a single-species population with a constant food input, as are the parasitoids in the above model. In the original gain models, the roughly constant adult population (caused by the invulnerability of the long-lived adult) substitutes for the experimenter to supply the constant rate of supply of hosts. We discuss the semi-chemostat analogy further in chapter 11.

In chapter 4 we discovered that the Nicholson-Bailey model with Ricker hosts can collapse to effectively a single-species system in which the host cycles, with period twice its development time, whereas the parasitoid is simply following these delayed-feedback cycles in the host. Above, we again discovered delayed-feedback cycles, but this time the cycles are driven by the parasitoid, whose development time is the main determinant of the period.

In fact, our gain models can give both classes of single-species cycles: delayed-feedback and single-generation cycles. We turn next to the latter.

FIGURE 6.9. Simulation of the simplified sex-allocation model at the point marked with the star in Fig. 6.8. Juvenile host stage is divided into 10 sub-stages, each 0.1^*T_J days broad. Shown through time over the course of two cycles are: (a) per-head host death rate ($aP(t)$); (b) the age structure within the juvenile host stage; (c) per-parasitoid birth rate of immature parasitoids ($aJ_2(t)$); (d) total birth rate of immature parasitoids ($aJ_2(t)P(t)$); and (e) densities of juvenile and adult parasitoids. In (b), hosts shown by the line labeled "1" are 0 to 0.1^*T_J days old; the line labeled "2" shows hosts 0.1^*T_J to 0.2^*T_J days old, etc. Dotted lines in (b) show stages in J_1 that do not produce female parasitoid offspring when attacked by the parasitoid, and solid lines show stages in J_2 that do produce female parasitoid offspring when attacked. In (a), (c), and (d), dotted line shows the value of that rate at the equilibrium host and parasitoid densities. $T_1 = 0.6^*T_J$, $T_P = 0.4^*T_J$, $R = 100$, $a = 1$, $d_P = 8/T_J$. See text for description of points A through F.

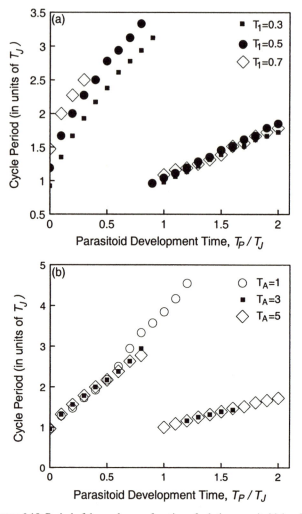

FIGURE 6.10. Period of the cycles as a function of relative parasitoid develop-ment time in (a) simplified model of Box 6.4 as measured from simulations (all parameters as in Fig. 6.8) and (b) for the sex-allocation model of Box 6.1 ($T_1 = 0.3*T_J$; all other parameters as in Fig. 6.3c). Cycles on the steeply rising points to the left are parasitoid delayed-feedback cycles; cycles on the slowly rising points to the right are parasitoid single-generation cycles. Modified with permission from Figs. 5 and 8 in Briggs et al. 1999b.

Single-Species-Like Dynamics: Single-Generation Cycles

Delayed-feedback cycles are seen in Fig. 6.8 when the parasitoid development time is less than that of the host. This is of course the rule in parasitoids. However, a second region with cycles begins in Fig. 6.8 when the parasitoid development time is almost equal to that of the host (i.e., T_P/T_J is close to 1). These cycles extend through much of the region with an even longer parasitoid development time. Fig. 6.10a shows that the period of these cycles starts with a value of 1, when both species have a development time of 1. Thereafter, as parasitoid development time, T_P, increases, the period increases with a slope close to 1. That is, the period is almost exactly the development time of the parasitoid. These are single-generation cycles in the parasitoid, unlike previous single-generation cycles, discussed at length in chapter 5, whose period is set by the host development time. They are fundamentally different types of cycles. Parasitoid single-generation cycles are also observed in the full sex-allocation model of Box 6.1 for relatively long parasitoid development times. The lower boundary of the parameters yielding these cycles is shown at the top of Fig. 6.3c and occurs for values of T_P/T_J greater than about 1 (Fig. 6.10b).

Real parasitoids generally do not take longer to develop than their host, so these cycles are not particularly relevant to parasitoid-host interactions. However, models in this chapter are stage-structured extensions of the Lotka-Volterra model, whose main application has been predators and prey and which is the archetype of the generic consumer-resource model; in fact, Hastings (1984) introduced the version with invulnerable immatures as a predator-prey model. We therefore use the gain models here to get insight into consumer-resource systems where the consumer, like many predators, takes longer to develop than its living resource. We refer to the protagonists as predator and prey. Gain to a predator from a meal also typically increases with prey size.

The origin of the predator single-generation cycles is slightly different from that of the delayed-feedback cycles. As before, a high density of predators suppresses the density of the juvenile predator stage (by suppressing prey density). Because development time is long, however, the offspring of the predators from one peak become adults after the prey population has recovered from attack by the initial predator peak.

As before, this is made possible by the constant supply rate of prey which is generated by the more or less constant population of invulnerable adults.

In the previous section we borrowed from a model by Hastings (1987) the substitution of constant recruitment of larvae for a roughly constant adult stage. The analogy with that single-species model of cannibalism in the flour beetle *Tribolium* is particularly apt for the single-generation cycles in Fig. 6.10b.

In the Hastings (1987) model, a large cohort of *Tribolium* larvae entering the old juvenile stage causes a high mortality on young juveniles through cannibalism. This drives down the density of young juveniles. The young juveniles remain suppressed until the large cohort matures out of the old juvenile stage. Once the dominant cohort matures, cohorts in the younger age class can have a high survival rate, and large cohorts of individuals can begin to progress through the age classes until they reach the old juvenile stage. Once the large cohorts reach the old juvenile stage, they start to suppress younger cohorts, and the cycle repeats. Thus, just as in our predator-prey model, the result is cycles that have a period approximately equal to the insect's development time. In our case, however, suppression is by the adult predators (rather than by old juveniles), since only the adults feed (i.e., parasitize). And in our case, the suppression is via suppressing the prey population, rather than via cannibalism. The origin of the cycles is also analogous to that of other consumer laboratory populations generating single-generation cycles, such as *Plodia*, where food is again supplied at a constant rate (chapter 5). They occur because suppression by the dominant cohort is direct rather than delayed.

Another similarity between Hastings' (1987) cannibalism model and the age-structured host-parasitoid model presented here is that the dynamics of both models include multiple attractors. As in our model, Hastings found that throughout much of the locally stable region of his model, limit cycles could occur following a large perturbation.

CONCLUDING REMARKS

The main point of this chapter is that most of the ways in which parasitoids respond to differences in host individuals are simply different manifestations of a single phenomenon: gain to the future female parasitoid population changes (usually increases) with host age. This large range of lifestyles has a uniform effect on population dynamics that fits nicely into a single quantitative framework, with the Basic model and the sex-allocation model being the limiting cases. We return in the final chapter to the issue of generality in a yet wider framework.

The gain mechanism can induce either of two types of cycles: delayed-feedback or single-generation cycles, with only the former likely in parasitoid-host interactions. These cycles are more likely to appear, and consumer-resource cycles to be suppressed, when the invulnerable adult host (resource) stage is long-lived. Long-lived adult hosts induce rather constant adult host density, and this in turn causes a rather constant recruitment of young hosts. This process produces semi-chemostat–like dynamics, in which the parasitoid (consumer) population acts like a single-species population.

The cycles are caused by what are effectively age-dependent density-dependent interactions in the consumer population, though the effects of one age class on another are mediated through resource density. When consumer development time is less than that of the resource (an almost universal condition in parasitoids and hosts), the consumer density dependence is delayed, and the cycles are delayed-feedback cycles corresponding with those seen in stage-structured single-species models with delayed feedback. When the consumer (now better considered a predator) takes as long or longer to develop than the resource (now prey), the consumer density dependence is direct and single-generation cycles occur. Note that in both types of cycles, the consumer development time is the main determinant of the period, unlike in the host-determined single-generation cycles in chapter 5.

The models in this chapter suggest that delayed-feedback cycles are possible in parasitoid-host interactions when dynamics are not too seasonal, the host has a long-lived invulnerable adult stage, and of course there is an increasing gain function. Since the period of delayed-feedback cycles can be as short as one host development time, the gain mechanism offers an alternative explanation for some examples of single-generation cycles (with the period equal to the host development time) seen in tropical pests (Godfray and Hassell 1989). These are cases in which the host adult is long lived (*Saccharosydne*, a sugar cane hopper, and two beetles, *Coelaenomenodera* and *Promecotheca*).

It would be useful to investigate in more detail cyclic host-parasitoid systems in which life histories fall into one or the other of these two patterns: brief and long-lived adult host stages. The dynamics of host stage structure are a potentially diagnostic probe. In cycles induced by the gain mechanism, adult host abundance should be much less variable than the later immature stages (Fig. 6.9b, where the density of the youngest host stage serves as a proxy for adult density). In cycles in hosts with a brief adult stage, induced by immediate density dependence in the parasitoid, adults should cycle as strongly as juveniles.

Other types of consumers, such as predators, also fit the gain mechanism—they typically get more resources by feeding on larger (usually older) individuals of their resource species. The dynamic effects discussed above may therefore arise also in these systems. Indeed, McNair (1987) developed a simple model for just such an interaction in predators.

In contrast to parasitoids, the development time of a real predator is typically longer than that of its prey. Predator-prey systems may therefore show cycles in which the period is equal to the predator development time—cycles to the right in Figs. 6.10a and b. We discuss real cycles in predators at some length in chapter 11.

CHAPTER SEVEN

State-Dependent Decisions

In chapter 6 we discussed how the response of a female parasitoid to an encountered host depends on the host's quality, especially its size or age. We noted that the overall pattern of responses is what we would expect based on evolutionary considerations. We showed that the range of responses, both within a parasitoid species and across parasitoids in general, exemplifies a single process: gain to the future female parasitoid population changes (usually increases) with host age. Finally, we showed that these responses have a consistent effect on the predicted population dynamics.

In this chapter we look at the dynamical consequences of two ways in which a parasitoid's state affects its response to an encountered host. The first effect is on the "decisions" the parasitoid makes. Here again, there is evolutionary theory to guide our expectations about what should happen, and real parasitoids behave approximately as the theory leads us to expect: parasitoid state does influence parasitoid decisions. In this case, however, we will discover that the state-dependent decisions examined, per se, sometimes can have no dynamical consequences.

State-dependent decisions can have dynamical consequences when they interact with host-age- or host-size-dependent decisions. As we will show, their effect is entirely analogous to the pure host-age-dependent attacks we discussed in chapter 6, and fits nicely into the general framework developed there.

The second type of effect consists of the purely physiological limits arising from parasitoid state: parasitoids cannot lay eggs if they have none, and host-feeders cannot host-feed if they are already at capacity. In their simplest form these limits effectively induce a type 2 functional response. But they can also have surprising effects on population dynamics.

EFFECTS OF EGG LOAD ON PARASITOID DECISIONS

Key Parasitoid State: Egg Load

We can assume that a parasitoid's goal is to maximize its fitness by the optimal placement of her eggs in hosts. Dynamic state variable models (Mangel and Clark 1988; Clark and Mangel 2000) have been used extensively to make predictions about optimal oviposition strategies of parasitoids. In these models, the current best strategy can depend on the parasitoid's current physiological state. The strategy that a parasitoid takes can in turn influence its future physiological state. These models make predictions about the optimal behavior at every step to maximize the parasitoid's lifetime fitness.

We might expect a parasitoid's decisions about placement of eggs to be affected by the number of mature eggs she has available—her egg load. Broadly, we would expect, and optimality models predict, the parasitoid to be more sparing and careful in ovipositing an egg if she has few available, and less choosy if she has an abundance (Iwasa et al. 1984; Mangel 1989; Chan and Godfray 1993; Collier et al. 1994; Mangel and Heimpel 1998).

The dynamic state variable models have inspired a growing number of experimental studies that mostly agree with the model expectations. Probability of host-feeding decreases with increasing egg load (Collier et al. 1994; Collier 1995a; Heimpel et al. 1996; Ueno 1999); probability of superparasitism increases with increasing egg load (Fletcher et al. 1994; Hughes et al. 1994; Sirot et al. 1997); clutch size increases with increasing egg load (Rosenheim and Rosen 1991); and behavioral interference and aggression increase with decreasing egg load (Hughes et al. 1994; Stokkebo and Hardy 2000).

Other aspects of state can also affect the parasitoid's decisions. For example, some parasitoids have been shown to be less choosy as they age and therefore can expect fewer egg-laying opportunities (van Baalen 2000). In the particular case of host-feeding parasitoids, we might expect the decision to parasitize a host, rather than eat it, to depend on gut fullness as well as egg load, though this has not been investigated.

ADULT PARASITOID PHYSIOLOGY

Several recent studies have contributed to a growing understanding of the details of the physiology of resource allocation and egg production in parasitoids (Ellers and van Alphen 1997; Rivero-Lynch and Godfray 1997; Ellers et al. 1998; Rivero and Casas 1999a, 1999b). These physiological studies have been in part motivated by, and contribute to, recent discussions about whether parasitoids are egg limited or time limited in the field (i.e., are they limited by the number of eggs they have available or by the time they have to search out hosts on which to lay those eggs) (Rosenheim 1996, 1999; Heimpel and Rosenheim 1998; Heimpel et al. 1998; Ellers et al. 2000; Rosenheim et al. 2000). We do not discuss the evolutionary arguments about whether parasitoids are likely to experience periods of egg limitation here, but instead concentrate on the effects on the population dynamics of some of the behaviors that have been found empirically to depend on egg load. We present here only a simple caricature of the physiology of nutrient allocation in parasitoids.

Adult parasitoids in general need to obtain energy to survive for more than a few days (e.g., Heimpel et al. 1997) and typically get it from nectar or honeydew. The need for energy may have dynamical consequences, which we do not pursue here. Instead, we assume adult parasitoids have all the energy they need and that protein is the key resource.

Adult parasitoids presumably need protein and other nutrients to replace natural breakdown. Pro-ovigenic species may emerge as adults with a lifetime's supply of protein, but host-feeding species need protein to mature new eggs, and may also use it for maintenance (Collier [1995b] showed that *Aphytis melinus* live somewhat longer when allowed to host-feed). The adult parasitoids probably have a source of protein in the haemolymph and/or "yellow bodies," but they also resorb eggs to supply maintenance (Collier 1995b; Heimpel and Rosenheim 1995; Heimpel et al. 1997; Rivero-Lynch and Godfray 1997; Rosenheim et al. 2000; Jervis et al. 2001), and we assume the eggs are effectively a protein store that drains at a steady rate to provide maintenance. Adult females quickly die when they cannot meet their maintenance needs, and in particular, Collier (1995b) found that *A. melinus* died within a day or so when they ran out of eggs and were not allowed to host-feed. Synovigenic species obtain protein mainly from feeding on hosts, but in some cases they may also feed on non-host prey or on plant proteins, and we explore the effects of such alternative resources.

In this section we concentrate on the dynamical effects of host-feeding to produce eggs, and of a decision to host-feed versus oviposit on an encountered host that depends on the parasitoid's egg load. We look first at models with an unstructured host population and then at the situation in which the parasitoids' decisions also depend on the stage of the host.

Model of State-Dependent Host-Feeding versus Parasitism, Unstructured Host

Kidd and Jervis (1989; 1991a; 1991b) investigated the dynamical effect of host-feeding in a simulation model. In their model, all host individuals were the same, and the probability a parasitoid host-fed rather than parasitized an encountered host increased as her egg load decreased. They found that population dynamics were unaffected by introducing such state-dependent behavior. We present here two simpler models from Briggs et al. (1995) which confirm Kidd and Jervis's results under some circumstances. The first model keeps track of the egg-load structure of the parasitoid population (the number of parasitoids in the population carrying each number of eggs); the second keeps track of only the average egg load in the population.

INTEGER EGG-LOAD MODEL

In this model we follow the density of female parasitoids in the different "egg-classes," i.e., those with 0 eggs, 1 egg, 2, 3 ... n eggs. We set the maximum number of eggs at $n = 12$, reflecting the biology of *Aphytis*. The structure of this female parasitoid population and the processes that move individuals among egg classes are diagrammed in Fig. 7.1a. A parasitoid moves into a lower egg class following oviposition and moves up one or more egg classes following host-feeding; the number moved depends on how many eggs are produced from a host meal. In *A. melinus* the number ranges from 1 to 3 depending on the size of the host eaten (S. Swarbrick, pers. comm.). (Briggs et al. [1995] show that assuming meals come in integer-egg equivalents does not affect the results.) In *Aphytis* it takes less than a day for a host meal to be reflected

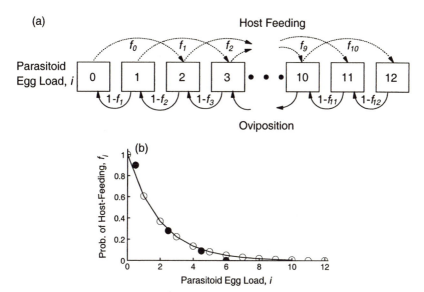

FIGURE 7.1. Integer egg-load model. (a) Movement of parasitoids between egg-load classes due to host-feeding (gain assumed to be 2 eggs per meal) and parasitism. Redrawn with the permission of Cambridge University Press from Fig. 2.5 in Briggs et al. 1999a. (b) Function describing how probability of host-feeding decreases with increasing egg load (the function is defined only at integer egg-load values, shown with the open circles). Solid circles represent data from Collier et al. (1994), in which each individual parasitoid was exposed to hosts and observed until she either host-fed or oviposited. The parasitoid was then dissected to determine egg load at the time the decision was made.

in an increase in egg load (Collier 1995b), and to begin with we assume no lag between a host meal and the maturation of the appropriate number of new eggs.

In *A. melinus*, non-feeding adult females develop what appears to be the maximum egg load for their size (egg load increases with size, a wrinkle we ignore here) within a day or so of emergence (Collier 1995b); i.e., this egg load is developed from nutrients gained in the larval stage. In the model we investigate the effect of variation in initial egg load.

We assume that all encountered hosts are attacked. The parasitoid either oviposits or feeds. When she has no eggs she always host-feeds; so we are assuming gut capacity is never limiting. When she is at her maximum egg load, or will exceed her maximum egg load through host-feeding, she always parasitizes. The host is of the pure Lotka-Volterra type (chapter 3), in which all individuals are identical and are reproductive from birth.

Finally, we assume that the probability a parasitoid will host-feed, rather than oviposit, declines exponentially with her egg load, being 1 when there are no eggs and 0 when the female has a complete egg load (Fig. 7.1b). This is an approximation of the data for *Aphytis* shown by the solid circles in Fig. 7.1b (from Collier et al. 1994).

We explore the effect of egg load on parasitoid death rate by assuming either that the adult female parasitoid death rate increases gradually as egg load declines, or that it is constant and then greatly increases at zero egg load, as was seen in *Aphytis*. Briggs et al. (1995) investigated the effect of a continuous drain of protein for maintenance in an analogous partial differential equation model in which protein gained from host-feeding and used for egg production was a continuous variable and not packaged only into discrete integer egg values.

The model is laid out in Table 7.1. The main results are as follows.

1. State-dependent host-feeding on its own has no effect on stability. The main effect of host-feeding is that it reduces the efficiency with which the parasitoids convert attacked hosts into new parasitoids (the conversion efficiency). Some hosts are used for the production of eggs rather than immediate parasitoid offspring. In the basic Lotka-Volterra model, reducing the conversion efficiency has the effect of increasing the host equilibrium density, but it has no effect on the stability of the equilibrium. If the death rate is constant and does not depend on egg load, the equilibrium in the integer egg-load model in Table 7.1 remains neutrally stable (Fig. 7.2a) just as in the basic Lotka-Volterra model. This occurs because the parasitoids rapidly reach a stable egg-load distribution, meaning that even though total numbers of parasitoids are fluctuating, the fraction of the parasitoids in each egg-load class remains constant (Fig. 7.2b), and the egg-load dynamics have no effect on the population dynamics.

Host-feeding does alter the actual equilibrium densities. In contrast with the simple Lotka-Volterra model, we now need more than one host to produce a single parasitoid—the parasitized host plus some part of a host-as-meal. Host density is thus higher at equilibrium than in the Lotka-Volterra model. The resources to produce an adult parasitoid

TABLE 7.1. Integer egg-load model

Populations

$H(t)$ density of hosts at time t
$P_i(t)$ density of parasitoids with i eggs at time t
$P(t)$ total density of parasitoids at time t

Properties of Individuals

r host rate of increase
a parasitoid attack rate
n maximum parasitoid egg capacity
μ_i fraction of adult parasitoid recruits with i eggs
z protein gain (in egg equivalents) per host-feeding
k_a constant setting rate of decline of f_i with increasing egg-load
f_i probability of host-feeding in an attack by parasitoid with i eggs
      $f_i = 1$ if $0 \le i < 1$
      $f_i = \exp\{-k_a i\}$ if $1 \le i \le n - z$, for $i = 0$ to n
      $f_i = 0$ if $i > n - z$
k_d constant setting rate of decline of d_i with increasing egg-load
d_i per-head death rate of parasitoids with i eggs; $d_i = d_{min} + \Delta \exp\{-k_d i\}$

Vital Processes

$$V(t) = \sum_{i=0}^{n}[aH(t)(1 - f_i)P_i] \quad \text{total parasitoid oviposition rate at time } t$$

$$R_i(t) = \mu_i V(t) \quad\quad\quad\quad \text{recruitment (birth) rate of parasitoids with } i \text{ eggs}$$

Balance Equations

$$dH(t)/dt = rH(t) - aH(t)P(t) \quad\quad\quad \text{hosts}$$

$$\begin{aligned} dP_i(t)/dt = R_i(t) &+ aH(t)\{(1 - f_{i+1})P_{i+1}(t) \\ &+ f_{i-z}P_{i-z}(t) - P_i(t)\} - d_i P_i(t) \end{aligned} \quad \text{parasitoids with } i \text{ eggs}$$

$$P(t) = \sum_{i=0}^{n} P_i(t) \quad\quad\quad\quad \text{total parasitoid density}$$

---

must of course come from somewhere, so pro-ovigenic parasitoids presumably produce a smaller adult from a given amount of host resource, since their eggs are smaller.

The result that state-dependent host-feeding has no effect on population stability, however, may be a consequence of the Lotka-Volterra structure of the model, in which altering the parasitoid conversion

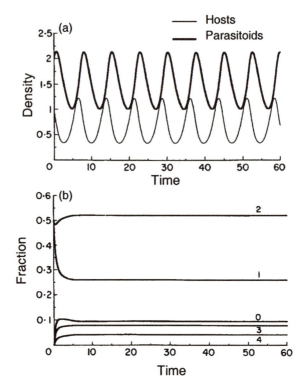

FIGURE 7.2. Simulation of integer egg-load model with a constant parasitoid death rate. (a) Host and total parasitoid populations show neutrally stable cycles. (b) Fraction of parasitoids in each egg-load class rapidly reaches a stable equilibrium (shown for egg-load classes 0–4). Numbers next to lines are egg-load classes. $r = 1.5, a = 1, k_a = 1, z = 2, d_i = 0.5$ for all $i, \mu_2 = 1, \mu_i = 0$ for all $i \neq 2$. From Briggs et al. 1995 with permission from the British Ecological Society.

efficiency has no effect on stability. In more complicated models, such as the Rosenzweig-MacArthur model (Rosenzweig and MacArthur 1963; Rosenzweig 1971) presented in chapter 3, reducing the parasitoid conversion efficiency has a stabilizing effect because it increases the host equilibrium. Increasing the host equilibrium causes the host population to be nearer its carrying capacity where intraspecific competition has a greater stabilizing effect. Therefore host-feeding is likely to be stabilizing in such models. It remains to be seen whether the state-dependent component of host-feeding (i.e., the increased probability of host-feeding at low egg loads) has any effect in such models, or whether the parasitoid population again rapidly reaches a stable egg-load distribution.

2. An egg-load-dependent death rate in the adult female parasitoid is stabilizing (discussed below), but the drain of eggs for maintenance is destabilizing (Briggs et al. 1995). These two related processes thus have opposing effects, and whether the equilibrium is stable or unstable depends on their relative strengths. There does not appear to be any information on which might be more important in real parasitoid-host systems.

3. The presence of an alternative source of protein for egg development (and hence also for maintenance) is stabilizing (Briggs et al. 1995). This result is not unexpected in light of the stabilizing effect of refuges, steady immigration, or other processes that provide a source of recruits to the population that is independent of the main parasitoid-host interaction (chapter 3).

Mean Egg-Load Model

The full version of the above model is fairly realistic in its portrayal of the movement of parasitoids between egg-load classes. But it can be analyzed only through simulations. Our second, simpler type of model gives the same answer, even though it contains a biologically impossible approximation: a parasitoid's decision depends on the mean egg load in the population. We later present other circumstances in which we can get away with this improper biology.

In this unstructured model there are no egg classes. Instead, we keep track of the total number (or density) of eggs in the parasitoid population, and since we know the number (or density) of females, we can also follow the average number of eggs per female (Table 7.2). Of course, real individual parasitoids cannot know the mean egg load. Nevertheless, this analytical model can give us insight into the importance of knowing the distribution of egg loads. Once again, if parasitoid death rate is unrelated to egg load, the equilibrium is neutrally stable, and again this occurs because the mean egg load becomes constant and uncoupled from population dynamics.

Unlike the structured model, however, the unstructured model retains a neutrally stable equilibrium when adult parasitoid death rate declines

TABLE 7.2. Mean egg-load model

Populations

$H(t)$ density of hosts at time t
$P(t)$ total density of parasitoids at time t
$E(t)$ total density of eggs in parasitoids at time t
$Q(t)$ average number of eggs per parasitoid at time t

Properties of Individuals

r	host rate of increase
a	parasitoid attack rate
z	protein gain (in egg equivalents) per host-feeding
E_b	eggs at birth
b	rate of protein drain
k_a	constant setting rate of decline of $f(t)$ with increasing egg load
$f(t)$	probability of host-feeding $f(t) = 1$ if $Q(t) < 1$ $f(t) = \exp\{-k_a Q(t)\}$ if $Q(t) \geq 1$
k_d	constant setting rate of decline of $d(t)$ with increasing egg load
$d(t) = d_{\min} + \Delta \exp\{-k_d Q(t)\}$	parasitoid death rate

Balance Equations

$dH(t)/dt = rH(t) - aH(t)P(t)$	hosts
$dP(t)/dt = [1 - f(t)]aH(t)P(t) - d(t)P(t)$	parasitoids
$dE(t)/dt = [zf(t) + (E_b - 1)(1 - f(t))]aH(t)P(t)$ $\quad -d(t)P(t)Q(t) - bP(t)$	eggs
$Q(t) = E(t)/P(t)$	average eggs per parasitoid

with mean egg load—this is stabilizing in the model with egg classes. Thus the stabilizing effect of a state-dependent death rate depends on the distribution of eggs among individuals and the response of these different individuals to their own egg load. In the structured model the different classes fluctuate out of phase with each other, egg-load distribution is not uncoupled from population dynamics, and neutral stability cannot ensue (Fig. 7.3). We will see later, however, that such

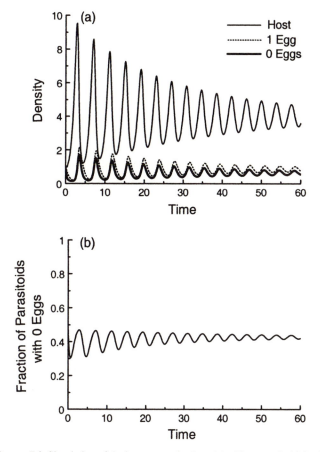

FIGURE 7.3. Simulation of the integer egg-load model with a parasitoid death rate that increases with decreasing egg load. Shown is a version of the model with only 2 egg-load classes (0 or 1). (a) Dynamics of host and parasitoids with 0 or 1 egg. (b) Fraction of parasitoids with 0 eggs. $r = 1.5$, $a = 1$, $z = 1$, $d_1 = 1$, $\Delta = 3$, $\mu_0 = \mu_1 = 0.5$. From Briggs et al. 1995 with permission from the British Ecological Society.

an unstructured model is a good guide to the dynamics of the structured model in some cases where the equilibrium is not neutrally stable.

The stabilizing effect of structuring the parasitoid population by its egg load is yet another example of how asynchronous fluctuations in different components of the population are potentially stabilizing. We saw this general effect when we compared spatially and stage-structured models in chapter 3, and further when we looked at properly stage-structured models in chapter 5. The effect reappears in chapter 10.

Combining Parasitoid-State and Host-Size Dependence

In chapter 6 we explored how the response of a searching female parasitoid depends on the size (age) of the encountered host. In this chapter we have seen that her behavior also depends on her own state, especially her egg load. Host size and parasitoid egg load both vary in the field, so we need to consider their joint effect. Fig. 7.4 shows the outcome for *Aphytis melinus*. As we might expect, when a host size-class can be either host-fed or parasitized, the probability it is host-fed decreases with egg load, but does so faster for larger hosts.

We can easily incorporate this effect into our model of two host age classes, which was discussed in detail in chapter 6 (Fig. 7.5a). The functions describing, for large (old) and small (young) hosts, the shift from host-feeding to parasitism with increasing egg load are in Fig. 7.5b. The (exponential) rate of decline is k_1 for the young age class and k_2 for the old age class, and k_1/k_2 measures the difference in response to the two age classes. Parasitoid death rate does not depend on egg load in this model.

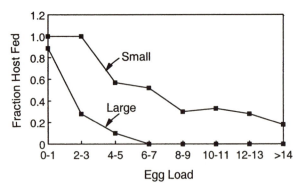

FIGURE 7.4. Fraction of attacks by *Aphytis melinus* that results in host-feeding rather than oviposition decrease faster, as egg load increases, on instar 3 (large) than on instar 2 (small) hosts (California red scale). Data from unpublished laboratory experiments (S. Swarbrick and W. Murdoch).

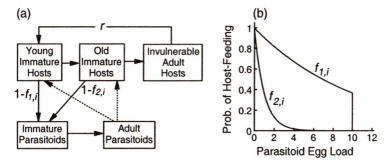

FIGURE 7.5. Model combining parasitoid-state- and host-size-dependent behavior. (a) Diagram showing structure of the model. (b) Functions describing how probability of host-feeding on the two stages of host decreases with increasing egg load, $f_{1,i}$, and $f_{2,i}$. Redrawn from Murdoch et al. 1997 with permission from the British Ecological Society.

We first look at the result when we follow the abundance of parasitoids in separate egg-load classes (0–12 eggs). As before, parasitoids move down an egg-load class by ovipositing and move up by host-feeding (Box 7.1) (Murdoch et al. 1997). This model can be analyzed only through simulations. We can obtain stability boundaries from a simplified version of the model that follows only the average egg load (Box 7.1). In all cases investigated in Fig. 7.6 the simulations from the egg-load structured model (at points marked with the squares) show the types of dynamics predicted by the stability boundaries for the mean egg-load version of the model (boundaries in Fig. 7.6).

The resulting dynamics are precisely those seen throughout chapter 6. If there is no difference between the host age classes in the rate at which the probability of host-feeding declines with egg load, we regain the Basic model ($k_1/k_2 = 1$ in Fig. 7.6). As we increase the rate at which the probability of host-feeding on old hosts falls off with egg load, relative to the rate on young hosts, we see first that the consumer-resource limit cycles are confined to a progressively smaller region of parameter space (e.g., Fig. 7.6 for the case where $k_1/k_2 = 0.1$); that is, the consumer-resource boundary moves to the left. As the difference between the two host age classes increases (k_1/k_2 gets smaller), delayed-feedback cycles appear and that boundary moves up and to the left ($k_1/k_2 = 0.01$ in Fig. 7.6). As k_1/k_2 approaches zero, we recapture the limit seen in the sex-allocation model of chapter 6.

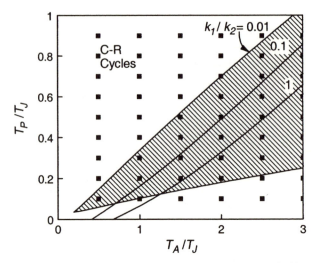

FIGURE 7.6. Stability boundaries for the model combining parasitoid-state- and host-size-dependent host-feeding. Stability boundaries shown were computed numerically for the mean egg-load version of the model. Black squares indicate points at which simulations of both the integer egg-load and mean egg-load versions of the model were run. In all cases, the two versions of the model gave virtually identical dynamics. For $k_1/k_2 = 1$ and $k_1/k_2 = 0.1$, the equilibrium is locally stable to the right and below the indicated boundaries. For $k_1/k_2 = 0.01$ the equilibrium is locally stable in the shaded region (with consumer resource [C–R] cycles dominating dynamics above the upper boundary and delayed-feedback cycles dominating dynamics below the lower boundary). $T_1 = T_2 = 0.5$. $T_J = 1$, $\rho = 33$, $d_P = 8$, $a_1 = a_2 = 1$, $k_2 = 1$, $s_1 = 1$. Redrawn from Murdoch et al. 1997 with permission from the British Ecological Society.

Thus, adding a state-dependent response has the same effect as a straightforward size-dependent response. The reason is quite simple and is the same as in the last chapter. For the average parasitoid, the gain from encountering an older (larger) host is greater because it is more likely to result in a female egg than in a host meal (or male egg—see chapter 6). The amount of delayed density dependence induced will be greater as the difference between young and old hosts, in the probability of host-feeding rather than parasitism, is increased.

We can see that variation in egg number among parasitoids has little effect on the result by looking at our simple model which has the odd

BOX 7.1.

MODEL COMBINING PARASITOID-STATE- AND HOST-SIZE-DEPENDENT
HOST-FEEDING

PARASITOID-STATE- AND HOST-SIZE-DEPENDENT
HOST-FEEDING: INTEGER EGG-LOAD VERSION

Populations

$J_1(t)$ number of young juvenile hosts at time t (receive male eggs)

$J_2(t)$ number of old juvenile hosts at time t (receive female eggs)

$A(t)$ number of adult hosts at time t (invulnerable)

$P_i(t)$ number of adult female parasitoids with i eggs at time t

$P(t)$ total number of parasitoids at time t

Properties of Individuals

T_1	duration of young juvenile host stage
T_2	duration of old juvenile host stage
$T_J = T_1 + T_2$	total juvenile host development time
T_P	duration of immature parasitoid stage
$T_A = 1/d_A$	average duration of adult host stage
ρ	per-individual lifetime adult host fecundity
a_1	parasitoid attack rate on young juvenile hosts
a_2	parasitoid attack rate on old juvenile hosts
s_I	immature parasitoid through-stage survival
d_{J1}	background death rate of young juvenile hosts
d_{J2}	background death rate of old juvenile hosts
d_A	death rate of adult hosts
d_P	death rate of adult parasitoids
n	maximum parasitoid egg capacity
z	protein gain (in egg equivalents) per host-feeding
μ_i	fraction of recruiting female parasitoids emerging with i eggs
k_1	constant setting rate of decline of $f_{1,i}$ with increasing egg load
k_2	constant setting rate of decline of $f_{2,i}$ with increasing egg load

(Box 7.1. continued)

$f_{1,i}$ prob. of host-feeding in an attack on J_1 by parasitoid with i eggs
$f_{2,i}$ prob. of host-feeding in an attack on J_2 by parasitoid with i eggs
$\quad f_{x,i} = 1$ if $0 \leq i < 1$ for $i = 0$ to n; $x = 1, 2$
$\quad f_{x,i} = \exp\{-k_x i\}$ if $1 \leq i \leq n - z$
$\quad f_{x,i} = 0$ if $i > n - z$

Vital Processes

$R_{J1}(t) = (\rho/T_A)A(t)$ recruitment rate of
                                                     young juvenile hosts
                                                     at time t

$R_{J2}(t) = R_{J1}(t - T_1)S_{J1}(t)$ recruitment rate of old
                                                     juvenile hosts at time t

$R_A(t) = R_{J2}(t - T_2)S_{J2}(t)$ recruitment rate into
                                                     adult host stage at
                                                     time t

$S_{J1}(t) = \exp\left\{-\int_{t-T_1}^{t} (a_1 P(x) + d_{J1})\,dx\right\}$ survival through J_1 stage
                                                     at time t

$S_{J2}(t) = \exp\left\{-\int_{t-T_2}^{t} (a_2 P(x) + d_{J2})\,dx\right\}$ survival through J_2 stage
                                                     at time t

$V(t) = a_1 J_1(t) \sum_{i=0}^{n} [(1 - f_{1,i})P_i(t)]$ total parasitoid
$\qquad + a_2 J_2(t) \sum_{i=0}^{n} [(1 - f_{2,i})P_i(t)]$ oviposition rate at
                                                             time t

$R_{Pi}(t) = \mu_i s_I V(t - T_P)$ recruitment rate of adult
                                                     female parasitoids
                                                     with i eggs at time t

Balance Equations

$dJ_1(t)/dt = R_{J1}(t) - R_{J2}(t) - a_1 P(t)J_1(t)$
$\qquad\qquad - d_{J1}J_1(t)$ young juvenile hosts

$dJ_2(t)/dt = R_{J2}(t) - R_A(t) - a_2 P(t)J_2(t)$
$\qquad\qquad - d_{J2}J_2(t)$ old juvenile hosts

$dA(t)/dt = R_A(t) - d_A A(t)$ adult hosts

(Box 7.1. continued)

$$dP_i(t)/dt = R_{Pi}(t) - d_P P(t)$$
$$+a_1 J_1(t)[f_{1,i-z}P_{i-z}(t)$$
$$+(1 - f_{1,i+1})P_{i+1}(t) - P_i(t)]$$
$$+a_2 J_2(t)[f_{2,i-z}P_{i-z}(t)$$
$$+(1 - f_{2,i+1})P_{i+1}(t) - P_i(t)]$$

adult female parasitoids
with i eggs

$$P(t) = \sum_{i=0}^{n} P_i(t)$$

total number of adult
female parasitoids

PARASITOID-STATE- AND HOST-SIZE-DEPENDENT HOST-FEEDING: MEAN EGG-LOAD VERSION

Same as integer egg-load version above, except:

Populations

$E(t)$ total density of eggs in parasitoids at time t
$Q(t)$ average number of eggs per parasitoid at time t

Properties of Individuals

E_b eggs at birth
$f_1(t)$ probability of host-feeding in an attack on J_1
$f_2(t)$ probability of host-feeding in an attack on J_2
$\quad f_i(t) = 1$ if $Q(t) < 1$
$\quad f_i(t) = \exp\{-k_i Q(t)\}$ if $Q(t) \geq 1$,   for $i = 1, 2$

Vital Processes

$$R_P(t) = \{[1 - f_1(t - T_P)]a_1 J_1(t - T_P)$$
$$+[1 - f_2(t - T_P)]a_2 J_2(t - T_P)\}$$
$$\times P(t - T_P)s_I$$

recruitment rate of adult
female parasitoids
at time t

Balance Equations

$$dE(t)/dt = E_b R_P(t)$$
$$+\{f_1(t)z - [1 - f_1(t)]\}a_1 J_1(t)P(t)$$
$$+\{f_2(t)z - [1 - f_2(t)]\}a_2 J_2(t)P(t)$$
$$-d_P P(t)Q(t)$$

eggs

$$Q(t) = E(t)/P(t)$$

average eggs
per parasitoid

assumption that each parasitoid responds to the average egg load, rather than her own ("Mean Egg-Load Version" in Box 7.1). Simulations with the integer egg-load and mean egg-load versions of the model are virtually indistinguishable provided the number of egg classes is not too small (Murdoch et al. 1997). This simplistic model thus provides a good approximation of the more complicated model, reinforcing a theme that runs through this book, that in many cases simple models give reliable guidance to the behavior of more detailed and realistic models.

This point in the text is the natural culmination of the series of models begun with the Basic model of chapter 5. Before drawing together the theory developed in that series, however, we need a detour into the population effects of limits to egg production, which we have ignored to this point. The next section makes that detour, and we return to a broader picture at the end of the chapter.

EFFECTS OF LIMITS TO EGG PRODUCTION

In this section we examine the effects of physiological limits to parasitoid egg production on the dynamics of host-parasitoid interactions. Like true predators, parasitoids spend time "handling" hosts, which can limit the rate at which parasitoids can attack hosts. Some parasitoids spend very little time handling a host once it is encountered. They rapidly stab it with the ovipositor, paralyzing it, and quickly oviposit. The entire sequence may take only a few minutes. In *Aphytis* the typical parasitism event takes about 4 minutes (S. Swarbrick, pers. comm.). Other species may spend much longer on an individual host. For example, the parasitoid *Sympiensis sericeicornis* attacks leaf miners. Once it detects a miner in the leaf (via its vibrations), it needs to find its precise location before stabbing it. This process produces vibrations the miner can detect, and it tries to escape by moving in the mine. There is thus a blindman's buff game of hide-and-seek, which can take up to 20 minutes per encounter (Casas 1989; Djemai et al. 2000). The effect of a handling time is straightforward. It changes the functional response from a type 1 to a type 2, just as in predators. This has a destabilizing effect, as we saw in chapter 3. In general, however, we expect egg limitation to be much more important than handling time, so we ignore the latter from now on.

Parasitoids also run out of eggs. We follow Shea et al. (1996) in defining egg limitation as any situation in which a parasitoid cannot oviposit on an encountered host because she lacks the egg(s) to do so. Egg limitation manifests itself differently in host-feeding versus non-host-feeding species. Our archetypal non-host-feeding species is a pro-ovigenic species that emerges as an adult with her full complement of eggs. (Other non-host-feeding species emerge with all of the nutrients required to produce all of their eggs, but production of the actual eggs continues for some time after she emerges as an adult). Egg limitation occurs in these species when an individual encounters enough hosts within her life to use up all of her eggs. At that point the female parasitoid is effectively dead from evolutionary and population dynamics perspectives.

Defining egg limitation is more complicated in host-feeding synovigenic species. These species ingest protein and nutrients from host-feeding, and it takes some finite amount of time for the material to be transferred from the gut and converted into new eggs. A parasitoid with no eggs can still host-feed, and a parasitoid with a full gut can still oviposit if she has eggs. The only case in which a host-feeding species must reject a host because of physiological limits is if she simultaneously has a full gut and no eggs.

In this section we investigate the dynamical effects of the various aspects of egg limitation, first in non-host-feeders and then in host-feeding species.

Egg Limitation in Non-Host-Feeding Species

We assume that adult parasitoids in this class emerge with their lifetime supply of mature eggs, or if they mature more eggs as time goes on, the rate of maturation is adequate to maintain a constant potential attack rate until all the eggs are used.

The model of egg limitation in pro-ovigenic species by Shea et al. (1996) uses a structure very similar to the integer egg-load models above. Parasitoids emerge with their full complement of eggs, and since the parasitoids do not host-feed, they move only down in egg-load classes through ovipositing on hosts (Table 7.3). When the egg load reaches zero, the parasitoids are effectively removed from the

TABLE 7.3. Egg limitation in non-feeding parasitoid species

Populations

$H(t)$ density of hosts at time t
$P_i(t)$ density of parasitoids with i eggs at time t
$P(t)$ total density of parasitoids at time t

Properties of Individuals

r host rate of increase
a parasitoid attack rate
n maximum parasitoid egg capacity
μ_i fraction of adult parasitoid recruits with i eggs
d per-head death rate of parasitoids

Balance Equations

$dH(t)/dt = rH(t) - aH(t)P(t)$ hosts

$dP_i(t)/dt = \mu_i aH(t)P(t) + aH(t)P_{i+1}(t)$
$\qquad\qquad - aH(t)P_i(t) - d_i Pi(t)$ parasitoids with i eggs, $i > 0$

$P(t) = \sum_{i=1}^{n} P_i(t)$ total parasitoid density

Notes: From Shea et al. 1996.

population. The rest of the structure of the model is the same as the integer egg-load model above: the host population is unstructured, and in the absence of parasitism the host would grow exponentially. A key assumption of the model is that hosts that are attacked by the parasitoids are immediately converted into new parasitoids.

This type of egg limitation has no effect on stability. The equilibrium is neutrally stable, just as in the basic Lotka-Volterra model. Thus merely having a limit to the number of eggs does not affect stability. The neutral stability comes about once again because the distribution of parasitoids among egg-load classes reaches a stable distribution and has no effect on dynamics. Thus, although the numbers in the populations are undergoing cycles, the egg-load distribution is constant.

It seems at first sight surprising that egg limitation has no effect on stability. The reason is that in this model, parasitized hosts are turned instantly into new parasitoids. This lack of a time delay actually results in no limit to the potential for parasitoid attack, even though the number of eggs carried by a parasitoid is limited. If the host population suddenly increases, the parasitoid population, and hence the rate of attack on hosts, increases immediately—new parasitoids are produced instead of new eggs per parasitoid. Adding a developmental delay in the parasitoid population is known to be destabilizing. It would be informative to look at the effects of different egg capacities in a model in which there is a parasitoid developmental delay. This crucial model, however, has not been investigated.

In the present model, the number of eggs does affect the host equilibrium density. As n decreases, host equilibrium increases. This is yet again an example of Lotka-Volterra insights persisting into more complicated models: lower n is the analog of a higher parasitoid death rate, since parasitoids effectively die when they run out of eggs, and we know that in the simple Lotka-Volterra model the resource density is set by the consumer death rate (chapter 3).

Egg Limitation in Host-Feeding Species

Whereas non-host-feeding species may be limited by their total number of eggs, host-feeding species may instead be limited by the rate at which eggs can be produced. Host-feeding parasitoids feed on host haemocoele or tissues. The nutrients pass through the gut to the haemocoele, are chemically transformed, and are used to mature new eggs. Some material may also be stored in an intermediate state. All of these processes are potentially rate limited via:

- gut capacity;
- gut emptying;
- intermediate storage capacity;
- chemical transformation;
- egg capacity.

We lump all but the last under the category "gut limitation." Thus a parasitoid can be in a situation where it encounters potential meals but is "satiated" and cannot eat them, or it can encounter potential hosts but has no mature eggs and cannot parasitize them.

Gut limitation is like satiation in predators (and like handling time—chapter 3) and induces a type 2 functional response in the host-feeding rate. In addition, there is a time delay between a host meal and egg maturation. The delay is short relative to the parasitoid developmental delay; for example, in *Aphytis* the development delay is 20 times longer than the egg-maturation delay. Nevertheless, it is a minor source of instability.

We model gut and egg limitation next. This model assumes that the maximum gut capacity is the nutrients from a single host meal (based on *Aphytis*) that can be converted to a fixed number of eggs. (In *Aphytis*, the number of eggs produced per host-feeding actually depends on the stage of the host that is attacked, and ranges from 1 egg/host-feeding for attacks on first-instar scale to 3 eggs/host-feeding for attacks on large second instars; S. Swarbrick, pers. comm.). The model assumes that there is a single time delay between host-feeding, which fills the gut, and emptying of the gut and production of new eggs. That is, new eggs are produced at the same moment the gut becomes empty. This results in a "latent" period after each host-feeding during which the parasitoid cannot host-feed again, but she can oviposit if she has eggs.

In the full model presented by Shea et al. (1996), the egg capacity is again $n = 12$ eggs, and z eggs are gained from host-feeding, now after a latent period of τ time units. However, the addition of this latent period, during which the parasitoids can oviposit, adds additional complexity to the model, which cannot be easily handled by the delay differential equation approach presented thus far in this book. It is not possible to keep track of which egg-load class the parasitoid is in throughout the latent period, τ. For this reason we do not present the details of the full model here, and refer interested readers to Shea et al. 1996 for a description of the technical details. Instead we present a simplified version of the model (Table 7.4), which contains the key feature of a latent period, and which gives the same results as the full model.

In this simplified model, there are two possible gut states (empty or full) and two possible egg loads (0 or 1). That is, the maximum egg load is 1 egg, each host-feeding results in the gut state going from empty to full, and following each host-feeding after a fixed latent period, τ, the gut becomes empty and the egg load is incremented from 0 to 1 (Fig. 7.7). This model is a simple caricature of the full model. There are only three possible states that a female parasitoid can be in: no eggs and empty (with density P_{00}), one egg and empty (P_{10}), and no eggs and full (P_{01}). It is not possible for the parasitoid to be in the one

TABLE 7.4. Simplified model of egg limitation in a host-feeding parasitoid

Populations

$H(t)$ density of hosts at time t
$P_{00}(t)$ density of parasitoids with 0 eggs and empty gut at time t
$P_{10}(t)$ density of parasitoids with 1 egg and empty gut at time t
$P_{01}(t)$ density of parasitoids with 0 eggs and full gut at time t

Properties of Individuals

r host rate of increase
a parasitoid attack rate
μ_i fraction of adult parasitoid recruits with i eggs
d per-head death rate of parasitoids
τ latent time required for a parasitoid to digest a host meal and produce
     a new egg

Balance Equations

$$dH(t)/dt = rH(t) - aH(t)[P_{00}(t) + P_{10}(t)]$$

$$dP_{00}(t)/dt = \mu_0 aH(t)P_{10}(t) - dP_{00}(t) + aH(t)P_{10}(t) - aH(t)P_{00}(t)$$

$$dP_{10}(t)/dt = \mu_1 aH(t)P_{10}(t) - dP_{10}(t) + aH(t-\tau)P_{00}(t-\tau)e^{-d\tau}$$
$$- aH(t)P_{10}(t)$$

$$dP_{01}(t)/dt = aH(t)P_{00}(t) - dP_{01}(t) - aH(t-\tau)P_{00}(t-\tau)e^{-d\tau}$$

Notes: From Shea et al. 1996.

egg and full (P_{11}) category, because whenever a parasitoid has an egg and encounters a host (in the P_{10}), she will always oviposit (moving to the P_{00} class, rather than P_{11}). Parasitoids are assumed to emerge with an empty gut and either 0 or 1 egg (where μ_0 and μ_1 are the fractions of adult parasitoids recruiting with 0 and 1 egg, respectively).

When the latent period, τ, is zero this model is again neutrally stable. Increasing τ to values greater than zero results in an unstable equilibrium. Thus, the time delay between host-feeding and the production of a new egg is destabilizing. This model includes no parasitoid developmental delay, so it has the strange assumption that it takes longer for eggs to be produced from host-feeding than for parasitoid offspring to be produced from oviposition. In reality, the parasitoid developmental delay is much longer (e.g., 20 times longer in *Aphytis*) than the latent period, and is likely to have a much stronger destabilizing effect than that found in this model.

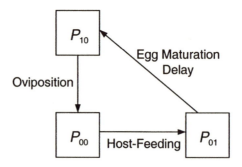

FIGURE 7.7. Diagram of the simplified model of egg limitation for a host-feeding species, incorporating an egg development delay. Shown are transitions by which parasitoids move between egg-load classes. Redrawn from Shea et al. 1996 with permission from the British Ecological Society.

In the full model with multiple egg-load classes, Shea et al. (1996) found the same result that long latent periods led to greater instability. They also found that the distribution of egg loads at birth affected stability. If parasitoids emerged with a large number of eggs, the equilibrium was more stable than if they emerged with few eggs.

A GENERAL DYNAMICAL THEORY OF PARASITOID BEHAVIOR

We make two main points in this chapter. First, the theory developed in this and the two previous chapters, like other theory based on the properties of individual organisms, is a vehicle for integrating different levels of biology and different disciplines in biology. Second, the theory achieves substantial generality.

The theory we have presented derives predictions about population dynamics from information on how individual parasitoids respond both to differences among the individual hosts they encounter and to variation in their own state. That is, it links behavior (and some physiology) to population dynamics. Since individual behavior is the subject of much evolutionary ecology, the theory also provides a framework for exploring the dynamical consequences of evolutionary predictions about parasitoid behavior in a variable world, though we have so far examined this link only cursorily.

Ecological modelers often restrict the term "individual-based model" to computer simulations that follow the fate of every individual, and its state, in the population (i-state representations). The models described so far are not individual-based models in this sense. Instead, for the most part they assume that the individual hosts in an age, stage, or developmental class have the same properties, and that parasitoids with the same egg load make the same decisions given an encounter with a host of a given class. The models are nevertheless based on the behavior of individual parasitoids in response to individual differences in themselves and their hosts. We suggest that the kind of framework we have used, where individuals with similar properties are lumped into a single class, is useful for exploring the links among behavior/physiology, evolution, and ecology.

The theory speaks to a fundamental ecological problem: how to achieve generality amidst the virtually infinite particularity of ecological systems. Each species is, by definition, unique. Indeed Thompson (1994; 1999) argues that the different populations of a species frequently are unique, each having evolved, and each continuing to evolve, to its local environment, especially to other key locally adapted populations in that local environment. Since there probably are tens of millions of species, each in many different local environments with different physical and biological properties, and since these properties and systems are also changing through time, there are countless variants and permutations. The behavioral and evolutionary ecologists, while reveling in this variety and attempting to explain it, have the theory of evolution to provide a unifying framework. But evolutionary ideas do not lead in the same way to theory for population dynamics, which must be developed from whole cloth.

The classical way to achieve general theory for population dynamics, in the face of biological complexity, is to omit the complexity. Thus, we write the Lotka-Volterra model or the Nicholson-Bailey model and claim that each captures the essential features of consumer-resource interactions, more or less regardless of the actual biology. Although one of our goals in this book is to show that much of the insight gained from these simple models does indeed carry through to models with more biology, these models are also far from telling us everything that is important. As we have shown, new dynamics appear when we add more realism.

The theory developed in these chapters, however, reassures us on several key points. First, for one of the most varied and complex group of organisms—parasitoids—a great variety of lifestyles boils down to a few major patterns. Second, the dynamical consequences of all these patterns can be explored in a single theoretical framework. Third, state-based responses by the parasitoid also fall nicely within this pattern. And finally, the resulting dynamics contain only four major patterns—stable equilibrium, consumer-resource cycles, delayed-feedback cycles, and single-generation cycles.

CHAPTER EIGHT

Competition between Consumer Species

In this chapter we examine interspecific competition between consumer species, with emphasis on competition between parasitoid species that attack the same host. Interspecific competition occurs between two species when individuals of each species have a negative effect on the population growth rate of the other species. The distinction is often made between exploitative and interference competition. In exploitative (or resource) competition, individuals interact only through suppression of their shared resource. In interference competition, individuals have a direct negative effect on each other through fighting, causing each other to waste searching time, or in the case of parasitoids, attacking and ovipositing on each other's offspring.

For much of this chapter we concentrate on parasitoid-host interactions, although most of the results are also relevant to competition between other types of consumers. Two aspects of parasitoid-host interactions make them excellent case studies of exploitative interspecific competition. First, parasitoids require hosts in order to produce offspring. In many cases, each host attacked potentially leads to at least one new searching parasitoid. The host is therefore an obvious potential limiting resource for the parasitoids. Second, parasitoids are often able to reduce dramatically the abundance of their host, which affects the availability of the host for use by competitors. These two facts make it likely that exploitative competition will be strong between parasitoid species that attack the same host.

These conclusions should be especially relevant in biological control of pests in agriculture and forestry, which is frequently accomplished by parasitoids. Competition ought to be especially severe since the resource, the pest, is driven to very low levels. In this chapter we therefore comment throughout on the effect on resource (host) density. In the next chapter we explore implications for biological control by bringing together and amplifying these results.

Interference competition is also rampant in host-parasitoid interactions. In many systems, host individuals are commonly encountered and attacked by more than one parasitoid individual. *Superparasitism* is the situation in which a parasitized host individual is re-encountered by another individual of the same parasitoid species and further eggs are laid on that host. *Multiparasitism* is the analogous situation in which the parasitized host individual is encountered and attacked by an individual of a different parasitoid species. When multiparasitism occurs, generally not all of the juvenile parasitoids laid on the host survive to adulthood. In some cases the adult female of the later attacking species can kill the eggs of other species during oviposition (*ovicide*). In other cases, the parasitoid larvae in or on the host fight directly (termed *larval competition*), with only one individual on the host surviving to emerge as an adult. In some systems, particular parasitoid species always win in larval competition over other species, whereas in other systems the outcome depends on the relative timing of oviposition by the various species.

Hyperparasitoids are species that attack juveniles of other parasitoid species. Hyperparasitism can be either obligate or facultative. Obligate hyperparasitoids can only attack hosts that have already been attacked by another parasitoid species. In facultative hyperparasitism, the hyperparasitoid can attack and lay eggs on either the host or on juveniles of other parasitoid species that have attacked the host. There is a fine distinction between multiparasitism and facultative hyperparasitism. In multiparasitism, theoretically only the host is attacked and used for food for the developing juvenile of the victorious parasitoid species, whereas in facultative hyperparasitism the developing hyperparasitoid juvenile can attack and eat either the host and/or the juvenile of the other parasitoid species. The outcome in terms of population dynamics, however, is the same: the host and the juvenile of one species are killed and the juvenile of the other species survives. As we discuss more fully later, facultative hyperparasitism in host-parasitoid systems is one example of a more general phenomenon termed *intraguild predation*.

Insect parasitoid interactions provide some of the most impressive examples of displacement of one species following the addition of its competitor. Perhaps the most famous and dramatic example comes from biological control of California red scale (chapters 2 and 9). Insect parasitoid systems also provide some of the most impressive examples of coexistence of multiple consumer species, apparently on a single resource. This occurs commonly in biological control and is also common in natural situations (chapters 2 and 9). For example, the midge *Rhopalomyia californica*, which forms galls on the shrub *Baccharis pilularis* in California, is commonly attacked by five or more parasitoid species in the field, several of which are thought to be specialists on the midge (Force 1970, 1974; Ehler 1985; Briggs and Latto 2000).

In this chapter we start with the familiar unstructured Lotka-Volterra competition model in which the resource is modeled only implicitly. We then move on to simple unstructured models of exploitative competition between species for an explicit resource. Many of the models of competition between parasitoid species have been discrete-time models, written as systems of difference equations. We review these and then look at the situation in which explicit age structure in the host can or cannot alter predictions about the outcome of competition. We discuss how temporal variability and spatial structure sometimes change the predictions about coexistence. The main conclusions of these models in terms of the outcome of competition are summarized in Table 8.1.

LOTKA-VOLTERRA COMPETITION MODEL: COMPETITION FOR AN IMPLICIT RESOURCE

The Lotka-Volterra model of interspecific competition (Volterra 1926), which appears in virtually every introductory ecology textbook, is based on the logistic model of population growth discussed in chapter 2:

$$\frac{dN}{dt} = rN\left(\frac{K-N}{K}\right) \tag{8.1}$$

where N is the density of the population, r is the maximum per-head growth rate that occurs at low population densities, and K is the carrying capacity, the density the environment can support over the long term.

The logistic model is a simple model of intraspecific competition. The resource over which the individuals in the population are presumably

TABLE 8.1. Summary of models in chapter 8 and their main conclusions regarding competition

Model	Main Conclusions
Lotka-Volterra competition model	Coexistence possible only when intraspecific competition is greater than interspecific competition.
Exploitative competition for an explicit resource	Species that leads to lowest abundance of resource wins; no coexistence.
Dependence of per-head rates on resource density	Does not qualitatively alter outcome of competition.
Dependence of per-head rates on consumer density	Promotes coexistence.
Facultative hyperparasitism (intraguild predation)	Coexistence possible if intraguild prey is superior at attacking resource: intraguild predation effect.
Competition in Discrete Time	
Competing parasitoid species aggregate independently of each other	Leads to density dependence in attack rate of each parasitoid species, promoting coexistence.
Competing parasitoid species aggregate to same host patches	Coexistence possible only through intraguild predation effect, as above.
Competing parasitoid species have different niches	Coexistence possible only if parasitoids have different niches, or through intraguild predation effect.
Obligate hyperparasitoids	Coexistence possible only if hyperparasitoid has higher attack rate than primary parasitoid.
Effects of Age Structure	
Parasitoids attack different host development stages, constant development times	Species that leads to lowest abundance of resource used by its competitor wins; no coexistence; earlier attacking species gains competitive advantage.

TABLE 8.1. (*continued*)

Model	Main Conclusions
Parasitoids have overlapping windows of vulnerability	Does not qualitatively alter outcome of competition.
Facultative hyperparasitism (intraguild predation) in a stage-structured model	Similar to unstructured case, but earlier attacking species gains additional advantage.
Variable host stage durations	Promotes coexistence by creating different host types.
Larval and adult stages attacked	Coexistence possible because attacks on adult hosts reduce duration of that stage, leading to different host types.

Non-Equilibrial Mechanisms of Coexistence

Seasonally varying parameters	Does not promote coexistence.
Species-specific non-linear responses to competition	Coexistence possible in a varying environment because of non-linearity: different species become superior competitor at different times.
Interactions between competition and environment: storage effect	Coexistence possible in a variable environment because of interaction between competition and environment that results in intraspecific competition greater than interspecific competition.

Spatial Structure

Competition/colonization trade-off	Promotes coexistence.
Spatial niche partitioning	Promotes coexistence.

competing, however, is modeled only implicitly. It is assumed that the net effect of resource supply and consumption by individuals of the species can be incorporated into an instantaneous per-head growth rate that decreases linearly as the density of the population increases. This is equivalent to having a per-head birth rate that decreases linearly with increasing population density, or a per-head death rate that increases linearly with increasing population density, or both.

The Lotka-Volterra competition model includes analogous equations for the rate of change of density of each of two competing species, N_1 and N_2, with r_i being the maximum per-head growth of species i at low densities of both species, and K_i being the density of species i that the environment can support in the absence of its competitor. The model is

$$\frac{dN_1}{dt} = r_1 N_1 \left(\frac{K_1 - N_1 - \alpha_{12} N_2}{K_1} \right) = f_1(N_1, N_2) \qquad (8.2a)$$

$$\frac{dN_2}{dt} = r_2 N_2 \left(\frac{K_2 - N_2 - \alpha_{21} N_1}{K_2} \right) = f_2(N_1, N_2). \qquad (8.2b)$$

The only change from the logistic model is that the reduction in the per-head growth rate of each species that is due to competition now depends not only on the density of that species but also on the density of its competitor. The per-head growth rate of each species decreases in a linear way with the weighted total density of both species. The constants determining this weighting are the competition coefficients, α_{12} and α_{21}. α_{12} is the relative effect of species 2 on the per-head growth rate of species 1—relative to the effect that an individual of species 1 would have on its own growth rate. For α_{12} less than 1, individuals of species 2 have a smaller per-head effect on the growth rate of species 1 than do individuals of species 1. For α_{12} greater than 1, individuals of species 2 have a greater per-head effect on the growth rate of species 1 than do individuals of species 1. Likewise α_{21} is the relative effect of species 1 on the per-head growth rate of species 2.

In basic ecology textbooks, the Lotka-Volterra competition model is analyzed most commonly through the use of phase-plane diagrams plotting the zero-growth isoclines for each species. We do not present this approach here because we do not use it in the rest of the book, and it is not useful when dealing with more than two equations (some authors use it for systems of three equations, but this requires interpreting three-dimensional plots, which can be cumbersome), or when dealing with the time lags inherent to stage-structured models.

Instead we use a single approach to determine the outcome of competition throughout the chapter: we calculate the invasion criterion for each species (MacArthur and Levins 1967; Turelli 1981). Coexistence between the two species is possible only if each species can increase in abundance when it is rare and its competitor is present at its long-term persistent density. It is most straightforward to calculate these invasion criteria when the persistent system that the competitor is attempting to invade is at a stable equilibrium. The concept is the same, however, if the long-term dynamics of the resident are a limit cycle or other complex persistent dynamics. For the next several sections of this chapter we assume that the resident species can persist at a stable equilibrium, and ask whether its competitor can invade this equilibrium. In the section "Non-Equilibrial Mechanisms of Coexistence" we look at the potential for coexistence in a varying environment.

Below we show for the Lotka-Volterra competition model that calculating the invasion criteria is equivalent to determining that each of the boundary equilibria (i.e., the equilibrium with the density of one species set to zero and the other species at its carrying capacity) is unstable to a small perturbation in density (i.e., to a small inoculum of the previously absent species). For the Lotka-Volterra competition model, this is also equivalent to the conditions for the stability of the equilibrium with the two species coexisting (however, this last result is not general in more complicated models).

To calculate the invasion criterion for species 2, we ask: what are the conditions in which species 2 can increase in abundance when it is rare and species 1 is present at its equilibrium value? This is the same as asking in what cases does N_2 have a positive growth rate, $dN_2/dt > 0$, when N_2 is very close to zero and N_1 is equal to K_1. The criterion is met when the term in parentheses in equation 8.2b is greater than zero for $N_1 = K_1$ and N_2 equal to zero, i.e., when $((K_2 - \alpha_{21}K_1)/K_2) > 0$, or after rearranging, when

$$\frac{1}{K_1} > \frac{\alpha_{21}}{K_2}.$$

Likewise, species 1 can invade the system when it is rare and species 2 is at its equilibrium when the term in parentheses in equation 8.2a is

TABLE 8.2. Invasion criteria for the
Lotka-Volterra competition model

Invasion by species 2 can occur when $\dfrac{1}{K_1} > \dfrac{\alpha_{21}}{K_2}$.

Invasion by species 1 can occur when $\dfrac{1}{K_2} > \dfrac{\alpha_{12}}{K_1}$.

greater than zero for $N_1 = 0$ and $N_2 = K_2$, i.e., when

$$\frac{1}{K_2} > \frac{\alpha_{12}}{K_1}.$$

These invasion criteria for the Lotka-Volterra model are listed in Table 8.2.

We next show that these invasion criteria are equivalent to asking under what conditions the boundary equilibria are unstable; e.g., for the equilibrium $(N_1^*, N_2^*) = (K_1, 0)$, following a small perturbation, such as the addition of small inoculum of species 2, does the system return to the boundary equilibrium or does the small inoculum increase in size through time?

We can determine the stability of an equilibrium by looking at the elements of the Jacobian matrix (see chapter 3 appendix):

$$A_{11} = \left.\frac{\partial f_1}{\partial N_1}\right|_* = r_1 - 2\frac{r_1}{K_1}N_1^* - r_1\frac{\alpha_{12}}{K_1}N_2^*$$

$$A_{12} = \left.\frac{\partial f_1}{\partial N_2}\right|_* = -r_1\frac{\alpha_{12}}{K_1}N_1^*$$

$$A_{21} = \left.\frac{\partial f_2}{\partial N_1}\right|_* = -r_2\frac{\alpha_{21}}{K_2}N_2^*$$

$$A_{22} = \left.\frac{\partial f_2}{\partial N_2}\right|_* = r_2 - 2\frac{r_2}{K_2}N_2^* - r_2\frac{\alpha_{21}}{K_2}N_1^*$$

where $\left.\frac{\partial f_i}{\partial N_j}\right|_*$ is the partial derivative of function f_i with respect to N_j, evaluated at the equilibrium.

For the equilibrium $(N_1^*, N_2^*) = (K_1, 0)$, with N_1 at its carrying capacity and N_2 at zero, the Jacobian matrix is

$$\begin{pmatrix} -r_1 & -r_1\alpha_{12} \\ 0 & r_2 - \dfrac{r_2\alpha_{21}K_1}{K_2} \end{pmatrix}.$$

This boundary equilibrium is stable if

Condition 1: $A_{11} + A_{22} < 0$: $\quad -r_1 + r_2 - r_2\alpha_{21}\dfrac{K_1}{K_2} < 0$

Condition 2: $A_{11}A_{22} > A_{12}A_{21}$: $\quad -r_1\left(r_2 - r_2\alpha_{21}\dfrac{K_1}{K_2}\right) > 0.$

Condition 1 will be true whenever condition 2 is true. Condition 2 will be true if the expression in parentheses is less than zero.

That is, the boundary equilibrium $(K_1, 0)$ is stable if

$$\frac{1}{K_1} < \frac{\alpha_{21}}{K_2},$$

and the boundary equilibrium is unstable to a small perturbation (e.g., the addition of a small inoculum of N_2) if

$$\frac{1}{K_1} > \frac{\alpha_{21}}{K_2}.$$

The second inequality is exactly the condition in which we found that species 2 can invade. The same procedure can be followed to determine that the boundary equilibrium, $(N_1^*, N_2^*) = (0, K_2)$, is unstable if

$$\frac{1}{K_2} > \frac{\alpha_{12}}{K_1},$$

which is the same condition in which we found that species 1 can invade.

If the invasion criterion for only one of the two species is met, then that species will win in competition and exclude the other species. If both are met, then the two species can coexist. It is also possible in the Lotka-Volterra competition model for neither criterion to be met. In this case, neither species can increase in abundance from low density when the other species is present at equilibrium. The outcome of competition then depends on the initial conditions. The species that gains an early advantage will win (this is often termed a *priority effect*). Thus there are four possible outcomes of competition in the Lotka-Volterra model, depending on the particular values of K_1, K_2, α_{12}, and α_{21}: (1) species 1 wins, excluding species 2, (2) species 2 wins, excluding species 1, (3) the two species coexist, and (4) either species 1 or species 2 wins, excluding the other species, but which species wins depends on the initial conditions. We will find that analogous outcomes occur in more complicated models.

Coexistence occurs only if both invasion criteria are true, i.e., if there is mutual invasibility. The invasion criteria can be rearranged to show that the necessary condition for coexistence is $\alpha_{12}\alpha_{21}$ less than 1. Thus, coexistence is possible when the per-head effect of species 2 on species 1 is less than the per-head effect of species 1 on itself ($\alpha_{12} < 1$) and the per-head effect of species 1 on species 2 is less than the per-head effect of species 2 on itself ($\alpha_{21} < 1$). In other words, coexistence is possible when individuals of each species have a greater per-head effect on the growth rate of their own species than they do on the growth rate of the competitor. That is, coexistence is possible when intraspecific competition is stronger than interspecific competition.

For the Lotka-Volterra competition model, the conditions for mutual invasibility are equivalent to the conditions in which the coexistence equilibrium is locally stable. The equivalence of these two criteria found in this model is not general to all competition models. In some models it is possible for both species to invade but for the resulting equilibrium to be unstable. In these cases the species may coexist and display non-equilibrium-persistent dynamics, such as cycles or chaotic dynamics, or one species may be driven extinct when the other species reaches densities away from equilibrium.

The two-species coexistence equilibrium, found by setting the right-hand sides of equations 8.2a and 8.2b equal to zero, is

$$\left(N_1^*, N_2^*\right) = \left(\frac{K_1 - \alpha_{12}K_2}{1 - \alpha_{12}\alpha_{21}}, \frac{K_2 - \alpha_{21}K_1}{1 - \alpha_{12}\alpha_{21}}\right).$$

The Jacobian matrix is

$$\begin{pmatrix} \dfrac{-r_1 N_1^*}{K_1} & \dfrac{-r_1 \alpha_{12} N_1^*}{K_1} \\[2ex] \dfrac{-r_2 \alpha_{21} N_2^*}{K_2} & \dfrac{-r_2 N_2^*}{K_2} \end{pmatrix}.$$

This is stable if

Condition 1: $A_{11} + A_{22} < 0$: $\qquad \dfrac{-r_1 N_1^*}{K_1} + \dfrac{-r_2 N_2^*}{K_2} < 0$;

this is always true whenever N_1^* and N_2^* are both positive;

Condition 2: $A_{11}A_{22} > A_{12}A_{21}$:

$$\left(\frac{-r_1 N_1^*}{K_1}\right)\left(\frac{-r_2 N_2^*}{K_2}\right) > \left(\frac{-r_1 \alpha_{12} N_1^*}{K_1}\right)\left(\frac{-r_2 \alpha_{21} N_2^*}{K_2}\right).$$

This second condition can be simplified to the same condition that we found to determine mutual invasibility: $\alpha_{12}\alpha_{21} < 1$.

EXPLOITATIVE COMPETITION FOR AN EXPLICIT RESOURCE

When we explicitly model the resource for which consumer species are competing, we gain insight into the effect of competition on resource density, as well as into the possibilities for coexistence. Such insight is of general ecological interest but is especially useful in relation to biological control of pest (resource) species in agriculture. Throughout the remainder of this chapter we therefore report on both coexistence and resource suppression, pointing ahead to the discussion of biological control in chapter 9.

We now turn to a model in which the resource for which competition is occurring is modeled explicitly. Consider two parasitoid (or predator) species, P and Q, which both attack the same host (or prey), H. A simple non-stage-structured model for this interaction is

$$\frac{dH}{dt} = rH\left(1 - \frac{H}{K}\right) - a_P H P - a_Q H Q = f_1(H, P, Q)$$

$$\frac{dP}{dt} = e_P a_P H P - d_P P = f_2(H, P, Q) \qquad (8.3)$$

$$\frac{dQ}{dt} = e_Q a_Q H Q - d_Q Q = f_3(H, P, Q)$$

where a_P and a_Q are the attack rates, e_P and e_Q are the parasitoid conversion efficiencies, and d_P and d_Q are the per-head death rates of parasitoids P and Q, respectively. We have included a logistic growth term for the host in order to stabilize the interaction, where r is the host intrinsic growth rate and K is the host carrying capacity.

Here, rather than being incorporated into competition coefficients as in the Lotka-Volterra competition model, the effect of species P on the growth rate of species Q is included through the effect that species P has on suppressing the density of H. Low densities of H lead to low per-head growth rates for Q.

We can ask under what conditions parasitoid Q can invade the system when parasitoid P and the host are present at an equilibrium, and vice versa. In Box 8.1 we show that, once again, the conditions for invasion of each species are equivalent to the conditions in which the boundary equilibria (i.e., the equilibria with one species absent and its competitor at equilibrium with the host) are unstable.

The two-species equilibrium with the host and parasitoid Q present is

$$H^*\big|_Q = \frac{d_Q}{e_Q a_Q}, \quad Q^*\big|_Q = \frac{r}{a_Q}\left(1 - \frac{H^*\big|_Q}{K}\right),$$

where $H^*\big|_Q$ is the host density at the equilibrium reached with parasitoid Q present but P absent. This notation is used extensively in this chapter.

At this equilibrium, P can invade from low densities when the per-head growth rate of P at low densities is greater than zero: $e_P a_P H^*\big|_Q - d_P > 0$. That is,

parasitoid P can invade when $\dfrac{d_Q}{e_Q a_Q} > \dfrac{d_P}{e_P a_P}$.

The two-species equilibrium with the host and parasitoid P present is exactly the same as that with Q present, but with the subscript Qs replaced with Ps. Therefore,

parasitoid Q can invade when $\dfrac{d_P}{e_P a_P} > \dfrac{d_Q}{e_Q a_Q}$.

It is obvious that these two criteria, for invasion of P from low densities and invasion of Q from low densities, are mutually exclusive. Therefore, the two parasitoid species will not be able to coexist on the single host species. Unless the combination of death rate, conversion efficiency, and attack rate happens to be exactly the same for the two parasitoid species (a highly unlikely situation), the species that is able

BOX 8.1.

INVASION CRITERIA IN THE SIMPLE MODEL OF EXPLOITATIVE
COMPETITION

The elements of the Jacobian matrix are:

$$A_{11} = \left.\frac{\partial f_1}{\partial H}\right|_* = r - \frac{2rH^*}{K} - a_P P^* - a_Q Q^*;$$

$$A_{12} = \left.\frac{\partial f_1}{\partial P}\right|_* = -a_P H^*; \qquad A_{13} = \left.\frac{\partial f_1}{\partial Q}\right|_* = -a_Q H^*$$

$$A_{21} = \left.\frac{\partial f_2}{\partial H}\right|_* = e_P a_P P^*; \qquad A_{22} = \left.\frac{\partial f_2}{\partial P}\right|_* = e_P a_P H^* - d_P;$$

$$A_{23} = \left.\frac{\partial f_2}{\partial Q}\right|_* = 0 \qquad A_{31} = \left.\frac{\partial f_3}{\partial H}\right|_* = e_Q a_Q Q^*;$$

$$A_{32} = \left.\frac{\partial f_3}{\partial P}\right|_* = 0; \qquad A_{33} = \left.\frac{\partial f_3}{\partial Q}\right|_* = e_Q a_Q H^* - d_Q.$$

First we investigate the conditions in which the equilibrium with the host present without the parasitoids is stable, and determine the conditions under which the parasitoid species can invade.

For the equilibrium $(H^*, P^*, Q^*) = (K, 0, 0)$, the Jacobian matrix is

$$\begin{pmatrix} -r & -a_P K & -a_Q K \\ 0 & e_P a_P K - d_P & 0 \\ 0 & 0 & e_Q a_Q K - d_Q \end{pmatrix}.$$

The characteristic equation is $(-r - \lambda)(e_P a_P K - d_P - \lambda)(e_Q a_Q K - d_Q - \lambda) = 0$. This equilibrium will be unstable whenever $e_P a_P K - d_P$ or $e_Q a_Q K - d_Q$ is greater than zero. It can be seen from inspection of equation 8.3 that these are the conditions for parasitoids P or Q to have positive growth rates when the host is at its carrying capacity.

For the equilibrium

$$\left(H^*|_P, P^*|_P, 0\right) = \left(\frac{d_P}{e_P a_P}, \frac{r}{a_P}\left(1 - \frac{H^*|_P}{K}\right), 0\right),$$

(Box 8.1. continued)

the Jacobian matrix is

$$\begin{pmatrix} r - 2r\dfrac{H^*\big|_P}{K} - a_P P^*\big|_P & -a_P H^*\big|_P & -a_Q H^*\big|_P \\ e_P a_P P^*\big|_P & 0 & 0 \\ 0 & 0 & e_Q a_Q H^*\big|_P - d_Q \end{pmatrix}.$$

The characteristic equation is

$$\left(e_Q a_Q H^*\big|_P - d_Q - \lambda\right)\left[(-\lambda)\left(r - 2r\dfrac{H^*\big|_P}{K} - a_P P^*\big|_P - \lambda\right) \right.$$

$$\left. + a_P H^*\big|_P \left(e_P a_P P^*\big|_P\right) \right] = 0.$$

The term in the square brackets in the characteristic equation defines the eigenvalues determining the stability of the two-species equilibrium $(H^*\big|_P, P^*\big|_P)$ in the two-equation system without Q. The $(H^*\big|_P, P^*\big|_P)$ equilibrium is always stable for this model for all parameters for which the two-species equilibrium is positive. In determining the invasion criteria here, we assume that H and P on their own can exist at a stable equilibrium, so the invasion criterion for Q depends only on the portion of the characteristic equation preceding the square brackets. It can be seen that the eigenvalue will be positive, and the $(H^*\big|_P, P^*\big|_P)$ equilibrium can be invaded by Q, if

$$\frac{d_P}{e_P a_P} > \frac{d_Q}{e_Q a_Q},$$

or in other words, if $H^*\big|_P$ is greater than $H^*\big|_Q$. Thus we arrive at the same invasion criteria by looking at the stability of the boundary equilibrium as we did by looking at the conditions in which the population of the second parasitoid species can have a positive growth rate when it is rare, and its competitor is at equilibrium with the host.

---

to invade from low densities, when its competitor cannot, will be able to displace its competitor.

Note that the parasitoid species that wins in competition is able to do so because it can suppress the abundance of the host to a level below the density required by its competitor to sustain itself. The condition for parasitoid P to invade can be rewritten as: P can invade when $H^*|_Q$ is greater than $H^*|_P$, and Q can invade when $H^*|_P$ is greater than $H^*|_Q$. This is an example of Tilman's R^* criteria (Tilman 1980) (which was foretold by Nicholson 1933), which states that in exploitative competition for a single resource, R, the species that can suppress the abundance of the resource to the lowest equilibrium level, R^*, will be able to win in competition and exclude its competitors (MacArthur and Levins 1967).

Dependence of Per-Head Rates on Host Density

The conclusion that the species that leads to the lowest host abundance is the one that wins in competition does not depend on the particular linear form of the functional response used in the model. If we make the per-head attack rate on hosts depend on host density, as in a type 2 functional response, we arrive at the same conclusion. (However, we will see next that a functional response that depends on parasitoid density does influence the outcome of competition.)

For example, consider the equations:

$$\frac{dH}{dt} = rH\left(1 - \frac{H}{K}\right) - \frac{a_P H P}{1 + a_P T_{hp} H} - \frac{a_Q H Q}{1 + a_Q T_{hq} H} \tag{8.4a}$$

$$\frac{dP}{dt} = \frac{e_P a_P H P}{1 + a_P T_{hp} H} - d_P P \tag{8.4b}$$

$$\frac{dQ}{dt} = \frac{e_Q a_Q H Q}{1 + a_Q T_{hq} H} - d_Q Q \tag{8.4c}$$

where T_{hp} and T_{hq} are the handling times for P and Q, respectively. As we saw in chapter 3, the two-species system with the host and one of the parasitoid species, either P or Q, can have either a stable equilibrium or can undergo limit cycles, depending on how far below the carrying capacity the parasitoid suppresses the host density. For now we consider only the case in which the two-species equilibrium is stable. (We discuss non-equilibrium coexistence below.)

If only parasitoid P and the host are present, the host equilibrium is

$$H^*|_P = \frac{d_P}{e_P a_P - d_P a_P T_{hp}},$$

and if only parasitoid Q and the host are present the host equilibrium is

$$H^*\big|_Q = \frac{d_Q}{e_Q a_Q - d_Q a_Q T_{hq}}.$$

From equation 8.4c, we can see that Q can have a positive growth rate and invade the system with parasitoid P and the host at equilibrium if $H^*\big|_P$ is greater than $H^*\big|_Q$, and from equation 8.4b, parasitoid P can invade the system with parasitoid Q and the host at equilibrium if $H^*\big|_Q$ is greater than $H^*\big|_P$. Therefore once again, the two parasitoid species cannot coexist. The species that wins in competition is the one that suppresses the density of the shared resource, the host, to the lower level.

Note that this conclusion sounds like good news for biological control (chapter 9). If we want to suppress the density of a pest to the lowest level, and we have more than one contending parasitoid species, the results of the models up to now suggest that we should simply release all possible parasitoid species, and the one that is most efficient at driving down the abundance of the host will prevail, and all lesser parasitoid species will be driven out of the system. If this were always true, there would be no harm in releasing all candidate natural enemy species. Unfortunately, however, the story becomes much more complicated.

Dependence of Per-Head Rates on Parasitoid Density

Above we saw that density dependence in the host population has no effect on the outcome of exploitative competition. This includes host dependence in the functional response (as in equations 8.4), and also any form of dependence on host density of the host birth or death rates (e.g., the logistic term in equation 8.4a above). Density dependence in the parasitoid population, in either the functional response or in the parasitoid death rates, however, can alter drastically our conclusions about the potential for coexistence and the impact on the host population. The reason for this goes back to the conclusions of the Lotka-Volterra competition model: if intraspecific competition is greater than interspecific competition, then coexistence is possible. We saw in chapter 3 that any density dependence in the per-head rate of increase of the parasitoid population has the effect of making the parasitoids less efficient at controlling the host population: they become self-limiting. Host-population densities tend to increase as the strength of density dependence in the parasitoid population increases.

For example, in chapter 3 we looked at a predator-prey model with a density-dependent predator death rate. We found that the predator density dependence had a stabilizing effect on the dynamics and increased prey density. Here we redefine the predators as parasitoids and add a second parasitoid species. The model is

$$\frac{dH}{dt} = rH - a_P HP - a_Q HQ \tag{8.5a}$$

$$\frac{dP}{dt} = e_P a_P HP - d_P P - m_P P^2 \tag{8.5b}$$

$$\frac{dQ}{dt} = e_Q a_Q HQ - d_Q Q - m_Q Q^2 \tag{8.5c}$$

where m_P and m_Q determine the strength of density dependence in the death rates of parasitoids P and Q, respectively.

The equilibrium with only parasitoid P and the host present is

$$H^*\big|_P = \frac{d_P}{e_P a_P} + \frac{m_P P^*\big|_P}{e_P a_P}, \quad \text{where} \quad P^*\big|_P = \frac{r}{a_P}.$$

Remember that the host equilibrium without density dependence in the parasitoid death rate is $d_P/(e_P a_P)$. Thus, the host equilibrium density increases linearly with increases in the strength of density dependence, m_P.

From equation 8.5c, parasitoid Q can invade this equilibrium if

$$e_Q a_Q H^*\big|_P > d_Q,$$

which can be written as

$$\frac{d_P}{e_P a_P} + \frac{m_P r}{e_P a_P^2} > \frac{d_Q}{e_Q a_Q}.$$

Likewise the equilibrium with only parasitoid Q and the host present is

$$H^*\big|_Q = \frac{d_Q}{e_Q a_Q} + \frac{m_Q Q^*\big|_Q}{e_Q a_Q}; \quad Q^*\big|_Q = \frac{r}{a_Q}.$$

From equation 8.5b, parasitoid P can invade this equilibrium if

$$e_P a_P H^*\big|_Q > d_P,$$

which can be written as

$$\frac{d_Q}{e_Q a_Q} + \frac{m_Q r}{e_Q a_Q^2} > \frac{d_P}{e_P a_P}.$$

These invasion criteria can be interpreted as follows. Parasitoid Q can invade the system, with parasitoid P and the host at equilibrium, if the equilibrium host density set by P alone, in the presence of density dependence in P's death rate, is greater than the equilibrium host density set by Q alone in the absence of density dependence in Q's death rate.

We can see from inspecting the two invasion criteria that it is now possible for P and Q to coexist, for some combinations of the other parasitoid parameters, provided that either m_P or m_Q (or both) is greater than zero. That is, coexistence is possible if there is density dependence in the death rate of at least one of the parasitoid species. This is illustrated in Fig. 8.1, which shows the outcome of competition in terms of the attack rates of the two species (here we assume that the two parasitoid species have equal death rates and equal conversion efficiencies). In Fig. 8.1a, there is density dependence only in parasitoid Q's death rate. In this case, whenever parasitoid P has a higher attack rate than parasitoid Q, P wins in competition and excludes Q from the system. However, if parasitoid Q has a higher attack rate than parasitoid P, there is now a region of parameter space in which the two species can coexist. For very high relative attack rates of parasitoid Q, Q can win and exclude P. In Fig. 8.1b, there is density dependence in the death rates of both parasitoid species, and the region of coexistence is greatly expanded.

When a single parasitoid species wins in competition (in the "P wins" or "Q wins" regions of Fig. 8.1), it is the species that can lead to the lower abundance of the host that wins in competition. When there is density dependence in the death rate of only one of the parasitoid species, the host equilibrium reached when the two parasitoid species coexist is equal to the host equilibrium achieved when only the parasitoid species without density dependence in its death rate is present. For example, in the hatched region of Fig. 8.1a, the two parasitoid species can coexist, and the host equilibrium reached is the same as if parasitoid P were present on its own. It is lower than if parasitoid Q were present on its own. That is, for parameters in the hatched region of Fig. 8.1a, parasitoid Q can invade the system with P and the host at equilibrium (although in

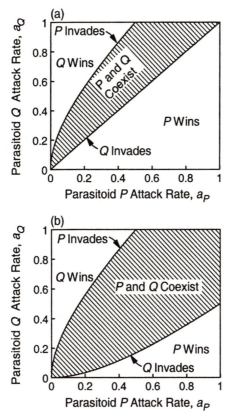

FIGURE 8.1. Outcome of competition for exploitative competition model with density-dependent parasitoid death rates. In each panel, parasitoid P can invade below and to the right of the line labeled "P Invades," and parasitoid Q can invade above and to the left of the line labeled "Q Invades." The two parasitoid species can coexist in the hatched region. In (a), density dependence is included only in parasitoid Q's death rate: $m_P = 0, m_Q = 0.5$. In (b), density dependence is included in the death rate of both parasitoid species: $m_P = m_Q = 0.5$. In both (a) and (b), $r = 2, d_P = d_Q = 1, e_P = e_Q = 1$.

this model without a carrying capacity for the host, this equilibrium is neutrally stable rather than stable), and the two parasitoid species can coexist, but the addition of Q has no impact on the host equilibrium (Fig. 8.2a). The addition of Q stabilizes the equilibrium because of its density-dependent death rate.

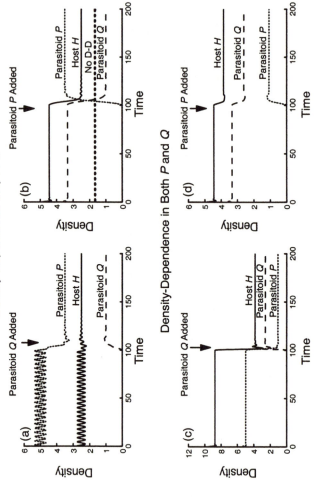

FIGURE 8.2. Simulations of the model of exploitative competition with density dependence in the death rate of one or both parasitoid species. In (a) and (b) only parasitoid Q's death rate is density dependent: $m_P = 0$, $m_Q = 0.5$. In (c) and (d) there is density dependence in the death rates of both species: $m_P = 0.5$, $m_Q = 0.5$. In (a) and (c), parasitoid Q invades the system with parasitoid P and the host present. In (b) and (d), parasitoid P invades the system with parasitoid Q and the host present. In (b) the line labeled "No D-D" shows the host equilibrium that would be reached by parasitoid Q alone if there were no density dependence in Q's death rate. $a_P = 0.4$, $a_Q = 0.6$, $r = 2$, $d_P = d_Q = 1$, $e_P = e_Q = 1$ for all simulations.

The addition of P, the species without density dependence in its death rate, into a system with Q and the host at equilibrium (this two-species equilibrium is stable because of the density dependence in Q's death rate) results in coexistence of the two parasitoid species and leads to further suppression of the host abundance (Fig. 8.2b). The final densities when the two species coexist are the same in Figs. 8.2a and b, meaning that the order of arrival of the two parasitoid species does not matter here.

The dotted line in Fig. 8.2b, labeled "No D-D," shows the host equilibrium that would be reached by parasitoid Q alone, for the same value of a_Q, if there were no density dependence in Q's death rate. This illustrates that for this combination of attack rates ($a_P = 0.4$, $a_Q = 0.6$), parasitoid Q on its own, with m_Q equal to zero, would lead to the lowest host density. But with density dependence ($m_Q = 0.5$ in this case), Q on its own leads to a higher host density, and a lower host density is achieved by parasitoid P either in the presence or absence of parasitoid Q.

When there is density dependence in the death rates of both parasitoid species, and the two parasitoid species coexist, as in the hatched regions of Fig. 8.1, the host equilibrium with both species present is lower than the host density with either species on its own (Figs. 8.2c and d). Once again, however, an even lower equilibrium would be achieved if the most effective parasitoid species (the one with the higher attack rate) did not have a density-dependent death rate. That is, density dependence in the parasitoid death rate increases the chances for coexistence, but only at the cost of reducing the effectiveness of control.

Box 8.2 shows that exactly the same conclusions are reached if intraspecific density dependence is included in the functional response of each parasitoid species rather than in its death rate. In each of these cases, coexistence is possible because introducing density dependence in the parasitoids' demographic rates is equivalent to introducing intraspecific competition. We know from the Lotka-Volterra competition model that coexistence is possible when the strength of intraspecific competition is greater than the strength of interspecific competition. In each of the above examples, the per-head growth rate of each of the parasitoid species depends not only on the density of its resource, the host, but also on the density of its own species. That is, there is self-limitation of each parasitoid species. If instead, the per-head growth rate of each species depends on the densities of the host and both parasitoid

BOX 8.2.

PARASITOID-DEPENDENT FUNCTIONAL RESPONSE

Consider a two-parasitoid version of the model with a parasitoid-dependent functional response that we presented in chapter 3. The model is

$$\frac{dH}{dt} = rH - \frac{a_P H P}{1 + z_P P} - \frac{a_Q H Q}{1 + z_Q Q} \tag{B8.1a}$$

$$\frac{dP}{dt} = \frac{e_P a_P H P}{1 + z_P P} - d_P P \tag{B8.1b}$$

$$\frac{dQ}{dt} = \frac{e_Q a_Q H Q}{1 + z_Q Q} - d_Q Q \tag{B8.1c}$$

where z_P and z_Q determine the strength of density dependence on parasitoid density in the functional responses of P and Q, respectively.

The equilibrium with only parasitoid P and the host present is

$$H^*\big|_P = \frac{d_P}{e_P a_P} \left(1 + z_P P^*\big|_P\right); \quad P^*\big|_P = \frac{r}{a_P - r z_P}.$$

From equation B8.1c, parasitoid Q can invade this equilibrium if

$$\frac{d_P}{e_P a_P} \left(1 + \frac{z_P r}{a_P - z_P r}\right) > \frac{d_Q}{e_Q a_Q}.$$

Likewise the equilibrium with only parasitoid Q and the host present is

$$H^*\big|_Q = \frac{d_Q}{e_Q a_Q} \left(1 + z_Q Q^*\big|_Q\right); \quad Q^*\big|_Q = \frac{r}{a_Q - r z_Q}.$$

From equation B8.1b, parasitoid P can invade this equilibrium if

$$\frac{d_Q}{e_Q a_Q} \left(1 + \frac{z_Q r}{a_Q - z_Q r}\right) > \frac{d_P}{e_P a_P}.$$

Once again, these invasion criteria can be interpreted as parasitoid Q can invade the system, with parasitoid P and the host at equilibrium, if the equilibrium host density set by P alone in the presence of density dependence in P's functional response is greater than the equilibrium host density set by Q alone in the absence of density dependence in Q's functional response.

species, the connection with the Lotka-Volterra model becomes more transparent.

For example, if the death rate of each parasitoid species increases linearly with increases in the density of both parasitoid species, the model becomes

$$\frac{dH}{dt} = rH - a_P HP - a_Q HQ$$

$$\frac{dP}{dt} = e_P a_P HP - d_P P - (m_{PP} P + m_{PQ} Q) P \qquad (8.6)$$

$$\frac{dQ}{dt} = e_Q a_Q HQ - d_Q Q - (m_{QP} P + m_{QQ} Q) Q$$

where m_{PP} measures the effect of P on its own death rate, m_{PQ} measures the effect of Q on P's death rate, m_{QP} measures the effect of P on Q's death rate, and m_{QQ} measures the effect of Q on its own death rate.

The equilibrium with each parasitoid species on its own is not changed from above (except m_P is replaced by m_{PP} and m_Q by m_{QQ}). Q can now invade the system with P and the host at equilibrium if

$$e_Q a_Q H^*\big|_P > d_Q + m_{QP} P^*\big|_P,$$

which can be written as

$$\frac{d_P}{e_P a_P} + \frac{m_{PP} r}{e_P a_P^2} > \frac{d_Q}{e_Q a_Q} + \frac{m_{QP} r}{e_Q a_Q a_P}.$$

Likewise P can invade the system with Q and the host at equilibrium if

$$e_P a_P H^*\big|_Q > d_P + m_{PQ} Q^*\big|_Q,$$

which can be written as

$$\frac{d_Q}{e_Q a_Q} + \frac{m_{QQ} r}{e_Q a_Q^2} > \frac{d_P}{e_P a_P} + \frac{m_{PQ} r}{e_P a_P a_Q}.$$

It can be seen that both of these criteria are more likely to be met if the intraspecific mortality terms, m_{PP} and m_{QQ}, are large relative to the interspecific mortality terms, m_{QP} and m_{PQ}. That is, coexistence is more likely to occur when intraspecific competition is stronger than interspecific competition.

Facultative Hyperparasitism Equals Intraguild Predation

Now consider what happens when there is "interference" as well as exploitative competition. For example, suppose the larva of one parasitoid species is able to kill that of its competitor when females of both species lay an egg in the same host individual. To the entomologist, this is facultative hyperparasitism (Briggs 1993a). To the community ecologist it is intraguild predation (Polis et al. 1989; Polis and Holt 1992) because the larval parasitoid is effectively an omnivore feeding on two trophic levels—the herbivore level represented by the host and the predator level represented by larvae of the first parasitoid species.

Consider the model of Holt and Polis (1997) in which H is the herbivore that can be consumed by either predator species. P is a predator but also the intraguild prey; it eats the herbivore but can also be consumed by the intraguild predator, Q. The intraguild predator can consume either the herbivore or the intraguild prey.

In parasitoid notation, H is the host that can be attacked by two parasitoid species, P and Q. P is a primary parasitoid, meaning it attacks only unparasitized hosts. Parasitoid Q is a facultative hyperparasitoid, which can attack either unparasitized hosts or hosts that have been parasitized by the primary parasitoid, P. In reality it is only the juveniles of parasitoid P that can be attacked by Q. We do not consider explicit age structure here but deal with it in the section "Effects of Age Structure on Competition." In parasitoid notation this model could also represent the situation in which Q is the superior larval competitor when P and Q attack the same host individual. As we noted in the introduction of this chapter, the outcome of multiparasitism and larval competition is the same, in terms of population dynamics, as facultative hyperparasitism.

We present here only one special case of a more general model investigated by Holt and Polis (1997). The model is

$$\frac{dH}{dt} = rH\left(1 - \frac{H}{K}\right) - a_{HP}HP - a_{HQ}HQ$$

$$\frac{dP}{dt} = e_{HP}a_{HP}HP - a_{PQ}PQ - d_PP \qquad (8.7)$$

$$\frac{dQ}{dt} = e_{HQ}a_{HQ}HQ + e_{PQ}a_{PQ}PQ - d_QQ$$

where r is the intrinsic rate of increase of the herbivore and K is its carrying capacity. a_{HP} and a_{HQ} are the attack rates of P and Q, respectively, on the herbivore, and a_{PQ} is the attack rate of Q on P. Likewise, e_{HP} and e_{HQ} are the conversion efficiencies of P and Q, respectively, on the herbivore (i.e., the numbers of new individuals of P or Q that result from eating one herbivore), and e_{PQ} is the conversion efficiency of Q on P. d_P and d_Q are the per-head death rates of the two predator species.

If neither predator species is present, the herbivore has a stable equilibrium at its carrying capacity, K. With only the intraguild prey, P, and the herbivore present, the equilibrium is

$$H^*\big|_P = \frac{d_P}{e_{HP}a_{HP}}; \quad P^*\big|_P = \frac{r}{a_{HP}}\left(1 - \frac{H^*\big|_P}{K}\right).$$

In any case in which the intraguild prey can persist (i.e., has a positive equilibrium), it suppresses the abundance of the herbivore below its carrying capacity, i.e., $H^*\big|_P$ is less than K. With only the intraguild predator, Q, and the herbivore present, the equilibrium is

$$H^*\big|_Q = \frac{d_Q}{e_{HQ}a_{HQ}}; \quad Q^*\big|_Q = \frac{r}{a_{HQ}}\left(1 - \frac{H^*\big|_Q}{K}\right).$$

The intraguild predator can have a positive equilibrium and persist on the herbivore if $H^*\big|_Q$ is less than K.

We now look at the invasion criterion for each of these two-species equilibria by the other predator species. The intraguild predator, Q, can invade the system with the intraguild prey and the herbivore at equilibrium if

$$e_{HQ}a_{HQ}H^*\big|_P + e_{PQ}a_{PQ}P^*\big|_P > d_Q.$$

This can be interpreted as: the per-head birth rate of the intraguild predator from attacking both the herbivore and the intraguild prey must be greater than its per-head death rate. Thus, P serves as an additional resource for Q, making it easier for Q to invade.

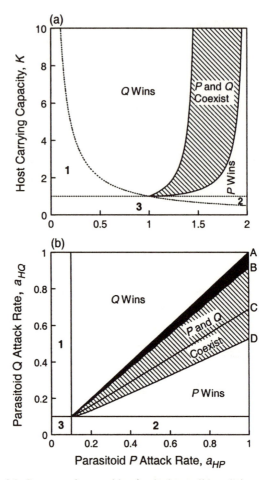

FIGURE 8.3. Outcome of competition for the intraguild predation model. In both (a) and (b): in hatched regions P and Q can coexist; in the black region either species can win, depending on initial conditions; in the region labeled "1" parasitoid P cannot persist; in region "2" parasitoid Q cannot persist; and in region "3" neither parasitoid species can persist. In (a) the outcome is shown in terms of the host carrying capacity and the attack rate by P, the intraguild prey, a_{HP}. $a_{HQ} = 1$, and coexistence is possible only when a_{HP} is greater than 1. $\alpha = 0.5$, $\beta = 2$, $r = 1$. In (b) the outcome is shown in terms of the attack rates of the two consumer species on the herbivore. The outcome is shown for four different values of ε, the relative conversion efficiency from intraguild prey versus the herbivore (or parasitized vs. unparasitized hosts). For all values of ε, parasitoid P can invade below and to the right of the line B. For $\varepsilon = 1$, parasitoid Q can invade above and to the left of line B,

The intraguild prey, P, can invade the system with the intraguild predator and the herbivore at equilibrium if

$$e_{HQ}a_{HQ}H^*\big|_Q > d_P + a_{PQ}Q^*\big|_Q.$$

This can be interpreted as: the per-head birth rate of the intraguild prey from attacking the herbivore must be greater than its per-head death rate that is due to background causes and attacks by the intraguild predator. Thus, direct attack by Q makes it harder for P to invade.

For convenience we let

$$\alpha = \frac{a_{PQ}}{a_{HQ}} \quad \text{and} \quad \varepsilon = \frac{e_{PQ}}{e_{HQ}}.$$

α is Q's relative attack rate on the intraguild prey versus the herbivore. Or in parasitoid terms, α is Q's relative attack rate on parasitized versus unparasitized hosts. If α is equal to 1, there is no preference or difference in ease of attack between the two prey types, and both are consumed in direct proportion to their abundance. If α is greater than 1, the intraguild predator prefers to attack the intraguild prey (or the intraguild prey is easier to capture and consume) relative to the herbivore. ε is the relative benefit to the intraguild predator in terms of production of new intraguild predator offspring (or biomass) from attacking an individual intraguild prey versus attacking a unit of the herbivore.

Figs. 8.3a and b show that it is possible for the two invasion criteria to be met simultaneously, and for the two predator species to coexist. The necessary conditions are: (1) the intraguild prey must be more efficient at attacking the herbivore than the intraguild predator, such that the intraguild prey on its own must lead to a lower equilibrium density of

---

so coexistence is not possible. For $\varepsilon = 5$, parasitoid Q can invade above and to the left of line C, so coexistence is possible in the region between lines B and C. For $\varepsilon = 10$, parasitoid Q can invade above and to the left of line D, so coexistence is possible in the region between lines B and D. For $\varepsilon < 1$, coexistence is not possible, and there is a region of parameter space in which the outcome of competition depends on initial conditions. This is shown in the black region for $\varepsilon = 0$, for which parasitoid Q can invade only above and to the left of line A. $r = 0.1$, $K = 10$, $c_{HP} = c_{HQ} = 1$, $d_P = d_Q = 1$, $\alpha = a_{PQ}/a_{HQ} = 1$.

the herbivore than the intraguild predator on its own:

$$\frac{d_P}{e_{HP}a_{HP}} < \frac{d_Q}{e_{HQ}a_{HQ}};$$

and (2) the intraguild predator must realize some benefit from attacking the intraguild prey,

$$\varepsilon = \frac{e_{PQ}}{e_{HQ}} > 0.$$

A popular theme in intraguild predation models is looking at the effect of altering the productivity of the basal species (the herbivore in this case) on the outcome of competition. We can examine this effect by inspecting the invasion criteria. The criterion for Q to invade can be rewritten as

$$H^*\big|_P + \left\{\frac{r\alpha\varepsilon}{a_{HP}}\left(1 - \frac{H^*\big|_P}{K}\right)\right\} > H^*\big|_Q.$$

A productive environment for the herbivore can be represented by high values of r and/or K. Increasing the value of either of these parameters leads to increases in the term in the brackets { } in the invasion criterion. This makes it easier for the intraguild predator to invade. Likewise, the criterion for P to invade can be rewritten as

$$H^*\big|_Q - \left\{\frac{r\alpha}{c_{HP}a_{HP}}\left(1 - \frac{H^*\big|_Q}{K}\right)\right\} > H^*\big|_P.$$

Once again, increasing r and/or K leads to higher values of the bracketed term in the invasion criterion. Since this term is subtracted from the left-hand side in this case, increasing r or K makes it harder for the intraguild prey to invade. The effects of increasing K on the invasion criteria are illustrated in Fig. 8.3a. Over some range of parameter space a highly productive environment should favor the intraguild predator, and a low productivity environment should favor the intraguild prey. At intermediate levels of productivity, where both criteria can be met, the two predators potentially coexist.

The equations for the equilibrium with both predator species present are not particularly revealing (see Holt and Polis 1997). When the two predators do coexist, the equilibrium of the herbivore reached is intermediate between the equilibrium density with the intraguild prey alone and the equilibrium density with the intraguild predator alone. Thus adding an intraguild predator that coexists with an intraguild prey (or a facultative hyperparasitoid that coexists with a primary parasitoid) is predicted to lead to an increase in the host equilibrium density, above that which would be achieved by the intraguild prey (or the primary parasitoid) on its own.

We call this the *intraguild predation effect*, where two consumer species can coexist by one being superior at attacking the herbivore, and the other species being able to use its competitor as an additional resource. When the two species do coexist through this mechanism, the equilibrium of the herbivore is higher than if the superior consumer species were present on its own.

In some cases it is possible for the intraguild predator to persist in the presence of the intraguild prey, where it could not persist on the herbivore alone (i.e., where $d_Q/(e_{HQ}a_{HQ}) > K$). This can occur when the intraguild prey is easier than the herbivore for the intraguild predator to catch, or when the intraguild predator gains a larger benefit from attacking the intraguild prey than the herbivore.

Direct interference without some benefit to one species ($\varepsilon = 0$) does not promote coexistence (Fig. 8.3b). If the intraguild predator simply kills the intraguild prey but does not gain an increased per-head growth rate from doing so, then this simply makes it harder for the intraguild prey to invade and has no effect on the intraguild predator. This results in one species or the other always winning in competition, but also results in a region of parameter space in which neither invasion criterion is met. In this case, the outcome of competition depends on which species gains an initial advantage. For different starting conditions, either P or Q will win in competition and exclude the other species. The species i with the lower value of $d_i/(e_{Hi}a_{Hi})$ will be the species that leads to the lower density of the herbivore. Thus, depending on the initial conditions, for ε equal to zero in the narrow region shown in black between lines A and B in Fig. 8.3b, Q may win, even though P on its own would be able to lead to a lower density of the herbivore.

COMPETITION IN DISCRETE TIME

We now use the above conclusions to interpret the results of discrete-time models of competition between parasitoid species. As we saw in chapter 4, these are more appropriate for temperate insect systems with a single generation of hosts and parasitoids per year. These models have contributed to the debate on whether a single or multiple parasitoid species would be best able to control the abundance of their common host (chapter 9).

Competing Parasitoid Species Aggregate Independently of Each Other

May and Hassell (1981) modeled one-host/two-parasitoid systems using discrete-time difference equations. In all of their models they used a negative binomial parasitism function, in which increases in the density of a given parasitoid species reduce the per-head efficiency with which that species attacks the host. For example, for parasitoid P, the fraction of hosts that escape parasitism is

$$f_P\left(P_t\right) = \left[1 + \frac{a_P P_t}{k_P}\right]^{-k_P},$$

where P_t is the density of parasitoid P in generation t, a_P is parasitoid P's attack rate, and k_P is the clumping parameter of the negative binomial distribution. In chapter 4 we discussed how the negative binomial distribution has been used in discrete-time host-parasitoid models as a phenomenological description of the effects of aggregated attacks by parasitoids. The negative binomial model can be interpreted as follows: the parasitoids distribute themselves among patches of hosts once per generation according to the negative binomial distribution. The parasitoids remain in their chosen patch for the entire generation, even as the density of hosts becomes severely depleted through time in patches with high densities of parasitoids. Therefore parasitoids in the high parasitoid density patches have a relatively low per-head efficiency, because the parasitoids are continuing to search there after the hosts have been depleted. Density dependence in the parasitoid population occurs because as the density of parasitoids increases, more of them are in the high-density patches, and therefore their per-head efficiency at attacking hosts is reduced. The distribution of parasitoids becomes more clumped at lower values of k_P, and therefore the strength of intraspecific density dependence increases at lower values of k_P. (See chapter 10 for further discussion of this subject.)

Thus, May and Hassell's (1981) models all incorporate intraspecific competition within each of the parasitoid species. We know from the section "Dependence of Per-Head Rates on Parasitoid Density," above, that the addition of intraspecific competition to the consumer species has the following effects:

1. It increases the host equilibrium density reached when each single parasitoid species is present on its own.
2. It makes coexistence of multiple consumer species possible on a single host species, because the growth rate of each species is limited by its own density as well as by the density of the shared host.
3. When two or more self-limited parasitoid species coexist, their combined action reduces the host density below the level imposed by the single self-limited parasitoid species. But this density is greater than that which would be achieved if the single most efficient parasitoid species were not self-limited.

In May and Hassell's (1981) first model, two parasitoid species, P and Q, attack the single host species, H. Only hosts that escape parasitism by Q can be successfully attacked by P. This could describe the following three situations: (a) Q attacks first in the season and renders the host unsuitable for subsequent parasitism by P; (b) Q is superior in larval competition to P; in this case, the two parasitoid species may attack the same stage of the host, but the larvae of P survive only in hosts that have not been attacked by Q; (c) Q is a facultative hyperparasitoid, attacking and successfully parasitizing either unparasitized hosts or hosts that have been previously parasitized by P. For brevity here we refer to P as the inferior larval competitor and Q as the superior larval competitor.

The model for this is

$$H_{t+1} = RH_t\phi(H_t)f_P(P_t)f_Q(Q_t)$$

$$P_{t+1} = H_tf_Q(Q_t)[1 - f_P(P_t)] \qquad (8.8)$$

$$Q_{t+1} = H_t[1 - f_Q(Q_t)]$$

with

$$f_P(P_t) = \left[1 + \frac{a_PP_t}{k_P}\right]^{-k_P} \quad \text{and} \quad f_Q(Q_t) = \left[1 + \frac{a_QQ_t}{k_Q}\right]^{-k_Q},$$

where H_t, P_t, and Q_t are the densities of hosts, parasitoid P, and parasitoid Q, respectively, in generation t; a_P and a_Q are the attack rates of parasitoid P and Q, respectively; and k_P and k_Q are the clumping parameters for the negative binomial distribution for P and Q, respectively. R is the yearly per-head net reproductive rate of the host. $\phi(H_t)$ is an optional function that incorporates host density dependence in the host's fecundity (in different versions of the model this is either 1 or $\exp(-gH_t)$, where g is a constant that determines the strength of host density dependence).

As we would predict from the discussion above, May and Hassell (1981) found that coexistence is impossible if there is no intraspecific density dependence in the parasitism function for either species, i.e., if there is no aggregation by either species ($k_P = k_Q = \infty$; the fractions of hosts that escape parasitism by the two species become: $f_P(P_t) = \exp\{-a_PP_t\}$ and $f_Q(Q_t) = \exp\{-a_QQ_t\}$). Coexistence is possible when there is relatively strong aggregation by one or both parasitoid species. Coexistence is more likely if the inferior larval competitor, P, has a higher attack rate than the superior larval competitor Q. This is consistent with the ideas of coexistence through the intraguild predation effect discussed above (Holt and Polis 1997) and was termed "counterbalanced competition" by Zwölfer (1973).

Hogarth and Diamond (1984) modified the May and Hassell (1981) model to include the situation in which one species is not always the winner in multiparasitism. When individuals of two parasitoid species attack the same host individual, species Q wins a proportion s of the time and species P wins the remaining $1 - s$ of the time. The result of varying s from zero to 1 is to simply shift the competitive advantage from P to Q, as would be expected.

May and Hassell's (1981) results about suppression of the host equilibrium density are understandable in light of the discussion in the section "Dependence of Per-Head Rates on Parasitoid Density" above. The addition of the inferior larval competitor, P, which has a higher attack rate on the host, leads to improved suppression of the host over that achieved by the superior larval competitor, Q, with a lower attack rate on the host. If P has a much greater attack rate than Q, P excludes Q and reduces the host equilibrium. If P has only a moderately higher attack rate than Q, the two can coexist and the resulting equilibrium is lower than with Q alone.

If the density dependence in P is relatively strong ($k_P < 1$) then the host equilibrium with both parasitoid species present is lower than that with P alone. But if k_P is greater than 1 (weak density dependence in P), then P on its own would lead to better host suppression. This final conclusion agrees with the results of models of intraguild predation (Holt and Polis 1997): if there is not strong intraspecific competition in the consumer species, then the species that has the higher attack rate on the host on its own also suppresses the host more. The addition of a second species that has a lower attack rate on the host, but that can attack hosts that are parasitized by the other species, can interfere with control of the host. May and Hassell (1981) found, however, that this region of parameter space in which Q interferes with P's superior control of the host is small in this model. Except for this intraguild predation effect, all of May and Hassell's (1981) conclusions support the idea that release of multiple parasitoid species would be the best strategy to achieve optimal control of the host (chapter 9).

Competing Parasitoid Species Aggregate to Same Host Patches

Kakehashi et al. (1984) pointed out that the above conclusions hinge on the fact that May and Hassell used parasitism functions for each species that depend only on the density of that parasitoid species. In the negative binomial model, each parasitoid species has a clumped distribution that is independent of the density of the other species (and independent of host density).

Kakehashi et al. (1984) modified the parasitism function in the model from a negative binomial distribution to a negative multinomial

distribution, which assumes that the two parasitoid species are aggregating in the same way, to the same patches of hosts (although again the
aggregation is independent of host density). For the situation described
in equations 8.8, in which Q is the superior larval competitor, the model
becomes

$$
H_{t+1} = RH_t\phi(H_t)\left[1 + \frac{a_P P_t}{k} + \frac{a_Q Q_t}{k}\right]^{-k}
$$

$$
P_{t+1} = H_t\left[\left(1 + \frac{a_Q Q_t}{k}\right)^{-k} - \left(1 + \frac{a_Q Q_t}{k} + \frac{a_P P_t}{k}\right)^{-k}\right] \quad (8.9)
$$

$$
Q_{t+1} = H_t\left[1 - \left(1 + \frac{a_Q Q_t}{k}\right)^{-k}\right]
$$

where k is the clumping parameter and is the same for the two species.

In this case, the addition of the superior larval competitor, Q, can
disrupt the level of control of the host achieved by P. When the two
species coexist, species P must be better at attacking the host ($a_P > a_Q$),
and the addition of Q leads to an increase in the host equilibrium density.
Thus, this model argues that the release of the single most effective
parasitoid species is the best strategy for successful biological control
(chapter 9).

By making the distribution of attacks depend on the densities of
both species, Kakehashi et al. removed the intraspecific competition
within each species present in the May and Hassell (1981) model. In
the Kakehashi et al. model, the two parasitoid species coexist only
through the intraguild predation effect, in which one species is superior
at attacking the host (the intraguild prey, P), and the other species, Q, is
an intraguild predator that effectively attacks juveniles of the intraguild
prey. When this occurs, the successful addition of the intraguild predator
always leads to an increase in the host equilibrium density.

Kakehashi et al. discussed their results in terms of "niche overlap."
In their terminology, Hassell and May assumed that the two parasitoid
species have different niches by aggregating independently of each
other, and the model in equations 8.9 assumes that the niches completely
overlap.

Competing Parasitoid Species Have Different Niches

Kakehashi et al. (1984) investigated a model in which the degree of niche overlap of the two parasitoid species can be varied directly. The model assumes that some of the hosts are vulnerable to attack by only one parasitoid species or the other, some are vulnerable to both parasitoid species, and others are in a refuge from attack by either species. These are defined as follows:

  S = fraction of hosts vulnerable only to Q;
  T = fraction of hosts vulnerable to both P and Q;
  U = fraction of hosts vulnerable only to P;
  V = fraction of hosts invulnerable to both species.

Attacks on the fraction of hosts vulnerable to each species occur at random, e.g., all hosts in the fraction S are equally vulnerable to attack by Q. Hosts in the fraction T that are actually attacked by individuals of both parasitoid species become adults of parasitoid Q in the next generation (thus again Q is the superior larval competitor).

The equations for this model are

$$H_{t+1} = RH_t \phi(H_t)[S \exp(-a_Q Q_t) + T \exp(-a_Q Q_t - a_P P_t)$$
$$+ U \exp(-a_P P_t) + V]$$

$$P_{t+1} = H_t[T \exp(-a_Q Q_t) + U][1 - \exp(-a_P P_t)] \qquad (8.10)$$

$$Q_{t+1} = H_t(S + T)[1 - \exp(-a_Q Q_t)].$$

Kakehashi et al. find the following results in terms of host suppression (chapter 9).

1. As expected, if the two parasitoid species attack exactly the same fraction of the host population (i.e., their niches overlap completely, $S = U = 0$), then the two species can coexist only through the intraguild predation mechanism, and the release of only the single most effective parasitoid species is the best strategy for effective biological control.
2. If the two parasitoid species attack completely different fractions of the host population ($T = 0$), then the two species can coexist

through niche partitioning, and the release of multiple parasitoid species is the best strategy for host suppression.

3. When there is a fraction of hosts that only the intraguild predator, Q, can attack ($S > 0$), but no fraction that only the intraguild prey, P, can attack ($U = 0$), then the multiple release strategy is better than single release for biological control. In this case, the parasitoid that is superior at attacking the host, P, cannot attack a fraction of the hosts, S, and therefore the addition of Q improves the degree of control.

4. When there is a fraction of hosts that only the intraguild prey, P, can attack ($U > 0$), but no fraction that only the intraguild predator, Q, can attack ($S = 0$), then the single release strategy is better than the multiple release strategy for biological control. In this case, P is better at attacking the host, and can attack all fractions of vulnerable hosts. The addition of Q simply interferes with the level of control achievable by P.

In general, the single release strategy is preferable when both parasitoids tend to attack the same hosts (i.e., high niche overlap), and the multiple release strategy is better when the two parasitoid species tend to attack different fractions of the host population (i.e., low niche overlap). Hochberg (1996) arrived at similar conclusions with a very different formulation of the model.

Obligate Hyperparasitoids

May and Hassell (1981) also investigated the situation in which parasitoid Q is an obligate hyperparasitoid on P. Parasitoid P attacks the host. Hosts that are attacked by P are then vulnerable to attack by Q. Only hosts that have been attacked by P that avoid attack by Q survive to become the next generation of parasitoid P. Only hosts that are attacked by both P and Q become the next generation of parasitoid Q.

The model for this is

$$H_{t+1} = RH_t f_P(P_t)$$

$$P_{t+1} = H_t[1 - f_P(P_t)]f_Q(Q_t) \qquad (8.11)$$

$$Q_{t+1} = H_t[1 - f_P(P_t)][1 - f_Q(Q_t)]$$

with $f_P(P_t)$ and $f_Q(Q_t)$ the same as in equations 8.8. Coexistence is possible if Q has a higher attack rate than P, or if there is strong intraspecific density dependence in one or both parasitoid species, i.e., if k_P and/or k_Q are less than 1 (the three-species equilibrium is stable only if the second of these two criteria is true). The addition of the obligate hyperparasitoid always leads to a higher equilibrium density of the host (this becomes simply a trophic cascade: addition of a third trophic level decreases the abundance of the second trophic level and increases the density of the bottom trophic level).

EFFECTS OF AGE STRUCTURE ON COMPETITION

If a host species is attacked by more than one parasitoid species, all of which attack exactly the same developmental stage of the host, the results in terms of competitive exclusion and host suppression carry over from the non-age-structured models. Unless there is some explicit mechanism that promotes coexistence (e.g., density dependence in the parasitoid death rate or attack rate), then the parasitoid species that suppresses the density of the host stage attacked by the parasitoids to the lowest level will be the species that wins in competition and excludes all competitors (Briggs 1993a).

Haigh and Maynard Smith (1972) were apparently the first to suggest through the use of a model that two consumer species may be able to coexist if they use different developmental stages of the same resource. We show here that the use of different juvenile developmental stages of the host is not enough for coexistence. However, if this is combined with some other process that causes there to be effectively more than one type of resource within the resource population, or that gives each species a low-density advantage, then coexistence is possible. We show that variability in the stage durations can have this effect, and that this type of variability was implicit in the model of Haigh and Maynard Smith (1972).

Parasitoids Attack Different Host Development Stages, Constant Development Times

Briggs (1993a) presented a model in which a stage-structured host population is attacked by two parasitoid species, each attacking a different juvenile life stage (see Fig. 8.4a). Adults of one parasitoid, P, attack the egg stage, E, and adults of the other parasitoid, Q, attack the larval stage, L. The adult stage is invulnerable. In this model, each juvenile stage of the host is fixed in duration. All host individuals spend T_E days in the egg stage before developing into the larval stage, in which they spent T_L days before maturing into the adult stage. The parasitoids also have fixed development times, T_{JP} and T_{JQ}, before maturing into adults.

The model for this is

host eggs: $\qquad \dfrac{dE(t)}{dt} = rA(t) - M_E(t) - a_P P(t) E(t) - d_E E(t)$

host larvae: $\qquad \dfrac{dL(t)}{dt} = M_E(t) - M_L(t) - a_Q Q(t) L(t) - d_L L(t)$

host adults: $\qquad \dfrac{dA(t)}{dt} = M_L(t) - d_A A(t)$

parasitoid P adults: (8.12)

$$\frac{dP(t)}{dt} = a_P P(t - T_{JP}) E(t - T_{JP}) s_{JP} - d_P P(t)$$

parasitoid Q adults:

$$\frac{dQ(t)}{dt} = a_Q Q(t - T_{JQ}) L(t - T_{JQ}) s_{JQ} - d_Q Q(t)$$

where r is the adult host per-head fecundity, a_P and a_Q are the attack rates of P and Q, respectively, and d_X is the constant background per-head death rate for each stage X. $M_E(t)$ is the maturation rate out of the host egg stage. As discussed in chapter 5, this is the recruitment rate into that stage T_E days ago, multiplied by the probability of surviving through that stage:

$$M_E(t) = rA(t - T_E) \exp\left\{ -\int_{t-T_E}^{t} [a_P P(x) + d_E]\, dx \right\}.$$

Likewise the maturation rate out of the host larval stage is

$$M_L(t) = M_E(t - T_L) \exp\left\{ -\int_{t-T_L}^{t} [a_Q Q(x) + d_L]\, dx \right\}.$$

s_{JP} and s_{JQ} are the probabilities of juveniles of each of the parasitoid species surviving through their respective juvenile stages. The juvenile

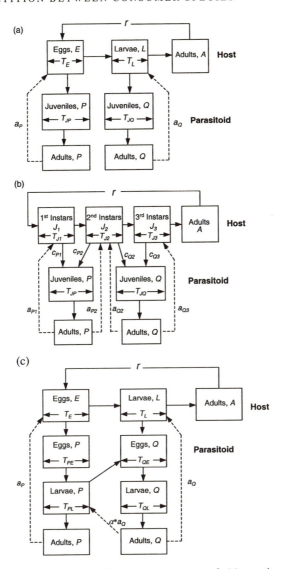

FIGURE 8.4. Diagrams showing the stage structure of: (a) equations 8.12 with fixed stage durations in which parasitoid P attacks only host eggs and parasitoid Q attacks only host larvae; (b) the model with overlapping windows of vulnerability; and (c) the facultative hyperparasitism model of Briggs (1993a). (a) and (c) are modified, with permission from University of Chicago Press, from figures in Briggs 1993b.

parasitoids are subjected to only a constant background mortality rate, d_{JP} or d_{JQ}, so these through-stage survival probabilities are simply $s_{JP} = \exp\{-d_{JP}T_{JP}\}$ and $s_{JQ} = \exp\{-d_{JQ}T_{JQ}\}$. These terms could also be scaled by the clutch size if more than one egg were laid on a host by either species. As in chapter 5, we define the average lifetime host fecundity, $\rho = rT_A$, as the per-head host fecundity multiplied by the average duration of the adult host stage, $T_A = 1/d_A$.

We can again look at the invasion criterion determining whether each species can invade the system when the other species is at equilibrium with the host. With only the egg parasitoid P and the host present, the equilibrium is

$$E^*\big|_P = \frac{d_P}{a_P s_{JP}}; \quad P^*\big|_P = \frac{\ln(\rho) - d_E T_E - d_L T_L}{a_P T_E};$$

$$A^*\big|_P = \frac{y_1 E^*\big|_P}{\rho\, d_A\{1 - \exp(-y_1 T_E)\}};$$

$$L^*\big|_P = \frac{d_A A^*\big|_P\{\rho \exp(-y_1 T_E) - 1\}}{d_L};$$

where $y_1 = a_P P^*\big|_P + d_E$.

Parasitoid Q can invade this equilibrium if

$$L^*\big|_P > \frac{d_Q}{(a_Q s_{JQ})}.$$

With only the larval parasitoid Q and the host present, the equilibrium is

$$L^*\big|_Q = \frac{d_Q}{a_Q s_{JQ}}; \quad Q^*\big|_Q = \frac{\ln(\rho) - d_E T_E - d_L T_L}{a_Q T_L};$$

$$A^*\big|_Q = \frac{y_2 L^*\big|_Q}{d_A\{\rho \exp(-d_E T_E) - 1\}};$$

$$E^*\big|_Q = \frac{\rho d_A A^*\big|_Q\{1 - \exp(-d_E T_E)\}}{d_E};$$

where $y_2 = a_Q Q^*\big|_Q + d_L$.

Parasitoid P can invade this equilibrium if

$$E^*\big|_Q > \frac{d_P}{(a_P s_{JP})}.$$

Thus, each species wins by suppressing the stage used by its competitor below the density the competitor needs to sustain its population. This result can have implications for biological control (chapter 9).

These criteria can be rewritten as:

P can invade if:

$$\frac{a_Q s_{JQ}}{d_Q} < \frac{a_P s_{JP}}{d_P}\left[\frac{\rho(\exp\{d_E T_E\} - 1)(d_E T_E - \ln(\rho))}{d_E T_L(\exp\{d_E T_E\} - \rho)}\right];$$

Q can invade if:

$$\frac{a_Q s_{JQ}}{d_Q} > \frac{a_P s_{JP}}{d_P}\left[\frac{d_L T_E(\rho - \exp\{d_L T_L\})}{(\exp\{d_L T_L\} - 1)(\ln(\rho) - d_L T_L)}\right].$$

This model illustrates that simply attacking different juvenile stages of the host does not promote coexistence of more than one parasitoid species. This is because attacks by one parasitoid species on host eggs suppress the density of future host larvae, but attacks by the other parasitoid species on host larvae suppress the density of host eggs in the next generation. Therefore different stages of the same host cannot be thought of as independent resources.

Fig. 8.5 shows that it is impossible for the two parasitoid species to coexist. We define scaled attack rates for the two parasitoid species, $\hat{a}_P = a_P s_{JP}/d_P$ and $\hat{a}_Q = a_Q s_{JQ}/d_Q$. For each species this is the attack rate scaled by the through-stage juvenile survival probability and the average adult parasitoid lifespan ($1/d_i$, for each species, i). The outcome of competition depends simply on the ratio of these scaled attack rates and on the characteristics of the host population: ρ, T_E, T_L, d_E, and d_L.

We can see from Fig. 8.5 that parasitoid P gains a competitive advantage by attacking the earlier host stage. If the parasitoids have equal scaled attack rates ($\hat{a}_Q/\hat{a}_P = 1$), then parasitoid P always wins. The later attacking species, Q, must have a substantially higher scaled attack rate than P to outcompete P. Increasing the relative duration of the egg stage attacked by P shifts the boundaries in Fig. 8.5 to favor P winning

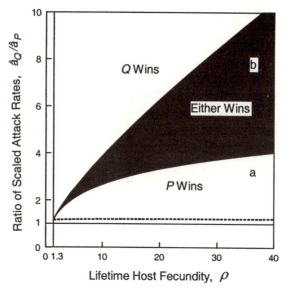

FIGURE 8.5. Outcome of competition for the stage-structured model with fixed stage durations in which parasitoid P attacks only host eggs and parasitoid Q attacks only host larvae, in terms of the ratio of the scaled attack rates of the two parasitoid species, and the lifetime host fecundity. On the line $\hat{a}_Q/\hat{a}_P = 1$, the two parasitoids have equal scaled attack rates, and parasitoid P always wins. In the region below the dotted line, parasitoid P wins and leads to a lower density of adult hosts than parasitoid Q, but in region "a" parasitoid P wins but parasitoid Q would lead to the lower density of adult hosts. In region "b" either species wins, depending on initial conditions. For values of ρ below about 1.3, the host has a negative per-head growth rate and goes extinct. $T_E = T_L = 0.5, d_E = 0.5, d_L = 0.1$.

in competition, and increasing the relative duration of the larval stage attacked by Q shifts the boundaries to favor Q winning. (This illustrates a point stressed in chapter 9: in stage-structured models, the demography of the host affects the interaction between the competing parasitoids.) In this model there is also a region of parameter space in which neither of the invasion criteria is met, and the competitive outcome depends on the initial conditions.

The above invasion criteria are strictly valid only if the host and resident parasitoid populations being invaded persist at a stable equilibrium. We showed in chapter 5 that the stability of a stage-structured host-parasitoid model of this form depends strongly on the duration of the adult host stage (in addition to other parameters). The two-species equilibrium tends to be stable if the invulnerable adult stage is long-lived (low d_A). Simulations suggest that these equilibrial invasion criteria also provide reasonable, but not exact, approximations to the invasion criteria when the resident species and the host instead persist in a limit cycle. In simulations of this model coexistence of the two parasitoid species was never observed, even in the non-equilibrial situation (Briggs 1993a).

We now consider the effects on host suppression. When parasitoid P wins in competition it does so by leading to a lower density of host larvae (the host stage used by Q) than parasitoid Q needs to sustain its population. Likewise, when parasitoid Q wins it does so by suppressing the density of host eggs below the level required for P to sustain its population. That is, by definition, the species that wins in competition is the one that leads to the lower density of the host stage required by its competitor. However, if it is desired to reduce the total density of hosts, or to reduce the density of the adult stage that is invulnerable to parasitism by both species, then it is possible for parasitoid P to win in competition but for parasitoid Q to be the superior species at host suppression.

The dotted line in Fig. 8.5 delineates the regions in which parasitoid P or parasitoid Q on its own would lead to the lower adult host density. In the region labeled "a", P wins but Q could lead to the lower host density. This can occur because parasitoid P wins by reducing larval hosts more than Q does. If parasitoid Q were present, however, it could attack the hosts throughout the larval stage and cause fewer larval hosts to make it to the age of maturation into the adult stage. That is, P leads to a lower total density of host larvae, but Q would lead to a lower density of host larvae of age $T_E + T_L$. In the region labeled "b", either species may win, depending on the initial conditions, but if only Q were released, it could lead to a lower host density than if P were also released and ended up being the winner.

Parasitoids Have Overlapping Windows of Vulnerability

In the above model, one host stage could be attacked only by P and another could be attacked only by Q. If the windows of vulnerability of the host to the two parasitoid species instead overlap, such that there is a host stage that can be attacked by both species, then the main conclusions are not changed.

Consider, for example, the holometabolous host species diagrammed in Fig. 8.4b, in which parasitoid P can attack first and second instars and parasitoid Q can attack second and third instars. We assume here that attack by one species renders the host unsuitable for subsequent attack by the other parasitoid species (we consider multiparasitism and facultative hyperparasitism in stage-structured models below).

Coexistence of the two parasitoid species is once again impossible in this model. The species that leads to the lower combined density of the stages used by its competitor wins in competition. Parasitoid P gains a competitive advantage from attacking the earlier host stage.

APHYTIS MELINUS/LINGNANENSIS MODEL

Murdoch et al. (1996a) used a model very similar to the above model to help explain the rapid displacement of the parasitoid *Aphytis lingnanensis* by its congener *A. melinus*; both attack California red scale on citrus (Box 8.3). In that model, the entire juvenile stage of the host can be attacked by both of the parasitoid species. However, the two parasitoids make somewhat different use of an attacked host, depending on its size (age). Both species host-feed on very young hosts, lay male eggs in hosts of intermediate age, and lay female eggs on old juvenile hosts. But *A. melinus* can produce female offspring on slightly younger hosts than can *A. lingnanensis*, and this gives *melinus* a large enough competitive advantage to explain the competitive exclusion of *lingnanensis* from inland regions of southern California within a few years of *melinus*'s release (see Fig. 9.3).

Facultative Hyperparasitism (Intraguild Predation) in a Stage-Structured Model

Briggs (1993a) investigated the stage-structured model of facultative hyperparasitism (intraguild predation) modified from equations 8.12 above, illustrated in Fig. 8.4c. Parasitoid P is a primary parasitoid attacking only host eggs. Parasitoid Q is a facultative hyperparasitoid attacking either unparasitized host larvae (at rate a_Q) or larvae that have

BOX 8.3.

MODEL OF DISPLACEMENT OF *APHYTIS LINGNANESIS* BY *A. MELINUS* ON CALIFORNIA RED SCALE

In this model there are three juvenile host (red scale) stages. Both parasitoid species host-feed or lay male eggs on the smallest stage (J_1). On the intermediate stage (J_2) *lingnanensis* again host-feeds or lays male eggs, but *melinus* can lay female eggs. Both parasitoid species can lay female eggs on the largest juvenile host stage (J_3).

Populations

$J_1(t)$	number of small juvenile hosts at time t
$J_2(t)$	number of intermediate juvenile hosts at time t
$J_3(t)$	number of large juvenile hosts at time t
$A(t)$	number of adult hosts at time t
$M(t)$	number of adult female *A. melinus* parasitoids at time t
$L(t)$	number of adult female *A. lingnanensis* parasitoids at time t

Properties of Individuals and Parameter Values

$T_{J1} = 0.33$	duration of small juvenile host stage
$T_{J2} = 0.22$	duration of intermediate juvenile host stage
$T_{J3} = 0.45$	duration of intermediate juvenile host stage
$T_M = 0.22$	duration of immature *A. melinus*
$T_L = 0.22$	duration of immature *A. lingnanensis*
$d_A = 1/T_A = 0.56$	adult scale death rate
$\rho = 33$	per-individual lifetime adult host fecundity
$d_{J1} = 1$	background death rate of small juvenile hosts
$d_{J2} = 1$	background death rate of intermediate juvenile hosts
$d_{J3} = 1$	background death rate of large juvenile hosts
$a_M = 1$	*A. melinus* attack rate
$a_L = 1$	*A. lingnanensis* attack rate
$s_M = 1$	*A. melinus* through-stage survival
$s_L = 1$	*A. lingnanensis* through-stage survival
$d_M = 8$	death rate of adult *A. melinus*
$d_L = 8$	death rate of adult *A. lingnanensis*
$c = 1$ or 2	number of adult female *A. melinus* from stage J_3 hosts

(Box 8.3. continued)

(All durations scaled relative to the total host development time, $T_J = T_{J1} + T_{J2} + T_{J3}$.)

Vital Processes

$R_{J1}(t) = (\rho/T_A)A(t)$ recruitment rate of small juvenile hosts at time t

$R_{J2}(t) = R_{J1}(t - T_{J1})S_{J1}(t)$ recruitment rate of intermediate juvenile hosts at time t

$R_{J3}(t) = R_{J2}(t - T_{J2})S_{J2}(t)$ recruitment rate of large juvenile hosts at time t

$R_A(t) = R_{J3}(t - T_{J3})S_{J3}(t)$ recruitment rate into adult host stage at time t

for $i = 1, 2, 3$:

$X_{i,M}(t) = a_M J_i(t)M(t)$ total death rate by *A. melinus* on J_i at time t

$X_{i,L}(t) = a_L J_i(t)L(t)$ total death rate by *A. lingnanensis* on J_i at time t

$S_{Ji}(t) = \exp\left\{ - \int_{t-T_{Ji}}^{t} \left(a_M M(x) + a_L L(x) + d_{Ji} \right) dx \right\}$ survival through J_i stage at time t

$R_M(t) = [X_{2,M}(t - T_M) + cX_{3,M}(t - T_M)]s_M$ recruitment rate of adult female *A. melinus* at time t

$R_L(t) = [X_{3,L}(t - T_L)]s_L$ recruitment rate of adult female *A. lingnanensis* at time t

Balance Equations

$dJ_1(t)/dt = R_{J1}(t) - R_{J2}(t) - X_{1,M}(t) - X_{1,L}(t) - d_{J1}J_1(t)$ small juvenile hosts

$dJ_2(t)/dt = R_{J2}(t) - R_{J3}(t) - X_{2,M}(t) - X_{2,L}(t) - d_{J2}J_2(t)$ intermediate juvenile hosts

$dJ_3(t)/dt = R_{J3}(t) - R_A(t) - X_{3,M}(t) - X_{3,L}(t) - d_{J3}J_3(t)$ large juvenile hosts

---

(Box 8.3. continued)

Balance Equations

$dA(t)/dt = R_A(t) - d_A A(t)$ adult hosts

$dM(t)/dt = R_M(t) - d_M M(t)$ adult *A. melinus*

$dL(t)/dt = R_L(t) - d_L L(t)$ adult *A. lingnanensis*

---

been previously attacked by the egg parasitoid (at rate $\alpha^* a_Q$). Each attack by Q on a parasitized host produces on average β parasitoid Q juveniles, whereas each attack on an unparasitized host produces 1 parasitoid Q juvenile (this allows for differential yield from the two types of hosts). The results from this model are similar to those obtained from the unstructured models of intraguild predation presented above (see Fig. 8.6). The two parasitoid species can coexist through a balance of parasitoid P being more efficient at attacking unparasitized hosts and parasitoid Q being able to attack parasitized hosts. The only difference from the non-stage-structured models is that parasitoid P gains an extra advantage from attacking an earlier host stage, and through this advantage coexistence is possible even in some cases in which P and Q have equal scaled attack rates on the host, $\hat{a}_Q = \hat{a}_P$. When the two parasitoids do coexist, the equilibrium densities of all of the host stages are always intermediate between the densities that would be achieved with parasitoid P present on its own and the densities with parasitoid Q present on its own. Thus again it would be preferable to release only the single most effective parasitoid species.

Fig. 8.6 shows that altering host productivity, which here is assumed to affect the lifetime host fecundity, ρ, has qualitatively the same effect on competition as it does in the unstructured models. For given values of the scaled attack rates, P is more likely to win at low values of ρ, coexistence is more likely at intermediate values of ρ, and Q is likely to win at high values of ρ.

Mylius et al. (2001) explored the effects of adding two types of stage structure to an intraguild predation model similar to Holt and Polis's (1997) model (but with a type 2 functional responses for each species). First they looked at the effect of an adult stage for the intraguild prey species, which could not be attacked by the intraguild predator. They

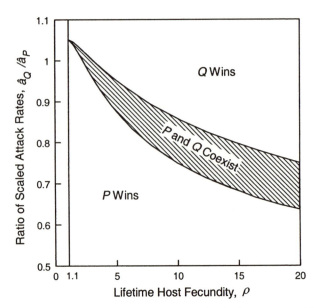

FIGURE 8.6. Outcome of competition for the facultative hyperparasitism model in terms of the ratio of the scaled attack rates of the two parasitoid species and the lifetime host fecundity. For values of ρ below about 1.1, the host has a negative per-head growth rate and goes extinct. $\beta = 1, T_E = T_L = 0.5, d_E = d_L = 0.1, T_{PL} = 0.5, d_{PE} = 0, d_{PL} = 0.1, \alpha = 1$.

found that, as in the Briggs (1993a) model, this had little qualitative effect on the predictions of the basic intraguild predation model. Second, they investigated the effect of a juvenile stage for the intraguild predator. This juvenile stage could feed on only the basal resource, whereas the adult stage could feed on either the resource or the intraguild prey. The effect of this is to allow for persistence of the intraguild prey, and coexistence of intraguild prey and predators, over a wider range of productivity values.

Variable Host Stage durations

In the above stage-structured models, each juvenile host stage was fixed in duration, and the use of different developmental stages by the two parasitoid species did not lead to coexistence. This was because the two parasitoid species imposed conflicting constraints on the densities and age structure within the various host stages. We show now that this conclusion can be altered when there is variability in the host stage durations.

Consider a modification of equations 8.12 in which there is a constant maturation rate out of each host developmental stage rather than a constant maturation time (Briggs et al. 1993):

host eggs: $\dfrac{dE(t)}{dt} = rA(t) - m_E E(t) - a_P P(t)E(t) - d_E E(t)$

host larvae: $\dfrac{dL(t)}{dt} = m_E E(t) - m_L L(t) - a_Q Q(t)L(t) - d_L L(t)$

host adults: $\dfrac{dA(t)}{dt} = m_L L(t) - d_A A(t)$

parasitoid P adults: (8.13)

$$\frac{dP(t)}{dt} = a_P P(t - T_{JP})E(t - T_{JP})s_{JP} - d_P P(t)$$

parasitoid Q adults:

$$\frac{dQ(t)}{dt} = a_Q Q(t - T_{JQ})L(t - T_{JQ})s_{JQ} - d_Q Q(t)$$

where m_E is the constant per-head rate of maturation out of the host egg stage and m_L is the constant per-head maturation rate out of the host larval stage.

With only the egg parasitoid P and the host present, the equilibrium is

$$E^*\big|_P = \frac{d_P}{a_P s_{JP}}; \quad L^*\big|_P = \frac{m_E E^*\big|_P}{(m_L + d_L)}; \quad A^*\big|_P = \frac{m_L L^*\big|_P}{d_A};$$

$$P^* = \rho \delta_A \frac{A^* - (m + d_E)E^*}{a_P E^*}, \quad \text{where } \rho = r/d_A.$$

Parasitoid Q can invade this equilibrium if $L^*\big|_P$ is greater than $d_Q/(a_Q s_{JQ})$. With only the larval parasitoid Q and the host present, the equilibrium is

$$L^*\big|_Q = \frac{d_Q}{a_Q s_{JQ}}; \quad A^*\big|_Q = \frac{m_L L^*\big|_Q}{d_A}; \quad E^*\big|_Q = \frac{\rho\, d_A A^*\big|_Q}{(m_E + d_E)};$$

$$Q^*\big|_Q = \frac{m_E E^*\big|_Q - (m_L + d_L) L^*\big|_Q}{a_Q L^*\big|_Q}.$$

Parasitoid P can invade this equilibrium if $E^*\big|_Q$ is greater than $d_P/(a_P s_{JP})$. These invasion criteria can be rewritten as: P can invade if

$$\hat{a}_Q < \hat{a}_P \left[\frac{\rho\, m_L}{(m_E + d_E)} \right]$$

and Q can invade if

$$\hat{a}_Q > \hat{a}_P \left[\frac{(m_L + d_L)}{m_E} \right],$$

where once again

$$\hat{a}_P = \frac{a_P s_{JP}}{d_P} \text{ and } \hat{a}_Q = \frac{a_Q s_{JQ}}{d_Q}$$

are the scaled attack rates.

Coexistence is now widely possible (e.g., Fig. 8.7). For example, if $d_E = d_L = 0$, then a region of coexistence is possible whenever the lifetime host fecundity, ρ, is greater than 1. Fig. 8.7 shows that increasing m_E, which reduces the average time that individual hosts spend in the egg stage, shifts the outcome to favor the larval parasitoid, Q. Likewise, increasing m_L favors the egg parasitoid, P.

We discuss the mechanism of coexistence in this model next, but first we point out another difference from the model with constant development times. In this model with constant maturation rates, the parasitoid that wins in competition always leads to the lowest density of all host stages, including the invulnerable adult stage. When the two parasitoids coexist, the adult host density equals the density that would be achieved by the larval parasitoid on its own. Thus, this model predicts that there is no risk, in terms of loss of host suppression, of releasing all possible parasitoid species.

The different predictions about the effects on adult host density arise because in the model with constant stage durations, the invasion criteria

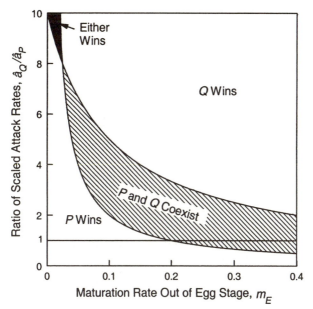

FIGURE 8.7. Outcome of competition for the stage-structured model with constant host maturation rates. P attacks only host eggs, Q attacks only host larvae. On the line $\hat{a}_Q/\hat{a}_P = 1$, the two parasitoids have equal scaled attack rates. The parasitoids coexist in the hatched region, and the outcome depends on initial conditions in the black region. $\rho = 10, m_L = 0.1, d_E = d_L = 0.1$.

depend on the total abundance of hosts in a stage, but the adult host density depends on the number of individuals that survive until the end of the larval stage. It is therefore possible for parasitoid P to lead to a lower total abundance of larval hosts, but for parasitoid Q to lead to a lower abundance of hosts that reach the adult stage. In the model with constant maturation rates, however, both the invasion criteria and the number of individuals maturing into the adult stage depend on the total density of hosts in the larval stage, rather than the number that reach a particular age. This is because all individuals in the larval stage have the same probability of maturing into the adult stage, regardless of age. Therefore this conclusion, that the species that wins in competition is always the one that could lead to the lowest abundance of adult hosts, rests on the unrealistic assumption that the probability of maturing into the adult stage is completely independent of larval age. This reinforces a message we drew in chapter 3, that this unrealistic assumption about development can produce misleading results. Thus, although making this assumption is often a useful ploy in an initial exploration of stage structure, the results need to be treated with caution.

The mechanism allowing for coexistence in this model can be understood in one of two different ways. First, in the earlier model in which host developmental time is constant (equations 8.12), a host with a relatively long egg stage favored the egg parasitoid in competition, and a host with a relatively long larval stage favored the larval parasitoid. The model with constant juvenile host maturation rates, by contrast, has a distributed developmental delay: in the absence of parasitism or background mortality, the time spent in a given stage i has an exponential distribution (with an average stage duration of $1/m_i$). This allows there to be a mixture of stage durations. In particular, some hosts pass through a long egg stage and then a short larval stage (so favor the egg parasitoid), whereas others have a short egg stage and long larval stage and thus favor the larval parasitoid. The existence of multiple host types allows coexistence.

Second, when the host developmental stages are not fixed in duration, mortality by each parasitoid species can reduce the average duration of the stage they attack. This gives each parasitoid species a relative advantage when it is relatively rare (because then the stage it attacks is relatively longer lived) and imposes a disadvantage when it is relatively abundant. Consider the egg stage attacked by parasitoid P in equations 8.13. The average duration of the egg stage is $1/(a_P P(t) + d_E + m_E)$. When parasitoid P is rare, the average duration of the egg stage is relatively long, and when parasitoid P is abundant it is relatively short. The abundance of parasitoid Q affects the average duration of the larval host stage in an analogous manner. Therefore, for some combinations of parameters, the stage durations of the host population are relatively favorable to the egg parasitoid when the egg parasitoid is rare, and to the larval parasitoid when the larval parasitoid is rare. In these circumstances, the population of each parasitoid species can have a positive growth rate at low density, and coexistence is possible.

This situation does not arise when both juvenile host stages are fixed in duration. Attack by parasitoid P on host eggs reduces the density of host eggs but does not affect the relative durations of the stages. Hosts must survive to the end of the egg stage to become larvae and must survive through the entire larval stage to become adults in order to produce new eggs.

Briggs et al. (1993) investigated a range of different models with more realistic assumptions about the distribution of times spent in the juvenile host stages. For example, one highly unrealistic feature of the model with a constant maturation rate is that the modal time spent in a stage is zero. That is, some individuals leave a stage as soon as they enter it. A more realistic modification made by Briggs et al. (1993) was to have a fixed minimum duration of each stage, followed by a portion of each stage that was variable in duration.

The general conclusion from these models was that coexistence was possible only if there was sufficient variability in the duration of one or both host developmental stages. If the stages were relatively fixed in duration, the results approximated those of the constant development time model above. If the stages were relatively variable in duration, the results approximated those of the constant maturation rate model. One result that did not carry over from the constant maturation rate model, however, was the conclusion about the effect on adult host density. If there is a minimum time that each individual must spend in each stage, when coexistence occurs, the density of adult hosts is intermediate between the densities that would be achieved if each of the two parasitoid species were present on its own. That is, the addition of the less efficient parasitoid species, resulting in coexistence, leads to an increase in the adult host density. Within-individual correlation between the duration of the egg and larval stages (i.e., if hosts with a long egg stage also tended to have a long larval stage) made coexistence less likely.

Larval and Adult Stages Attacked

Haigh and Maynard Smith (1972) presented a consumer-resource model in which one consumer species attacked the juvenile stage of the resource and a second consumer species attacked the adult stage (Fig. 8.8). This is a much less common situation in parasitoid-host interactions than attack on different portions of the juvenile stage by different parasitoid species. We will see that this can result in coexistence by a similar mechanism to that in the model with variable stage durations.

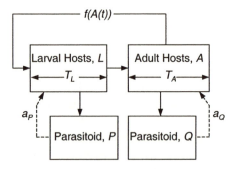

FIGURE 8.8. Diagram of the Haigh and Maynard Smith model in which parasitoid P attacks juvenile hosts and parasitoid Q attacks adult hosts.

The equations for this model are (modified slightly from the original, and expressed in terms of parasitoids and hosts)

host larvae:
$$\frac{dL(t)}{dt} = A(t)f(A(t)) - M_L(t)$$
$$- a_P P(t)L(t) - d_L L(t)$$

host adults:
$$\frac{dA(t)}{dt} = M_L(t) - M_A(t) - a_Q Q(t)A(t) - d_A A(t)$$

(8.14)

parasitoid P:
$$\frac{dP(t)}{dt} = s_P a_P L(t)P(t) - d_P P(t)$$

parasitoid Q:
$$\frac{dQ(t)}{dt} = s_Q a_Q A(t)Q(t) - d_Q Q(t)$$

where $M_L(t)$ is the maturation rate out of the host larval stage:

$$M_L(t) = A(t - T_L)f(A(t - T_L)) \exp\left\{-\int_{t-T_L}^{t} [a_P P(x) + d_L]\, dx\right\}.$$

Hosts that mature out of the larval stage immediately enter the host adult stage. $M_A(t)$ is the "maturation" rate out of the host adult stage:

$$M_A(t) = M_L(t - T_A) \exp\left\{-\int_{t-T_A}^{t} [a_Q Q(x) + d_A]\, dx\right\}.$$

Hosts that "mature" out of the adult die (i.e., they senesce).

This differs from equations 8.12 in several ways. First, there is no developmental delay in populations of either of the parasitoids. Hosts that are attacked by one of the parasitoid species immediately become new searching parasitoids of that species. s_P and s_Q can be interpreted here as the number of new parasitoids of each species produced per attack on the host. This does not qualitatively affect the outcome of competition. Second, we have added a maximum duration to the adult host stage, as represented by the $M_A(t)$ term in the model. The reason for this will become clear below. It will turn out that this has little effect. Third, the adult host is no longer invulnerable to parasitism, so that stabilizing mechanism is no longer present. Therefore, density dependence has been added to the host reproduction term in order to stabilize the host-parasitoid interaction. We showed above in the section "Dependence of Per-Head Rates on Host Density" above that density dependence in the host per-head rates has no effect on the conclusions about competitive exclusion and coexistence but does affect the stability of the resulting dynamics. For illustration here we used a Ricker form for the density-dependent fecundity term (see chapter 4): $f(A(t)) = \exp(-cA(t))$.

With only parasitoid P and the host present, the equilibrium can be calculated by numerically solving the following transcendental equations:

$$A^*\big|_P f(A^*\big|_P)\{1-\exp(-(a_P P^*\big|_P+d_L)T_L)\}-(a_P P^*\big|_P+d_L)L^*\big|_P = 0,$$

where

$$L^*\big|_P = \frac{d_P}{s_P a_P}$$

and

$$P^*\big|_P = \frac{1}{a_P T_L}\left\{\ln\left[f(A^*\big|_P)\left(\frac{1-\exp(-d_A T_A)}{d_A}\right)\right] - d_L T_L\right\}.$$

Parasitoid Q can invade this equilibrium if $A^*\big|_P$ is greater than $d_Q/(s_Q a_Q)$. With only parasitoid Q and the host present, the host equilibrium is

$$A^*\big|_Q = \frac{d_Q}{s_Q a_Q}; \quad L^*\big|_Q = A^*\big|_Q f(A^*\big|_Q)\left(\frac{1-\exp(-d_L T_L)}{d_L}\right).$$

The parasitoid equilibrium, $Q^*\big|_Q$, can be determined by numerically solving the following transcendental equation, but this is not necessary for determining the invasion criteria:

$$f(A^*|_Q) \exp(-d_L T_L)$$

$$\{1 - \exp(-(a_Q Q^*|_Q + d_A) T_Q)\} - (a_Q Q^*|_Q + d_A) = 0.$$

Parasitoid P can invade this equilibrium if $L^*|_Q$ is greater than $d_P/(s_P a_P)$.

Fig. 8.9 shows that coexistence is possible in this model. When the two parasitoid species do coexist, the three-species equilibrium becomes

$$L^* = \frac{d_P}{s_P a_P}; \quad A^* = \frac{d_Q}{s_Q a_Q}.$$

P^* can be found only by numerically solving the equation

$$A^* f(A^*)\{1 - \exp[-(a_P P^* + d_L)T_L]\} = (a_P P^* + d_L)L^*,$$

and Q^* can then be found by numerically solving the equation

$$f(A^*) \exp[-(a_P P^* + d_L)T_L]\{1 - \exp[-(a_Q Q^* + d_A)T_A]\}$$

$$= (a_Q Q^* + d_A).$$

The mechanism for coexistence in this model is similar to that in the model with variable juvenile host stage durations above. This model in which the adult stage is attacked and the model with constant juvenile stage durations in equations 8.12 share the feature that each host stage attacked by the parasitoids has a fixed maximum duration. The important difference, however, is that in the model with the two juvenile stages attacked, host individuals must survive all of the way to the end of the second attacked stage (the larval stage in that case) before they can reproduce and contribute to the next generation. In the model with the larval and adult stages attacked, adult hosts can reproduce continuously during the adult stage, and do not need to survive to the end of that stage to contribute to the next generation. Therefore, attacks by parasitoid Q on adult hosts can shorten the average duration of that stage, and the two parasitoid species can coexist through an analogous mechanism to that in the model with variable juvenile host stage durations above. A relatively long juvenile host stage favors parasitoid P, and a relatively long

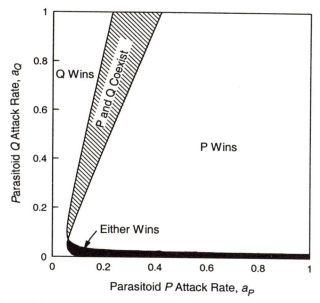

FIGURE 8.9. Outcome of competition in the Haigh and Maynard Smith model with upper limit to duration of the adult stage. $d_L = d_A = 0.1, r = 5, c = 0.1, T_L = T_A = 1, d_P = d_Q = 1.$

adult host stage favors parasitoid Q. High densities of Q lead to a short adult host stage, and low densities of Q lead to a long adult host stage. Therefore, Q can realize a relative low-density advantage, and for certain combinations of parameters the two parasitoid species can coexist.

NON-EQUILIBRIAL MECHANISMS OF COEXISTENCE

In this chapter so far, all of the models have examined only equilibrial mechanisms of coexistence, in which we ask whether the stable equilibrium achieved with one competitor can be invaded by a second competitor. The mechanisms of coexistence discussed thus far did not require temporal fluctuations either in population densities or in parameters driven by environmental variability, and we have ignored any effects that environmental variability may have. We call these *classical coexistence mechanisms*. We now examine the effects that temporal variability, either in the form of random or seasonal variation in parameter values caused by the environment, or variation in population densities caused by the interaction itself, may have on the outcome of competition.

The classical coexistence mechanisms comprise only one of three categories of coexistence mechanisms outlined by Chesson (1994) that are possible when we allow densities and/or parameters to vary through time. In all three categories more than one species can coexist because each can have a positive growth rate in the system when that species is at low density. In the classical equilibrium mechanisms, this can occur when intraspecific competition is greater than interspecific competition, or when the species in some way partition their resources. The two other types of coexistence mechanisms require some type of temporal variability (either externally driven or arising internally from the interaction itself) in order for each species to gain a low-density advantage. These coexistence mechanisms involve either (a) species-specific non-linear responses to the level of competition or (b) species-specific responses to environmental fluctuations and an interaction between competition and the environment (the *storage effect*).

Seasonally Varying Parameters

First we will show that temporal variability does not necessarily alter the outcome of competition or promote coexistence. For example, consider a situation in which one parasitoid species has an advantage during the summer months (e.g., through a higher attack rate or a lower death rate) and another parasitoid species has an advantage during the winter months. We can show in a simple model that this on its own is not sufficient to promote coexistence of the two species.

We can alter equation 8.3 in "Exploitative Competition for an Explicit Resource" above such that the parasitoid attack rates and/or death rates are no longer constant but vary through time:

$$\frac{dH}{dt} = rH\left(1 - \frac{H}{K}\right) - a_P(t)HP - a_Q(t)HQ$$

$$\frac{dP}{dt} = e_P a_P(t)HP - d_P(t)P \qquad (8.15)$$

$$\frac{dQ}{dt} = e_Q a_Q(t)HQ - d_Q(t)Q.$$

A seasonal system could be represented by making the attack rate (or death rate) of each parasitoid species vary sinusoidally, with a period of 1 year:

$$a_P(t) = a_{Pave} + a_{Pamp} \cos(\omega t + \phi_P)$$

$$a_Q(t) = a_{Qave} + a_{Qamp} \cos(\omega t + \phi_Q)$$

where a_{Pave} and a_{Qave} are the average values of the attack rates, a_{Pamp} and a_{Qamp} are the amplitudes of the sinusoidal variation, $\omega = 2\pi/1$ year is the radian frequency, and ϕ_P and ϕ_Q are the phase shifts. If we let ϕ_P equal zero and ϕ_Q equal π, then the attack rates of the two species will vary 180 degrees out of phase with each other.

We find that the two parasitoid species cannot coexist. The species i that has the higher temporal average $a_i e_i/d_i$ will be the species that wins in competition and excludes the other species. The important message here is that two parasitoid species cannot coexist simply by having different seasonal tolerances to the environment. In order for seasonality to allow for coexistence in a system in which coexistence is otherwise not possible, seasonal variability (or other form of temporal variability) must be combined with either (a) species-specific non-linear responses to the level of competition or (b) species-specific responses to environmental fluctuations and an interaction between competition and the environment. These two types of non-equilibrial coexistence mechanisms will be described in turn next.

Species-Specific Non-Linear Responses to Competition

This first class of mechanisms of coexistence through variability is illustrated by the model of Armstrong and McGehee (1980). In this model a single resource, R, is consumed by two species, P and Q. Consumer Q has a linear type 1 functional response with attack rate a_Q, but consumer P has a non-linear type 2 functional response with attack rate a_P and handling time T_h. The consumer species have density-independent, constant per-head death rates, d_P and d_Q, and conversion efficiencies c_P and c_Q. In the absence of the consumers, the resource population would grow logistically with intrinsic growth rate r and carrying capacity K.

The equations are

resource:

$$\frac{dR(t)}{dt} = rR(t)\left(1 - \frac{R(t)}{K}\right) - \frac{a_P P(t) R(t)}{1 + a_P T_h R(t)} - a_Q Q(t) R(t)$$

consumer P:

$$\frac{dP(t)}{dt} = \frac{c_P a_P R(t) P(t)}{1 + a_P T_h R(t)} - d_P P(t)$$

consumer Q:

$$\frac{dQ(t)}{dt} = c_Q a_Q R(t) Q(t) - d_Q Q(t).$$

As we discussed in chapter 3, in the absence of consumer P the two-species system with only R and Q present would always have a stable two-species equilibrium at

$$R^*|_Q = \frac{d_Q}{c_Q a_Q}; \quad Q^*|_Q = \frac{r}{a_Q}\left(1 - \frac{R^*|_Q}{K}\right)$$

because of the stabilizing effect of the logistic growth term in the resource.

The two-species system with only R and P present is equivalent to the Rosenzweig and MacArthur model (1963) discussed in chapter 3, which has an equilibrium at

$$R^*|_P = \frac{d_P}{a_P(c_P - d_P T_h)}; \quad P^*|_P = \frac{r}{a_P}\left(1 - \frac{R^*|_P}{K}\right)(1 + a_P T_h R^*|_P).$$

This equilibrium can be stable or unstable, because of the interaction between the stabilizing effect of the logistic growth term in the resource and the destabilizing effect of the type 2 functional response. As discussed in chapter 3, whether or not the equilibrium is stable depends on how low the resource is suppressed below its carrying capacity by the consumer. The equilibrium is stable for relatively low values of K and unstable for relatively high values. When the equilibrium is unstable the system undergoes limit cycle dynamics.

In this system, coexistence of P and Q is possible on the single resource, R, only when the three-species system displays limit cycle

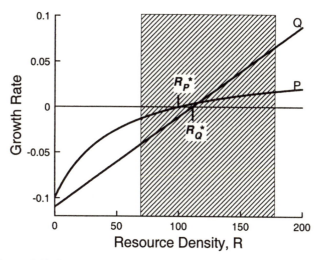

FIGURE 8.10. Per-head population growth rates of each consumer species as functions of the resource density, R, in the Armstrong and McGehee model. R_P^* is the equilibrium density with consumer P alone, and R_Q^* is the equilibrium density with consumer Q alone. Hatched region shows the range of densities over which the resource fluctuates in the limit cycle. $a_P = 0.01, a_Q = 0.003, d_P = 0.1, d_Q = 0.11, c_P = 0.03, c_Q = 0.33, T_h = 2$.

dynamics. It is never possible at a stable equilibrium. This can be illustrated by looking at the per-head growth rate of each consumer species as a function of resource density as in Fig. 8.10. The per-head growth rate of consumer Q increases linearly with resource density, whereas that of consumer P increases in a non-linear way. For each consumer species, the two-species equilibrium with only that consumer and the resource present is the point at which the per-head growth rate is equal to zero. These are marked with R_P^* and R_Q^* for P and Q respectively, in Fig. 8.10. If K is sufficiently small, each of the two-species equilibria is stable and the species with the lowest R^* value (i.e., the species that reduces the equilibrium density of the resource to the lowest level) will win in competition and exclude the other species. In the case shown in the figure, consumer P will be the winner. A small density of Q introduced into a system with P and the resource present would experience a resource density below the level necessary for Q's population to have a positive growth rate, and Q would be unable to invade.

If K is large, however, then $R_Q{}^*$ is still a stable point equilibrium, but $R_P{}^*$ is unstable and R and P undergo limit cycles in the absence of Q. Now the density of R with P present fluctuates over approximately the range indicated in Fig. 8.10. This range includes densities of R at which Q would have a positive growth rate as well as densities at which it would have a negative growth rate. A small inoculum of Q into this fluctuating system would experience alternating periods of high and low resource densities. Q will be able to invade the system with R and P present if the average value of Q's growth rate over the limit cycle is greater than zero. This is equivalent to: invasion of Q is possible if the time average value of R over the limit cycle is greater than $R_Q{}^*$.

In the standard Lotka-Volterra consumer-resource model, the average value of the resource over the neutrally stable cycle is equal to the equilibrium value of the resource. In the limit cycle displayed by the Rosenzweig and MacArthur model, however, the average value of the resource over the course of the limit cycle is always greater than the equilibrium value, and it is possible for the two species to coexist.

Thus, in this case variable resource densities allow for coexistence in cases in which coexistence would not be possible if the resource densities were constant. The mechanism acting here is different non-linear responses of the two consumer species to competition. The different non-linear responses allow there to be a range of resource densities favorable to each of the species. As the densities fluctuate across this range, each of the species has a period of disproportionately high growth.

In the Armstrong-McGehee model discussed here, the temporal variability is deterministic and caused by the unstable interaction between the resource and consumer P with the type 2 functional response. Coexistence would also be possible if the temporal variability was externally driven by some environmental factor (e.g., Ellner 1987), as long as the densities varied over the range in which the non-linear responses favor each of the consumer species.

It is easy to see how this mechanism of species-specific non-linear responses may act in a host-parasitoid system to allow for coexistence of two parasitoid species on a single host species. However, it is difficult to get coexistence of more than two consumer species through this mechanism (Grover 1988, 1990, 1991; Chesson 1994, 2000). The next mechanism can allow for coexistence of multiple consumer species in a variable environment, but its applicability to host-parasitoid system is less apparent.

Interactions between Competition and the Environment: The Storage Effect

This second class of mechanisms of coexistence through variability is termed the storage effect (Chesson 1994, 2000). In this mechanism multiple species can coexist in a variable environment through an interaction between competition and the environment that results in intraspecific competition being stronger than interspecific competition for each species. Three elements are necessary for this mechanism to work. First, the different species must respond differently to the changing environment. Second, there must be a covariance between the environment and competition, or more specifically a covariance between the effects of the environment on the per-head growth rate and the effect of competition on the per-head growth rate of each species. This allows the impact of intraspecific competition to be strongest for each species when the environment is good for that species. Third, there must be some mechanism that buffers population growth, which limits the impact of competition (both intra- and interspecific) when environmental conditions are bad for a species. Mechanisms such as this appear to be common in competing plant species (e.g., a seed bank) (Chesson and Huntly 1988, 1997; Chesson 1994) and in marine fish and invertebrate populations (a long-lived adult stage) (Warner and Chesson 1985; Chesson 1986). The combination of strong intraspecific competition during good years and little competition (both intra- and interspecific) during bad years for each species means that the overall growth rate of each species is more affected by intraspecific than interspecific competition, and therefore coexistence is possible.

As an example, consider a simple model of the seed bank of an annual plant (Chesson 1990; 1994):

$$N_i(t+1) = G_i(E_i, C_i)N_i(t)$$

$$= [(1 - E_i(t))s_i + E_i(t)Y_i/C_i(t)]N_i(t).$$

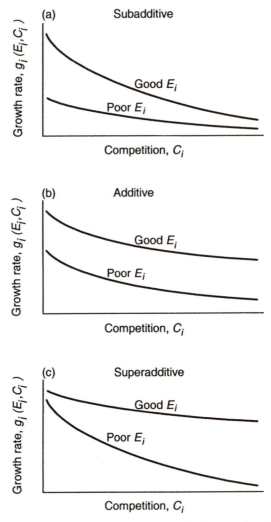

FIGURE 8.11. Three different forms of interaction between the effects of competition, $C_i(t)$, and the effects of the environment, $E_i(t)$, on population growth rate, $g_i(E_i(t), C_i(t))$, which equals the change in logarithmic population size. In (a), the subadditive case, population growth rate is relatively insensitive to the effects of competition during years in which the environment is poor for that species (poor Ei) and is sensitive to competition during good years (good E_i). In (b), the additive case, there is no covariance between the effects of competition and the environment. In (c), the superadditive case, the species is more sensitive to competition during the years in which the environment is poor for that species than during

Chesson's models are generally framed in discrete time, and here $N_i(t)$ is the number of seeds of species i in the seed bank at the beginning of year t prior to germination. $G_i(E_i, C_i)$ is the finite rate of increase of species i. $E_i(t)$ is the fraction of seeds of species i that germinate between t and $t+1$. $E_i(t)$ can be viewed as the effect of the environment on species i: when the environment is good for species i, more seeds germinate that year. s_i is the survival rate of ungerminated seeds between years. Y_i, the yield of species i, is the number of seeds of species i that would be produced by each germinated seed in the absence of competition (i.e., when $C_i(t) = 1$). $C_i(t)$ is the reduction in the yield of species i due to competition. $C_i(t)$ is assumed here to be determined by the weighted sum of the number of germinating seeds of each species:

$$C_i(t) = 1 + \sum_{j=1}^{n} \alpha_{ij} E_j(t) N_j(t),$$

where the α_{ij}'s are weighting constants.

If we take the natural log of the finite rate of increase we get

$$g_i(E_i(t), C_i(t)) = \ln G_i(E_i(t), C_i(t)) = \ln N_i(t+1) - \ln N_i(t),$$

and an important observation is that the population growth rate over any time period is the arithmetic average of g_i over that time. The ability of species i to invade from low density (the invasion criterion) is then determined by $\Delta_i = E(g_i[E_i(t), \hat{C}_i(t)])$, where $E[\ ]$ is the expected value (i.e., the mean), $g_i(E_i(t), \hat{C}_i(t))$ is the value of g_i when the density of species i is set equal to zero ($N_i(t) = 0$), and all other species are at their stationary distribution (i.e., have on average a zero population growth rate). If Δ_i is positive, species i can invade from low densities, and coexistence of two species is possible if the Δs for both species are positive.

Chesson (1990) shows that coexistence is possible if there is a certain form of *subadditive* interaction between the effects of competition and the effects of the environment on species i (Fig. 8.11a). In order for species i to invade, it must have a low-density advantage; that is, it must

the years when the environment is good. Only (a) promotes coexistence of competitors. Redrawn, with permission from the Royal Society of London, from Chesson 1990.

be able to have a positive growth rate when all other species are at their stationary distribution (and therefore have a zero population growth rate). If competition and the environment act in a subadditive way, then during years in which the environment is poor for species i, changes in the level of competition have relatively little impact on the population growth rate (i.e., the population growth rate is relatively insensitive to the effects of competition). During years when the environment is good for species i, however, the population growth rate is very sensitive to the level of competition.

This subadditive effect can occur in the annual plant example because during poor years for a species, a small fraction of the seeds germinate and most of the individuals of the species remain in the seed bank. Seeds in the seed bank are unaffected by competition with other species. During good years, a larger fraction of seeds germinate and are therefore affected by competition. Thus, the presence of the long-lived seed bank, which is invulnerable to competition, has a strong buffering effect on the population of each species. In order for this mechanism to work, there must be some difference between the species in their response to the environment. If all species respond to the environment through time in exactly the same way, then the competitively superior species (as determined by the α's) will exclude the other species.

This contrasts with the effects of additive (Fig. 8.11b) or superadditive (Fig. 8.11c) interactions between the effects of competition and the environment. Neither of these scenarios leads to a low-density advantage for species i, and therefore neither promotes competitive coexistence. In the additive case, a species is affected by competition in the same way, regardless of whether the environment is favorable or unfavorable to that species during a given year; that is, there is no covariance between competition and the environment. In the superadditive case, the species is more sensitive to competition during poor environmental years than in good years. This gives the species a low-density disadvantage, which is the opposite of what is required for coexistence.

EFFECTS OF SPATIAL STRUCTURE ON COMPETITION

None of the models presented so far in this chapter has taken into consideration the spatial configuration of the system in which the species are interacting, or the movement abilities of the individuals involved. They have all implicitly assumed a homogeneous, well-mixed, system. In chapter 10 we discuss the effects of spatial processes on the dynamics of single-consumer/resource systems. There we acknowledge that the literature on that topic is huge and cannot be covered exhaustively in a single chapter, so we cover only a few key topics. Likewise, there is an extensive and growing literature on the impacts of spatial structure and limited mobility of individuals on the potential for coexistence of competing species. We cannot provide a thorough review here so instead provide only highlights of a few models to make two fundamental points:

- Coexistence can occur through a trade-off between competitive ability in a local area and the ability to colonize new areas.
- Spatial heterogeneity can allow for the possibility of coexistence through spatial niche separation.

Coexistence through a Competition/Colonization Trade-Off

Hastings (1980), Nee and May (1992), and Tilman (1994) all present variations of a patch-occupancy model of competition (see also Levin 1974; Slatkin 1974). In a patch-occupancy model, the world is divided into (generally an infinite number of) discrete patches of suitable habitat where the organisms can live, surrounded by unsuitable habitat. Each patch is either occupied by a species or empty. A patch can be interpreted either as a site that can be occupied by a single individual of a species, or the location where a population of the species can live. If patches are locations for populations, it is assumed that the dynamics within a patch occur on a very rapid time scale, so once a patch is colonized by a species its population effectively immediately reaches its carrying capacity on that patch. Patch-occupancy models therefore do not follow the dynamics of populations within each patch, but instead keep track of only the fraction of patches occupied by each species.

The Hastings (1980), Nee and May (1992), and Tilman (1994) models build on the simple, but highly influential, model of a single species metapopulation by Levins (1969):

$$\frac{dp}{dt} = cp(1 - p) - mp,$$

where p is the fraction of patches occupied by a given species, c is the colonization rate, and m is the patch extinction rate (if patches are populations) or mortality rate (if patches are sites for single individuals). Each occupied patch produces c colonists per unit time which can lead to new occupied sites only when they land on unoccupied sites. The spatial configuration of patches is not considered, and colonization is a global process, meaning that any unoccupied patch can be colonized by propagules from any occupied patch.

The model has a stable equilibrium with the fraction $p^* = 1 - (m/c)$ of patches occupied. The implication of this for competition is that if this species is a superior competitor, at equilibrium the fraction m/c of patches remains empty and potentially available for colonization by other inferior competitor species.

We now consider two species (here we follow the presentation of Tilman 1994). Species 1 is a superior competitor that can successfully colonize any patch in the system regardless of whether the other species is present. Species 2 is an inferior competitor that can colonize only patches in which species 1 is absent, and subsequent colonization by species 1 leads to the immediate extinction of species 2 from the patch. p_1 is the fraction of patches occupied by species 1, and p_2 is the fraction of patches occupied by species 2. In this scenario, it is impossible for both species to be present in the same patch at the same time. For species $i = 1, 2$, c_i is the colonization rate and m_i is the patch extinction or mortality rate. The equations for the rate of change of fraction of patches occupied by each species become

$$\frac{dp_1}{dt} = c_1 p_1 (1 - p_1) - m_1 p_1$$

$$\frac{dp_2}{dt} = c_2 p_2 (1 - p_1 - p_2) - m_2 p_2 - c_1 p_1 p_2.$$

The dynamics of species 1, the superior competitor, are not affected by the presence of the inferior competitor in the system. Species 2, the inferior competitor, can invade, and the two species can coexist at a regional scale (although only one species can persist on each patch) only if

$$c_2 > \frac{c_1 (c_1 + m_2 - m_1)}{m_1}.$$

When both species have the same mortality rate, $m = m_1 = m_2$, the condition for coexistence becomes $c_2 > c_1^2/m$. Therefore two species can coexist spatially only if c_2 is greater than c_1, because c_1 must be greater than m for species 1 to persist, so $c_1/m > 1$. Thus the inferior competitor must have the superior colonization ability (i.e., there must be a competition/colonization trade-off).

If the mortality rates are different between the two species, the condition for coexistence really becomes a competition/colonization/ mortality rate trade-off. In particular, if m_2 is less than m_1 (the inferior competitor has a lower mortality rate), it is possible for the two species to coexist in some cases in which c_2 is less than c_1.

The model can be expanded to include multiple species. If species 1 and 2 can coexist, then at equilibrium a fraction of the environment still remains empty and potentially can be colonized by a species even more inferior in competition than species 2 with an even higher colonization rate. This pattern can theoretically continue for an infinite number of species; however, the criteria for coexistence (the combinations of c_i and m_i for which coexistence is possible) become more and more restrictive as more species are added.

Hassell et al. (1994) and Comins and Hassell (1996) also found coexistence through a competition/colonization trade-off in a spatially explicit lattice model of a single host species being attacked by two parasitoid species. In their model, space is divided into a two-dimensional grid of cells, with the dynamics of populations in each cell modeled explicitly using equations similar to equations 8.8. Each generation fractions of the populations of each species leave each patch and distribute themselves among the neighboring patches. Coexistence is found to be most likely when the species with the lower within-patch searching rate (and therefore the inferior competitor within a patch) has a higher dispersal rate. This type of model produces a range of interesting spatial dynamics and self-generated spatial patterning, including spiral waves (where waves of high and low densities of the different species spiral out from a relatively stationary central focus) and spatial chaos (see also Hassell et al. 1991a; Comins et al. 1992). When coexistence of the two parasitoid species does occur, it is generally accompanied by some degree of spatial separation of the species (e.g., with the less mobile species near the center of the spiral and the more mobile species in the spiral arm). Thus in these models the coexistence mechanism seems also to involve some aspect of spatial niche partitioning, discussed next.

Coexistence through Spatial Niche Partitioning

One assumption central to the patch-occupancy models described above is that the inferior and superior competitors cannot coexist within an individual patch. Amarasekare and Nisbet (2001), however, point to several examples from real systems in which competitors do coexist within a patch. They cite, for example, Lei and Hanski's (1998) study of the butterfly *Melitaea cinxia*, attacked by two parasitoid species, and Amarasekare's (2000a; 2000b) work on the harlequin bug, *Murgantia histrionica*, and its two parasitoid species. Both examples show the pattern of coexistence with both parasitoid species present in some patches and only the inferior competitor in other patches. This pattern is impossible in the patch occupancy models.

Amarasekare and Nisbet (2001) present a model of two patches connected by dispersal, in which the dynamics of the two species on each patch are modeled explicitly using the Lotka-Volterra competition equations with a dispersal term:

$$\frac{dX_i}{dt} = r_x X_i \left(1 - \frac{X_i}{K_{x,i}} - \phi_{x,i} \frac{Y_i}{K_{x,i}}\right) + d_x \left(X_j - X_i\right)$$

$$\frac{dY_i}{dt} = r_y Y_i \left(1 - \frac{Y_i}{K_{y,i}} - \phi_{y,i} \frac{X_i}{K_{y,i}}\right) + d_y \left(Y_j - Y_i\right)$$

for $i, j = 1, 2$ and $i \neq j$, where X_i is the density of species X in patch i, and Y_i is the density of species Y in patch i. $\phi_{x,i}$ and $\phi_{y,i}$ are the competition coefficients, such that, for example, $\phi_{x,1}$ is the relative per-head effect of Y on X in patch 1. $K_{x,i}$ and $K_{y,i}$ are the carrying capacities for each species in patch i. r_x and r_y are the per-head growth rates of each species, and d_x and d_y are the per-head emigration rates of each species. All individuals that leave one patch immediately enter the other patch.

In this model, unlike in the patch-occupancy models, dispersal and competition occur on the same time scale, so that dispersal can potentially rescue populations on individual patches from extinction.

Amarasekare and Nisbet make the two species identical, except for the competition coefficients and dispersal abilities. Thus in the absence of dispersal, there are three possible outcomes depending on the competition coefficients, as seen in the Lotka-Volterra competition model: coexistence, competitive exclusion of one species or the other depending on the initial conditions (the priority effect), and competitive exclusion of the inferior competitor by the superior competitor. They concentrate on this last case, where there is competitive exclusion that is due to asymmetrical competition in the absence of dispersal. This occurs when $\phi_{x,i}$ is less than 1 and $\phi_{y,i}$ is greater than 1 if species X is the superior competitor in patch i (or vice versa if species Y is superior).

Amarasekare and Nisbet (2001) find that if one species is the superior competitor throughout the landscape (in both patches in this case), then competitive exclusion of the inferior competitor cannot be avoided through the inferior competitor having a higher dispersal ability. That is, the competition/colonization trade-off does not result in coexistence in this case, because the superior competitor causes the inferior competitor to have a negative population growth rate in both patches. Movement between patches with a negative growth rate cannot rescue the inferior competitor from extinction.

If, however, the superior competitor is confined to patch 1, such that the inferior competitor has a refuge in patch 2, then the two species can coexist on patch 1 through source-sink dynamics, but only if there is a limited amount of dispersal. Patch 1 is the sink for the inferior competitor, and patch 2 is the source. Small amounts of movement from the source can rescue the inferior competitor from extinction in the sink patch. If the dispersal rate is too high, however, the population growth rate of the inferior competitor on patch 2 (which was otherwise the source) can become negative, and the inferior competitor can be driven extinct from both patches.

Coexistence can occur also if there is spatial heterogeneity in the competitive abilities of the two species, such that each species has a competitive advantage in at least part of the landscape (e.g., $\phi_{x,i} < 1$ and $\phi_{x,j} > 1$ for $i, j = 1, 2$ and $i \neq j$). Spatial heterogeneity is the key to coexistence here, not a competition/colonization trade-off. In fact, high levels of dispersal by the overall inferior competitor can sometimes lead to its extinction (again because of source-sink dynamics).

A variety of different biological mechanisms can lead to spatial niche partitioning. For example, in the neighborhood models of plant competition (e.g., Pacala 1986; Pacala and Levin 1997), limited dispersal abilities of competitors can naturally lead to clumping of each species through time. This is simply because the offspring for each species land near the adults. If the spatial neighborhood over which the effects of competition are felt (e.g., the spatial range over which each plant draws down the level of nutrients) is small, then individuals will compete mainly with nearby individuals, which will mainly be conspecifics. This can therefore make the net effect of intraspecific competition greater than interspecific competition.

Several models have described intraspecific aggregation of species that develop on patchy and ephemeral resources, such as carrion flies, dung beetles, and *Drosophila* (Atkinson and Shorrocks 1981, 1984; Hanski 1981; Ives 1991). If there is intraspecific aggregation such that different species tend to aggregate in different patches, then coexistence is again possible. Individuals of each species will compete mainly with conspecifics, so intraspecific competition is greater than interspecific competition.

Another interesting example that combines elements of spatial niche partitioning with a colonization/coexistence trade-off is the work by Richards et al. (2000) on competition between consumers with different feeding strategies. They consider a system such as benthic snails feeding on algae (e.g. Schmitt 1996; Wilson et al. 1999). One type of species in the model is the "digger," which stays in one place longer, feeds more intensively in each area, and depletes the local density of algae to a lower level. The other type of species is the "grazer," which moves more rapidly across the environment, only grazing off the top level of algae. In a single patch the digger is likely to outcompete the grazer, because it is more efficient at driving down the local density of the resource. In a spatial context, however, coexistence is possible between species using the two feeding strategies because they produce a heterogeneous distribution of the resource across the landscape. Coexistence requires that the grazer either move faster across the environment or be more efficient at converting the resource into new grazers.

CONCLUDING REMARKS

This chapter has reviewed how simple models of multiple consumer species attacking a single resource predict that only a single consumer species will be able to persist (the competitive exclusion principle; Hardin 1960). The species that drives down the abundance of the resource to the lowest level will exclude all other species. In order for coexistence of multiple consumer species to be possible, there must be some mechanism acting that either causes intraspecific competition to be greater than interspecific competition, or that introduces additional resources into the system. This chapter discussed several of these mechanisms, summarized in Table 8.1. Any factor that leads to intraspecific density dependence in the growth rate of the consumer species promotes coexistence. Under certain conditions intraguild predation (facultative hyperparasitism) promotes coexistence, because the intraguild prey is an additional resource for the intraguild predator. We discussed how attack by different consumer species on different ages or stages of an age-structured resource population can lead to coexistence in only a restricted set of circumstances (e.g., when there is significant variation in the stage durations). A variable environment promotes coexistence when it either is combined with species-specific non-linear responses to the environment or when there is an interaction between competition and the environment that leads to some form of the storage effect. Coexistence can be promoted through several spatial mechanisms, including a competition/colonization trade-off and spatial niche partitioning.

Throughout, we have discussed, where appropriate, the effects of the outcome of competition on the resource (or host abundance). This will become relevant in the next chapter on biological control, where the goal is to reduce the abundance of the resource through the addition of one or more consumer species.

Implications for Biological Control

Here we synthesize results, mainly from chapters 8 and 10, that relate to biological control, and where possible relate them to real systems. Mostly we consider "classical" biological control in which the pest is an alien species and the specialist natural enemies released to control it have typically been found in the pest's region of origin. Such control has been most successful in long-lived crops such as forests and orchards, and this is where most of our models, with their long-term view, are relevant. But we also consider, briefly, temporary environments such as seasonal crops, where indigenous enemies are typically important. We comment only briefly on generalist enemies since the book does not deal with relevant theory (Holt and Lawton 1993; Murdoch et al. 1998a; Chang and Kareiva 1999). Hawkins and Cornell (1999) provide a wide-ranging summary of current theory for biological control and a good entry into the literature.

In spite of its successes, and its more numerous failures, there are few, if any, general principles, or even rules of thumb, to guide the efforts biological control. Nor is there a general scientific basis for biological control. There seem to be several reasons for this. First, success or failure depends on the detailed natural history of each situation, and these details vary in uncountable ways from one system to another. Second, there has been astonishingly little effort to learn from more than 100 years of experience. If a natural enemy works to control a pest somewhere, the entomologist must move on to the next problem; if it fails, the entomologist tries another enemy or moves on to the next problem. There has been little support for analyzing in the field the conditions separating success and failure.

Biological control has thus been an unrelenting trial-and-error process over its entire history. Whether theory can inform this process and make it more "scientific" is still an open question. Here we continue a long tradition of mining ecological theory, some of it developed specifically for biological control, for insights. Some possible rules of thumb emerge, but they should be viewed as hypotheses awaiting empirical test.

Trade-Off between Suppression and Stability and between Suppression and Coexistence

Discussion of a theoretical or conceptual basis for biological control has traditionally focused on stability of the pest–natural enemy interaction as a central goal. This book points out, however, that stabilizing processes usually tend to result in a higher pest equilibrium (Murdoch 1990; Hochberg and Holt 2000), and factors that drive the pest density lower tend to be destabilizing (Murdoch and Stewart-Oaten 1989) (chapter 10). In fact, we do not know what stabilizes most consumer-resource systems; they seem frequently to be stable in the absence of any obvious stabilizing process (Murdoch et al. 1998a). We therefore frankly do not know how conclusions based on achieving low pest density ought to be constrained by considerations of stability. We therefore focus mainly on factors that reduce pest equilibrium, rather than on those that stabilize it. In the same vein, when considering multiple natural enemies, processes that enhance coexistence of competing enemy species tend to increase the pest equilibrium (chapter 8). It is of course quite possible that many successful cases of biological control do not involve stable equilibria, and may even involve local pest eradication, at least on a local spatial scale (see "Spatial Processes and Control," below).

The issue of self-limiting parasitoids needs special mention in this context. It has become standard in the last few decades to include density dependence in the parasitoid population in models for biological control. This is done routinely by assuming the distribution of attacks among hosts is negative binomial (May 1978; Godfray and Waage 1991) (chapter 8). Whether density dependence in the parasitoid should be included for a model of any particular system is an empirical question, to be determined by observation or experiment. But since natural enemies are often at low density when they are effective, density dependence in the enemy may be rare.

We know of two stabilizing mechanisms that can actually decrease pest equilibrium density. The first is an external source of parasitoids, which both reduces pest equilibrium and increases the parasitoid equilibrium (chapter 3, equation 3.20). Second, in the stage-structured framework a shorter parasitoid development time both improves control and is less destabilizing. The equilibrium density of vulnerable immature pests in the Basic model of chapter 5 is

$$J^* = \frac{d_P}{as_I}, \qquad (9.1)$$

where $s_I = e^{-d_I T_P}$ is the fraction of immature parasitoids that survive to become adult when their death rate is d_I and their development time is T_P. Substituting for s_I in equation 9.1 shows the immature pest population increases exponentially as the parasitoid time lag increases:

$$J^* = \frac{d_P e^{d_I T_P}}{a},$$

as does total pest density. A shorter parasitoid development time is also stabilizing (chapter 5; see also the hybrid models at the end of chapter 4). Results of a field experiment suggest that a short parasitoid development time may indeed be a key feature of the successful stable suppression of California red scale by *Aphytis melinus* (W. Murdoch and S. Swarbrick, unpubl. data).

In summary, although there are exceptions, stability tends to be bought at the cost of higher pest density. This does not mean that stability is incompatible with control. It does mean, though, that where the trade-off exists, control is possible only if the pest equilibrium can be held below the economic threshold in spite of the stabilizing processes.

An analogous result holds, in theory, for coexistence of enemy species: coexistence generally requires a higher pest density. The exceptions are when both enemy species, or the potential winner, limit their own densities (intraspecific exceeds interspecific competition; chapter 8), or when the species attack different parts of the population.

A COMPARATIVE APPROACH TO EVALUATING
NATURAL ENEMIES

Ecologists and biological control workers have argued for decades about the features that define good natural enemies, and whether it is better to release only one (because there is a "best" species and others might interfere with its effectiveness) or multiple enemies (because collectively they may be more effective than any one). Here we synthesize the results from chapter 8 in response to these issues and use them to establish that the question of single versus multiple releases is too crude. It needs to be expanded to include:

- When can we release multiple enemy species at will?
- When should we be cautious about releasing more than one enemy species?
- Which enemy should we release?
- When should we release more than one species? (And what are good combinations?)

The answers are summarized in Table 9.1.

Life-History Trade-Offs

Until recently, students of biological control listed desirable attributes of natural enemies based on intuition and on parasitoid-host models with only one species of enemy. These guidelines may sometimes be generally useful. For example, pest suppression will be greater, all else equal, if the enemy's attack rate is higher, its death rate is lower, and it continually tracks patches that contain more hosts.

Such modeling exercises generally ignore the fact that a parasitoid's attributes come as a package, however, and that its evolution has involved various trade-offs between attributes. That is, "all else" is never equal. For example, as noted in chapter 8, a parasitoid species with one favorable attribute—attacking younger stages of the host—may as a result unavoidably lower its instantaneous successful attack rate, since it will typically be smaller than parasitoids that emerge from older, and hence larger, host stages. A similar trade-off may

TABLE 9.1. Implications of competition theory for biological control

Theory	Result	Release
Simple models of exploitative competition	Best competitor gives most control.	All species: best wins
Segregation of resource use	Combination better.	Complementary species
Enemies interfere	Coexistence can increase pest density.	Best agent, not necessarily winner
Enemies are self-limiting: intrasp. greater than intersp. competition	Coexistence decreases pest density.	More than 1 species
Stage-structured pest	Winner may increase key pest stage.	Best agent, not necessarily winner

Theory based on different mechanisms of competition or coexistence yields different advice on the release of multiple natural enemies, especially whether the winner in competition is the best control agent. Details on models in each class of theory are in chapter 8.

exist between development time and attack rate. We therefore need comparative analyses using models with competing parasitoid species.

The point is perhaps made most forcefully in relation to a major lifestyle trait: whether the parasitoid is a host-feeder or not. It has often been suggested that host-feeders are potentially poorer natural enemies because they need more host individuals to produce a new adult parasitoid—an integrative measure whose importance we stressed in chapter 8. The situation, however, is not that simple. Non-host-feeding parasitoids need to accumulate, as larvae, most of the nutrients they will require for the adult's lifetime reproduction. Host-feeders do not need to do this, and so can use relatively smaller hosts. The eggs of host-feeding species may also be relatively larger (since these species can accumulate new resources for each egg), which will reduce the development time and the size of host needed. That is, a suite of changes accompanies host-feeding that, in total, may lead to a lower pest density than the feasible alternative suites. And, in fact, a substantial number of successful control agents are host-feeders.

We thus need to evaluate suites of attributes in an integrative way. In addition, the number of potential combinations of attributes is virtually limitless, and we cannot explore them all. We therefore need a comparative approach that looks at the relative effectiveness of actual or likely species (Murdoch 1990; Waage 1990). The models in chapter 8 do exactly this kind of analysis.

Numbers of Enemy Species in Real Systems

The evidence from real systems usually does not tell us whether it is better to have only one or more than one enemy species. Successful control seems typically to be associated with the coexistence of several natural enemies (e.g., Clausen 1978). But we usually do not know what the mechanisms of coexistence are, or whether control is better or worse because there is more than one species present.

There are a few exceptions. For example, the parasitoid *Aphytis paramaculicornis* was introduced to control olive scale in California. However, it attacks only the larger stages and does poorly in summer. Control was incomplete and outbreaks still occurred. Another parasitoid, *Coccophagoides utilis*, which attacks smaller stages especially in summer, was then introduced, and the scale was brought under control (Huffaker and Kennett 1966). Takagi and Hirose (1994) suggest that a combination of two parasitoids is able to control arrowhead scale because one species (*Aphytis anonensis*) is effective at driving down the scale from high densities whereas the other species (*Coccobius fulvus*) can regulate the population at low densities. Roland and Embree (1995) analyze an analogous situation, discussed below.

The effectiveness of multiple enemies is less clear in other examples. The following three cases are typical. In mulberry orchards in Europe, each of three introduced parasitoids contributes to overall parasitism of the peach scale (*Pseudaulacapsis pentagona*) during a given season (Pedata et al. 1995). Many predatory species seem to be involved in control of rice pests in Indonesia (Settle et al. 1996). Finally, although *Aphytis melinus* seems to be the main control agent of California red scale in southern California, there is always at least one other introduced parasitoid present in any particular grove, together with one or two predators (chapter 2). In none of these examples, however, do we know if the additional species contribute significantly to control, or if they are merely killing pests that would be killed in any case. Indeed, it is quite possible that pest density would be lower if only one enemy species were present, if some species interfere with the most effective enemy.

We also typically do not know whether enemy species that were introduced but have not persisted were simply not well adapted to the crop environment or were outcompeted by other species. *Aphytis* spp. on red scale is one of the few cases where we know the answer (see below), as is control of the bayberry whitefly (*Parabemisia myricae*) which became a pest on citrus in the late 1970s. In the case of bayberry, two parasitoid species from Japan were released for the whitefly, but a short time later a native or accidentally released species, *Eretocerus debachi*, appeared. Within a few years it had excluded the released parasitoids and depressed the whitefly to extremely low levels (Rose and DeBach 1992).

This cursory survey is enough to show that the history of biological control cannot by itself tell us how many or what kinds of species to release. (As noted above, previous experience might have been a good guide if the appropriate studies of successes and failures had been done.) We discuss other examples below that illustrate particular points but turn now to insights from theory. We organize these insights under headings corresponding to the series of questions raised above.

Insights from Theory

WHEN CAN WE RELEASE MULTIPLE ENEMY SPECIES AT WILL?

In theory, we can release natural enemy species at will when, no matter which species we release, the best control agent is always the winner. In models, this condition holds mainly when there is no stage structure in the pest (or the various natural enemies all attack the same stage) and competition between the enemies is purely exploitative. In these circumstances, the invasion criterion (chapter 8) is a reliable guide to pest control: the winning enemy is always the best control agent.

In some real biological control cases the more effective species do seem to have survived as winners. For example, of the eight introduced enemies of red scale that became established in southern California, only *Aphytis melinus* gave adequate control, and it dominates most areas. Even here, though, one or more additional enemy species usually coexist with *melinus*, and it is possible that control would be better in their absence. Control of black scale in Israel may be another case: although 17 parasitoid species were released, control has been attributed mainly to the action of one, *Metaphycus bartletti* (Argov and Rossler 1993).

WHEN SHOULD WE BE CAUTIOUS ABOUT RELEASING MORE THAN ONE ENEMY SPECIES?

Two features militate against releasing multiple natural enemies: facultative hyperparasitism (intraguild predation) and pest stage structure. In each case, we are concerned both with the situation where the winner in competition is not the best agent and where coexistence of competitors induces higher pest density.

Natural enemies interfere (Intraguild predation—IGP). In parasitoids, IGP takes the form of facultative hyperparasitism, in which parasitized pest individuals can be reparasitized successfully by the competing species. If the species that interferes is worse at pure exploitation—direct parasitism—it is best to release only the better exploiter. This species may win in competition anyway if pest productivity is not too high. Of course, if a parasitoid can be both a better exploiter of the pest and superior at interference, it will always win and will always suppress the pest more, and we can safely release both.

The potential problem posed by IGP is made worse, in theory, because of the feature that the interfering enemy is increasingly likely to win as pest productivity increases (see "Facultative Hyperparasitism Equals Intraguild Predation," chapter 8) and, as a consequence, pest equilibrium increases. Since increases in crop density and quality typically improve the environment for the pest, IGP should be a worse problem as crop productivity increases, i.e., where pests are worst.

An excellent example of reduction of control caused by interference by a poorer exploiter comes from weed control in Australia (Woodburn 1996). The nodding thistle (*Carduus nutans*) is native to Europe but was accidentally introduced to Australia in the 1940s and 1950s. It has since spread and become a pest of pastures in eastern Australia. Studies of the plant and its natural enemies in Europe suggested that several species of herbivores might be useful control agents, of which two are of interest here. The first is a weevil, *Rhinocyllus conicus*, which had already contributed to controlling the thistle in North America. The adults attack the rosette leaves of the plant in spring, when it is small; they lay eggs that hatch into larvae that mine the developing flowers. Each final-instar larva destroys almost 30 seeds. The second enemy is the seed-fly *Urophora solstitialis*, whose eggs are laid into the developing flower; the larva then eats its way down through the developing ovule into the receptacle tissue, where it induces the plant to form a gall. The woody gall becomes a metabolic sink, diverting plant nutrients that otherwise would have been used to produce seeds, and each final-instar larva is reckoned to prevent the development of about seven seeds.

Competition between the weevil and fly had been studied in Europe, and there was even discussion of whether the weevil would (Harris 1989) or would not (Zwölfer 1973) displace or interfere with the fly. However, the weevil's success in North America stimulated its release in Australia in 1988.

Unfortunately, the weevil turned out to be both the poorer exploiter and an interfering competitor. By 1994 it had spread somewhat in the region of release, but it has not been able to control the weed. The main reason appears to be the much longer flowering season of the thistle in the Southern Hemisphere. Whereas the plant flowers over 6 months (November–May), the weevil has but one generation, which attacks the developing flowers only in November and December. Thus, although the weevil can cause almost total seed mortality early in the season, it hardly attacks seeds that are produced thereafter.

The fly turned out to be the better exploiter, and potentially the more effective control agent in Australia. It has two generations per year. In an experimental area where it was the only species present, its first generation caused close to 100% seed loss early in the season, as did its second generation at its peak later in the season.

The fly was released in 1992 in the hope that it would augment the effect of the weevil, especially in the latter part of the flowering season. By 1994, however, it had spread very little from its area of origin. Field studies suggest it loses in competition with the weevil: in the presence of the weevil the fly has been unable to cause more than about 20–30%

seed mortality early in the season. Consequently, the second generation is small and peak seed mortality in the latter part of the season is only about 30–40%. These peaks are also narrow, so seeds escape mortality for most of the year, the overall reduction in seed production being less than 20%.

The mechanism of interference is straightforward. The fly needs to draw plant nutrients to the gall through the plant's vascular tissue, and in particular the nutrients need to flow through the receptacle tissue. Heavy attacks by the weevil, however, result in mining of the receptacle tissue, and plant nutrients cannot pass to the gall.

Snyder and Ives (2001) provide a second example. They show experimentally how a generalist predator (a carabid beetle) reduces control of pea aphids in alfalfa by a specialist parasitoid, by attacking both parasitized and unparasitized aphids. Indeed, in some circumstances the unparasitized aphids are able to resist carabid attack, whereas parasitized "mummies" are not. Interestingly, Snyder and Ives show that carabids can reduce aphid densities in the short run, and it is only when the longer term consequences are tracked that interference with control is detected.

Stage structure: which pest stage is most damaging? The results above hold whether or not we consider pest stage structure. An entirely new issue appears, however, with stage structure: which stage of the pest causes the most damage? Recall that in a host with different stages and only exploitative competition, a competitor wins by reducing the host stage attacked by its competitor below the density that allows the competitor to persist (equations 8.12). But that stage may not be the most damaging to the crop. Most obviously, it is of no direct advantage to suppress to a minimum either the egg or the pupal stage of the pest, or even the adult stage if the pest is, say, a butterfly or moth. The winning enemy species might thus make control worse by increasing the density of the most damaging pest stage.

All insect pests have a structured life history, so the effect on the crop always depends on the stage that is most destructive. Thus taking stage structure into account is essential unless differences in efficiency between parasitoids are very large.

We know of no real cases that illustrate the above result. This does not mean they do not exist. They have not been looked for.

WHICH ENEMY SHOULD WE RELEASE? PARASITOID LIFESTYLES,
HOST LIFE HISTORIES, AND PEST CONTROL

We covered several reasons, above, why we might want to release only one species. In addition, even when it appears safe to release many, there may be good reasons for releasing only the best species. For example, even supposedly host-specific natural enemies may attack non-target species, and it may be wise to reduce the number introduced. We therefore present an approach for evaluating potential enemies based on what is known of the host's demography and the parasitoids' lifestyles and vital rates.

Just as competition is decided by the combination of parasitoid and host attributes, so too is the relative degree of pest control that can be achieved by two competing parasitoid species. Consider a pest that is currently attacked by a resident parasitoid. How would its equilibrium density change if an invading species were to displace the resident enemy?

We first need to specify which pest stage(s) we most want to suppress. In what follows we imagine the pest is a sucking insect and that we want to minimize the adult density. We could of course consider total density or a weighted sum of the densities of all stages.

As an example, suppose as in chapter 8 that we have a parasitoid P that attacks the egg stage and a parasitoid Q that attacks the larval stage, and the pest adult is invulnerable to attack. Assume the parasitoids are otherwise identical—they have the same survival and attack rates ($a_P = a_Q$, $s_{JP} = s_{JQ}$, $d_P = d_Q$). This is the age-structured parasitoid competition model in equations 8.12. When only P is present, adult pest equilibrium density is $A^*|_P$; and when only Q is present, adult pest equilibrium is $A^*|_Q$. We ask how the ratio $F = A^*|_Q/A^*|_P$ changes in relation to various parasitoid and host attributes. If F is less than 1, then control is improved by using Q rather than P as the control agent. From chapter 8 we would expect that P would usually be better, so we are interested mainly in conditions that reverse this outcome. The ratio F can be calculated from the equilibria in chapter 8 as

$$F = \frac{A^*|_Q}{A^*|_P} = \frac{T_E[\ln(\rho) - d_E T_E][\rho - \exp(-d_L T_L)]}{T_L[\ln(\rho) - d_L T_L][\rho \exp(-d_E T_E) - 1]}$$

(Murdoch and Briggs 1996; note that the ratio shown here is the inverse of that given in that paper).

By taking pest stage structure into account, we discover that the pest's demography, as well as parasitoid properties and lifestyles, is crucial (chapter 8). First consider the case in which the two juvenile host stages are equal in duration ($T_E = T_L$). If $d_E = d_L = 0$, so that the parasitoid is the only source of mortality on the juvenile hosts stages, then F is equal to 1, meaning that the same degree of pest suppression is achieved regardless of which enemy is present. If the background death rate on either or both juvenile host stage(s) is greater than zero, then P, which attacks the earlier host stage, will always be the better control agent ($F > 1$) (Fig. 9.1a). The larval parasitoid, Q, however, can be the better control agent if the larval stage is much longer than the egg stage ($T_E < T_L$), especially if the egg stage suffers high mortality from other causes (Fig. 9.1b). But pest demography sets limits to Q's relative effectiveness: Q can never be the better control agent if the egg stage is longer than the larval stage (or more specifically, if the duration of the stage attacked by the early-attacking species is longer than the duration of the stage attacked by the late-attacking species) (Fig. 9.1c). Our results also appear to hold for self-limiting parasitoids (Murdoch and Briggs 1996).

Godfray and Waage (1991) used the comparative approach to suggest which of two parasitoid species would be more effective at reducing the density of the mango mealy bug (*Rastrococcus invadens*) (though the model was developed too late to be predictive). The model suggested that one parasitoid species, *Gyranusoidea tebygi*, would lead to considerably lower pest density. The main difference between the species was that *G. tebygi* attacked an earlier stage of the mealy bug than did the other parasitoid species (*Anagryus sp.*). In reality, *G. tebygi* was the species released in the biological control program, though not because of the predictions of the model: it was the easier species to be raised in culture. Nevertheless, in conformity with the model, it has proven to be remarkably successful at reducing densities of the mealy bug.

This framework can of course deal with more detailed or more subtle aspects of parasitoid and pest ecology. We illustrated this in chapter 8, with the famous case of competitive displacement of *Aphytis lingnanensis* by *A. melinus*, and here we merely add the implications for biological control.

California red scale almost destroyed the citrus industry in southern California. Unchecked, the population on a single grapefruit or lemon

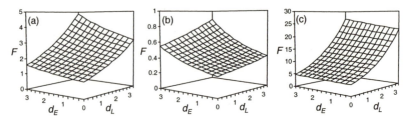

FIGURE 9.1. Effects on adult equilibrium host density of a parasitoid that attacks late in the host developmental period (larval stage attacked by Q) versus a parasitoid that attacks early in host development (egg stage attacked by P). The ratio of the equilibrium adult host density when Q is present alone relative to when P is present alone, $F = A^*|_Q / A^*|_P$, is shown as a function of the background death rates on host eggs (d_E) and host larvae (d_L). (a) The two juvenile host stages are equal in duration, $T_E = T_L = 0.5$. (b) Egg stage is shorter than larval stage, $T_E = 0.2$, $T_L = 0.8$. (c) Egg stage is longer than larval stage, $T_E = 0.8$, $T_L = 0.2$. When F is less than 1, which occurs only in (b), the parasitoid that attacks the larval stage leads to better suppression than the earlier attacking parasitoid species. In all cases, $\rho = 33$. The model is from equation 8.12.

tree can be in the millions and can kill the tree. Three species of the parasitoid *Aphytis* have been involved in control of red scale (Rosen and DeBach 1979). *A. chrysomphali* was present for most of the twentieth century, was relatively ineffective, and was displaced by *A. lingnanensis* (introduced in 1947), though this process was not studied in detail. *A. lingnanensis* also did not provide wholly adequate control, except along the coast, so *A. melinus* was introduced, mainly in 1959. It rapidly displaced *lingnanensis* and provided satisfactory control everywhere except along the coast (Luck and Podoler 1985) (Fig. 9.2).

A. *melinus* displaced *lingnanensis* very quickly. DeBach and Sundby (1963) showed that the process took only about a year in their study plots. The same had happened when *lingnanensis* displaced *chrysomphali*. This is only two to three generations of scale, and six to nine of the parasitoid. The apparently huge superiority of *melinus* is surprising. DeBach and Sundby (1963) stressed that the species are extremely similar and appeared to have identical lifestyles and properties. Indeed *lingnanensis* had a higher search rate, at least in the laboratory. Data from Kfir and Luck (1979) suggested *lingnanensis* might have a higher death rate in more extreme climates.

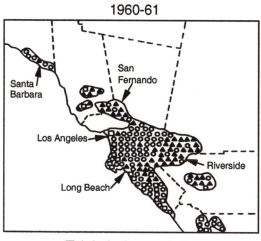

1960-61

□ Aphytis chrysomphali
○ Aphytis lingnanensis
▲ Aphytis melinus

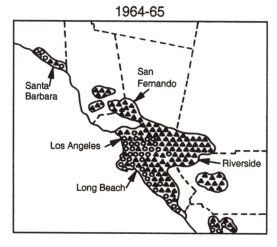

1964-65

○ Aphytis lingnanensis
▲ Aphytis melinus

FIGURE 9.2. Map of southern California showing distribution of *Aphytis melinus*, *lingnanensis*, and *chrysomphali* in the years 1960–1961 and again 4 years later in 1964–1965. In 1960 *chrysomphali* was still present at a few sites, but it was completely eliminated by 1964. *A. lingnanensis* dominated much of the inland area in 1960 but was displaced by 1964. Redrawn from Luck and Podoler 1985 with permission from the Ecological Society of America.

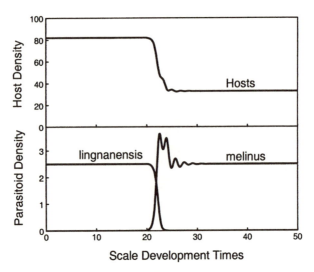

FIGURE 9.3. Simulation of the stage-structured model from Box 8.3, using default parameters (see also Murdoch et al. 1996a), showing rapid displacement of *A. lingnanensis* by *A. melinus*. The only difference between the two parasitoid species is that *melinus* can produce female offspring from a slightly younger host stage than can *lingnanensis*. The simulation starts with *lingnanesis* and the host (red scale) at equilibrium. A few individuals of *melinus* are introduced at time 20. Within a few scale development times *lingnanesis* is competitively excluded and *melinus* reduces host density. Redrawn, with permission of University of Chicago Press, from Fig. 3A in Murdoch et al. 1996a.

Luck and Podoler (1985) discovered a clue to what might be going on. They found, in both the laboratory and the field, that female offspring of *melinus* were able to develop and emerge from somewhat smaller scale than required by female *lingnanensis*. We showed in chapter 8, Box 8.3, that a well-parameterized model of the interaction (Murdoch et al. 1996a) could explain the rapid displacement of *lingnanensis* by *melinus*. Fig. 9.3 shows how the model can also account for the much greater suppression of scale by *melinus*.

Control of the cassava mealy bug (*Manihot esculenta*), in large areas of Africa is a recent example of highly successful biological control. Gutierrez et al. (1993) developed a stage-structured model that indicated that one parasitoid species (*Epidinocarsis lopezi*) gains a competitive advantage over its congener (*E. diversicornis*) through its ability to lay a female egg in smaller individuals of the mealy bug, and *E. lopezi* has been found to be the more successful of the two parasitoid species in the field.

WHEN SHOULD WE RELEASE MORE THAN ONE SPECIES?

We should release more than one species when coexisting enemies suppress the pest more than any single species would on its own. This can occur in models under two circumstances. (1) Pest density will be lower when the two species attack different parts of the pest population (niche separation). Pest density may increase, however, if they share some fraction of the pest and there is an exploitation-IGP trade-off (Kakehashi et al. 1984 and chapter 8). (2) Pest density will be lower when each enemy limits its own density through density dependence in the birth or death rate, i.e., coexistence is useful when each enemy is relatively ineffective (chapter 8). But recall that an enemy lacking self-limitation will, all else equal, be more effective than a combination of two self-limiting enemies.

UNRESOLVED ISSUES: GUILD THEORY, SEASONAL AND OTHER COMPLEMENTARITIES

We noted above that a few real examples show how enemy species may be more effective together when they are in some way complementary. At least three types of complementarity that might enhance control have been suggested. Guild theory (Ehler 1992, 1995; Mills 1994) suggests that control can be enhanced by choosing enemies that, for example, attack different stages of the pest. Or the enemies may dovetail seasonally, as in olive scale in California, which is controlled by *Aphytis* and *Coccophagoides*. *Aphytis* is most effective in cooler weather whereas *Coccophagoides* is more effective in summer (Huffaker and Kennett 1966). Finally, one enemy may be more effective at reducing very high pest densities whereas the other is more effective at maintaining a low pest density once that has been achieved (Takagi and Hirose 1994).

There is, unfortunately, no satisfactory theory underlying these three commonly stated ideas. It has been suggested, for example, that competition is minimized if individuals of the competing parasitoid species do not encounter each other in the same individual host: an egg parasitoid, for example, might always emerge before the parasitized host reaches the pupal stage and becomes vulnerable to reattack by a pupal parasitoid. But the analyses in chapter 8 ("Effects of Age Structure on Competition") clearly establish that, at least in theory, parasitoids cannot coexist merely by attacking different stages of the host. Two enemies that do not meet are nevertheless in exploitative competition; the more efficient exploiter wins, unless this is balanced by intraguild predation, and then coexistence comes at the expense of pest suppression (Briggs 1993a).

There is also no theoretical support for purely seasonal complementarity as a coexistence mechanism. Chapter 8 (equations 8.15) shows that superior performance in different seasons, cannot alone give coexistence. The species that is more effective on average over the year still wins by reducing the host to a density too low to support the less effective species. The problem is that the less effective species cannot accumulate gain that can carry it over its poor season. Finally, the only theory we know of that might relate to coexistence of enemies effective at high versus low pest densities requires a cyclic pest (chapter 8, "Species-Specific Non-Linear Responses to Competition").

Since natural enemies do coexist, and may indeed act in complementary ways, the failure obviously lies in the theory. But that means we do not understand the mechanisms.

One way, in models, to achieve coexistence of enemies on different stages, or in different seasons, is through sufficiently strong enemy self-limitation, also discussed in chapter 8. For example, Rochat and Gutierrez (2001) obtained coexistence of *Aphytis* and *Coccophagoides* in a seasonal model for olive scale. The parasitoids in their model, however, are both self-limiting via ratio-dependent functional responses (see chapter 4). Such self-limitation, which of course makes them less effective enemies, rather than the seasonal difference per se is what allows coexistence. This model, in addition, assumes the parasitoids immigrate into the grove from somewhere else, so does not give local coexistence. It thus nicely demonstrates the theoretical difficulty of simultaneously obtaining control and coexistence.

Rules of Thumb and Adaptive Management

Entomologists have developed some rules of thumb in the past. Some have been widely accepted. For example, a "good" natural enemy is thought to have a high search rate for the pest, can increase rapidly when the pest does, is synchronous with it, and needs few pest individuals to complete its life history (Huffaker and Messenger 1976). Others have been contentious: it is better to release one than many natural enemy species (Turnbull and Chant 1961; Huffaker 1971; Ehler and Hall 1982); specialist enemies are better than generalists (Murdoch 1975; Ehler 1977; Chang and Kareiva 1999). We reinforce a few of these rules, substantially expand their number, and suggest that to resolve some arguments we need to pose the problem differently.

Table 9.2 extracts the various conclusions of this and the following sections, and of models in chapter 8, and expresses them as rules of thumb. Care is needed, as with all such simplifications. They will never be adequate substitutes for a thorough understanding of each situation. But such understanding is often beyond economic or feasible reach, and rules of thumb may then provide a useful guide.

TABLE 9.2. Theory-based rules for biological control, ignoring effects on stability

---

Desirable attributes of single enemy species, "all else equal"
Choose the species that:
- is not self-limited at relevant densities
- has a higher rate of successful search
- has a shorter generation time
- preferentially attacks areas with higher host density
- leaves no refuge areas

Choose the species that attacks:
- an earlier host stage
- a host stage that lasts longer
- a host stage that suffers from lower mortality from other sources

Choose the species that produces:
- more adults per attacked host individual
- more fertilized females per host attacked (includes producing females from younger hosts)

Multiple-enemy situations to avoid
- hyperparasitoids
- facultative hyperparasitoids (intraguild predators)
- autoparasitoids

Rules of thumb, and detailed models, need to be tested. Biological control presents wonderful opportunities for such tests. Yet they are virtually lacking in the long history of biological control. Much work is often done on aspects of the enemy's natural history, host range, likely match to local climate, etc. But systematic effort is rarely directed, either before or after release, to those aspects of the population ecology of pest and natural enemy that theory suggests are important.

SPATIAL PROCESSES AND CONTROL

Here we draw on some results from chapter 10.

Aggregation within Populations

If we consider processes within a population, theory reinforces our intuition: pest suppression will be greater if the natural enemy continually responds to local variation in pest density by searching preferentially where the pest is more abundant (chapter 10). Once again, such efficient responsiveness is destabilizing. This result is reinforced by investigations of a spatially explicit individual-based model: one route to stability requires the enemy to ignore spatial differences in pest density, but the cost is higher mean pest abundance (chapter 10, "Route 2: Spatial Processes Induce Stability by Altering Spatially Averaged Properties"). Processes that induce persistent skewed risk among pest individuals, independent of local pest density, are also inefficient, but stabilizing.

Coupled Spatially Distributed Subpopulations

Our analyses to this point have assumed that the enemy-pest interaction persists, and is regulated, on a local spatial scale. Murdoch et al. (1996b) provide an example where this is a valid assumption. They show that a population of California red scale on a single tree isolated from the rest of the citrus grove is suppressed and stable. But surprisingly, there is little other hard evidence for spatially local stability in successful biological control, though it is often assumed that stability is the rule (Murdoch et al. 1985; Hochberg and Holt 2000). For example, although there has been argument about which natural enemies are responsible for maintaining winter moth on Vancouver Island, British Columbia, at a stable equilibrium (Roland 1994, 1995; Bonsall and Hassell 1995), in fact the data show that the population continued to decline exponentially, albeit at less than the initial rate, over the period of study.

There are some biological control situations where local regulation is not likely. Persistence of the interaction in these cases (if it occurs) must be regional, involving coupling of spatially distributed subpopulations. The most obvious examples are in short-lived crops, and especially those where harvesting is associated with severe habitat disruption such as plowing or burning. Although some fraction of the pest and/or enemy populations may carry over locally from one season to the next, perhaps in resting stages or in alternative plant hosts, there may well be local extinction with reinvasion from elsewhere at the start of the season. In areas with year-round cropping, "elsewhere" may be another field where the crop is growing, but it may also be a distant source. In these situations, dispersal properties of pest and enemy, as well as local dynamics, must be important.

The importance of enemy dispersal is illustrated indirectly in the study of pest control in rice in Indonesia by Settle et al. (1996). These authors show that predators are more common, especially early in the crop's development, in regions where rice is planted asynchronously rather than synchronously. Presumably predator populations build up in each paddy as the rice crop develops and then move on to other paddies as these are planted. Asynchronous planting was not an intentional pest-control strategy by the rice-growers, but it illustrates how such a strategy could be employed when a vagile natural enemy is available.

Regional rather than local regulation may also characterize some pests of long-lived crops, though this is difficult to demonstrate. Murdoch et al. (1985) presented evidence that in some cases local pest populations are unstable—indeed in a few cases the pest appeared to be driven locally extinct—and that persistence of the interacting populations is to be sought at a larger spatial scale. Additional evidence has accrued since then. Walde (1991; 1994) provided good evidence for local pest extinction in the successful control of apple mites by predatory mites. Walde showed that the predatory mite drives the prey mite locally extinct on several spatial scales: individual trees, groups of trees, and probably even occasionally entire orchards. Extinction at the orchard level suggests that the prey mite may even exist as a classical metapopulation, though this would be extremely difficult to establish. Walde and Nachman (1999) summarize evidence that control of mites in greenhouses by predatory mites also involves local extinction but prolonged interaction over a larger area.

Models of Coupled Subpopulations

There is little theory for coupled spatially distributed subpopulations that is useful to biological control. In chapter 10, we will note that coupling of such populations, especially in a heterogeneous environment, can lead to stability through asynchronous local fluctuations. But the models give little insight into factors determining the degree of pest suppression. There is clearly an opportunity for useful modeling in this area. We suspect that models in which individual movements couple submodels of local dynamics will be most useful, and we discuss several below.

DISPERSAL RATES AND BEHAVIORS

Some modeling reinforces our intuition that rapid enemy dispersal, relative to that of the pest, is a desirable attribute. Reeve (1988) analyzed a system of subpopulations of hosts and parasitoids in a stochastic environment and linked by random migration. The dynamics in a patch followed a Nicholson-Bailey model in which local stability was enhanced by a negative binomial parasitoid attack rate (chapter 4). His simulations showed that in systems that persisted for at least 500 generations, pest density was lower when parasitoid dispersal rate was greater than pest dispersal rate.

The cues that determine when natural enemies leave a local area of crop, and those that affect the next patch they choose, are likely to be at least as important as the actual rate of dispersal. We discuss two quite different approaches that illustrate the point.

Reeve (1988) looked at a version of his model in which both pest and parasitoid migrated in a density-dependent fashion. In persistent systems such behavior tended to reduce pest density. However, it is strongly destabilizing since it brings the patches into synchrony, so system extinction greatly increased. This model is hardly suitable for a crop system; in particular the parasitoid pays no cost, in time or other currency, of moving between subpopulations. It serves as a useful contrast, however, to the model discussed next, which was designed with a particular crop system in mind.

Ives and Settle (1997) developed a model for the rice-cropping system described above (Settle et al. 1996). They assume that the adult pest stage is both vulnerable and can migrate among fields. The immature stage is sedentary and is not attacked by the predator. Pest dynamics in a field are described by equation 4.7, which has a constant fraction of adults surviving between time units. The predators are described by the Nicholson-Bailey model (equation 4.12); they also have a constant fraction surviving from one time unit to the next, and in both populations a constant fraction survives in the field between harvesting and planting (harvesting occurs at the end of 8 time units). A fraction of both the predators and pest emigrate from the field at each time unit, and migrants mix and then immigrate equally among the eight fields in the system. At harvest all pests in a field die except for a fraction that persists until planting at the next time step.

Ives and Settle first show that asynchronous planting can enhance control when the pest and predators can migrate among fields. They then ask: how is final pest density at harvest affected when predator migration is density dependent? Specifically, the fraction of predators emigrating from a field is assumed to decrease with increasing pest density in the field.

The answer is that pest density increases the more strongly predator emigration rate responds to local pest density: that is, density-dependent movement by predators leads to less effective control. This is because the key to pest suppression is for the predators to be sufficiently abundant early in the life of the crop, when the pest is still at low density. The predator can then prevent rapid increase in the pest population. Predators are ineffective when they are attracted to and trapped by a pest population that is already peaking late in the season. They need to migrate to fields with incipient pest outbreaks, where they can do most good.

This model, of course, is also not a complete description of the real situation. Together with the other spatial models discussed, however, it serves to stress the importance of the cues to which enemy dispersal responds, and it illustrates a potentially useful approach that would repay further investigation.

van Lenteren and van Roermund (1999) have gone a step further by developing a spatially explicit individual-based simulation model for the parasitoid *Encarsia formosa* controlling whitefly in greenhouses. Most agricultural systems may be too extensive and heterogeneous, however, for this approach to be broadly useful.

NEED FOR EXPERIMENTAL TESTS

We have explored a variety of theoretical results that might yield useful insight or rules of thumb for managing real pests. These insights and rules gain occasional support from real cases of pest control. But as stressed at the start of this chapter, there have been discouragingly few efforts to test theory (whether mathematical or not) in real systems.

Active adaptive management is a potentially useful approach in which information from successes and failures, obtained from planned interventions, is used to inform general principles or theory and to guide future efforts (Parma et al. 1998). Active adaptive management views management decisions as experiments that test our understanding or underlying theory. Pest control is a much more amenable venue for such management than are many others. Because crops are replicated in space, it is actually possible to compare the efficacy of different management decisions (the Australian work on enemies of nodding thistle, described above, shows how this approach could work). We have tried to show that biological control is also an interesting laboratory for testing "pure" ecological ideas. Thus it is an area where basic and applied science can happily coexist and indeed reinforce each other.

Suppressing pest populations, like harvesting living resources and preserving biodiversity, is a problem in population management. It is thus ineluctably rooted in ecology, and for many decades there was great interest in placing biological control in an appropriate ecological framework. This interest has waned somewhat. But this chapter shows, we hope, that we are now in a better position to develop ecological theory that is potentially useful. We believe the time is ripe for a resurgence of research both in theory and in experimental tests of theory.

CHAPTER TEN

Dynamical Effects of Spatial Processes

Space has become a central focus in ecology over the past two decades, though 50 years ago Andrewartha and Birch (1954) recognized its general importance and Skellam (1951) wrote pioneering models. Throughout this chapter we concentrate on the effects of spatial processes, in particular spatial subdivision, spatial heterogeneity, and the movement of individuals, on the dynamics and stability of consumer-resource interactions. It is useful to distinguish two scales at which spatial processes can occur, though they may overlap in real systems. At the *among-subpopulation*, or metapopulation, scale, the population is divided into subpopulations that typically persist longer than one generation. At the *within-population* scale, individuals typically are redistributed among patches at least once per generation. The first two sections of the chapter correspond with these two scales.

SPATIAL PROCESSES AMONG SUBPOPULATIONS

Types of Spatially Structured Populations

Several approaches have been used to study spatially structured systems at the among-subpopulation scale, and here we briefly describe three of the main approaches. A central idea in all approaches is that coupling local interactions that would be unstable if isolated can, in some circumstances, yield a regional system that is dynamically stable; this is the idea of a metapopulation. Both consumer-resource interactions and single-species populations have been modeled through these approaches, yet the source of instability of the local populations generally differs in the two types of systems (Harrison and Taylor 1997). In single-species models, frequently it is assumed that a large, well-mixed population would be able to persist indefinitely. However, the population is instead fragmented into several small, isolated populations, each having a high probability of going extinct because of environmental and/or demographic stochasticity. High levels of movement among these local populations can rescue low-density populations from extinction and can recolonize patches from which the species has gone extinct.

In consumer-resource systems, however, the interaction itself is a source of instability. In a large, well-mixed system, or in a small local population, the consumer-resource interaction is assumed to be unstable (e.g., the Nicholson-Bailey model of chapter 4), and the consumer can potentially drive the resource extinct. Habitat fragmentation and limited movement of individuals between parts of the habitat can potentially keep the consumer from driving the resource extinct at a regional scale, and can potentially stabilize the interaction through one of the two routes to stability discussed below.

Approach 1. Cell-occupancy models. The simplest models, which we do not discuss further, assume that isolated subpopulations go extinct at some rate but can be rescued by immigrants from other subpopulations. These are extensions of Levins's reformulated logistic model (Levins 1969), but now the state variable is the fraction of patches that are occupied rather than the number of individuals in the total population (e.g., Gilpin and Hanski 1991; Hanski et al. 1996; Amarasekare 1998). The models do not usually describe the number of organisms in a subpopulation.

It has been difficult to demonstrate this type of metapopulation dynamics in natural populations driven by unstable consumer-resource interactions (Harrison and Taylor 1997), but a few clear examples have been found (Hanski and Gilpin 1997), and they may also characterize some biological control situations (chapter 9).

Approach 2. Patch models with explicit within-patch dynamics. The second approach also assumes the environment contains definable patches, each of which contains a distinct subpopulation that interacts with other subpopulations via the movement of individuals. In this case the metapopulation model follows the number of individuals in each subpopulation. Typically, only a small fraction of individuals leave or enter a subpopulation in each generation, so dynamics in a patch have a substantial local memory. Space, unlike local dynamics, is implicit rather than explicit. That is, a patch is not assigned a position in explicit space. Typically, the environment differs in some way among patches. Single-species examples include those in which the local density follows a fixed trajectory over time but is also subjected to random catastrophes (Hastings and Wolin 1989; Hastings 1991, 1995). We discuss consumer-resource examples below.

Approach 3. Spatially explicit individual-based models. The third approach assumes an effectively continuous environment, with no distinct patches, and follows the fate of individual organisms, including their movement, at each time step. Dynamics take on a markedly local aspect when individuals have restricted mobility.

Individual-based models offer the potential advantage of going directly from the behavior of individuals to population dynamics. Spatially explicit individual-based models in addition can examine the effect of interactions between individuals on an appropriate, local scale. They discard the "mass action" assumption that every individual can interact equally with every other individual in the population (or patch). In addition these are, by their nature, models with demographic stochasticity. For example, an individual either lives or dies during each time step, rather than having a death rate as does a group of individuals.

The models have the disadvantage, common to all complex models whose results are obtained via simulation, that it is often difficult to understand why they behave as they do. For this reason, they are most powerful when used in combination with simpler analytical models that can give insight into key processes; this is the approach followed below. For example, Keeling et al. (2002) suggest that the moment-closure approach can under some circumstances allow derivation of analytical models describing the stability properties of stochastic spatially explicit individual-based models.

In general, space in the models is homogeneous and is defined by a two-dimensional grid, or lattice, usually with no more than one individual occupying each point on the grid. Other spatially explicit models, which we do not discuss further, have a subpopulation at each point in the lattice. For example, Comins et al. (1992) simulate connected host and parasitoid subpopulations, each of which obeys a Nicholson-Bailey model (see also Hassell et al. 1991a; Hassell 2000). The results from the two approaches are broadly similar, at least in so far as they explore the effect of limited mobility. We also do not discuss here continuous-time, continuous-space reaction-diffusion models which describe the populations as density functions distributed in space (e.g., McLaughlin and Roughgarden 1991, 1992; Pascual 1993; Pascual and Caswell 1997).

Comment on Field Tests As mentioned above, the main notion in patch models with explicit within-patch dynamics and spatially explicit individual-based models is that an isolated subpopulation, or the individuals interacting in some small area, would have an unstable equilibrium but the dynamics of the collection of subpopulations can be stabilized, or at least well regulated (chapter 2), via movement among patches. This idea is difficult to test in the field. Walde (1994) has shown that movement among subpopulations is needed to maintain a predator-prey mite interaction in apple orchards. Extinction and reinvasion may even occur at the level of an individual tree. Such spatial processes, by contrast, do not contribute to the stability of the *Aphytis–* red scale interaction: temporal variability in population density did not increase when the populations in single trees were isolated (Murdoch et al. 1996b), and experimental outbreaks on this and an even smaller spatial scale were controlled by the local *Aphytis* population (W. Murdoch and S. Swarbrick, unpubl. data). In a similar experiment, Briggs and Latto (2000) showed in the *Baccharis* gall midge system (described in chapter 2) that parasitoid dispersal between bushes is important for maintaining parasitoid diversity, but not for midge population stability.

There is a vast literature on spatial processes, which we cannot cover here. Instead, in the rest of this first part of the chapter we bring together results that illustrate two routes by which spatial processes can stabilize otherwise unstable consumer-resource interactions, or at least reduce the fluctuations in density. The first revisits a mechanism examined in chapter 3: restricted movement of individuals linking subpopulations that have asynchronous trajectories. The second involves a combination of spatial variation and non-linearity. Recent sources for readers interested in spatial processes include Durrett and Levin (1994; 2000), Shigesada and Kawasaki (1997), Tilman et al. (1997), Dieckmann et al. (2000), and Okubo and Levin (2001).

Route 1 to Stability and Reduced Fluctuations: Asynchronous Trajectories in Linked Subpopulations

In chapter 3 we developed the following key insight regarding simple structured predator-prey models: differential predation in different patches in the environment, or on different stages of the prey, can stabilize the equilibrium via (indirect) density dependence in, respectively, per-head immigration or per-head recruitment. In this section we extend this insight to slightly more complicated spatial models.

EFFECT OF SPATIAL HETEROGENEITY IN MODELS OF LINKED SUBPOPULATIONS

This section reinforces the conclusion mentioned above. Spatial heterogeneity is necessary for stability in spatially distributed subpopulations linked by movement. Mere spatial structure can lead to reduced fluctuations in density, but it cannot induce true stability in the absence of spatial heterogeneity.

The earliest examples of linked subpopulations in a spatially heterogeneous environment extended the Lotka-Volterra model. They are similar to equation 3.22, except that they allow random movement in the predator as well as the prey. An early model (Murdoch and Oaten 1975; see also Crowley 1981) is

$$\frac{dH_1}{dt} = r_1 H_1 - a_1 H_1 P_1 - m_{h1} H_1 + m_{h2} H_2$$

$$\frac{dP_1}{dt} = c_1 H_1 P_1 - d_1 P_1 - m_{p1} P_1 + m_{p2} P_2$$

$$\frac{dH_2}{dt} = r_2 H_2 - a_2 H_2 P_2 - m_{h2} H_2 + m_{h1} H_1$$

$$\frac{dP_2}{dt} = c_2 H_2 P_2 - d_2 P_2 - m_{p2} P_2 + m_{p1} P_1$$

$$(10.1)$$

where the various per-head rates in patch i are r_i, prey births; a_i, predator attack; d_i, predator death; and m_{ji}, movement of species j out of patch i. We assume no deaths during movement between patches and that hosts and/or parasitoids that leave a patch immediately enter the other patch.

In this form, spatial heterogeneity can be introduced as any difference in parameter value between the two patches. Thus, for example, patch 1 might have higher prey birth rate, lower vulnerability, or a higher rate of predator emigration. A difference in any pair of parameter values can (but does not necessarily) lead to a stable equilibrium.

Murdoch et al. (1992a, Fig. 2) showed that random prey movement between the initially asynchronously fluctuating populations on the two patches creates density-dependent per-head immigration (because the number of immigrants is to some extent independent of the number in the patch; see chapter 3, "Simple Models of Stage and Spatial Structure"). These authors also extended the models to include predators that aggregated preferentially in areas of higher prey density. Weak aggregation was stabilizing, strong aggregation destabilizing. (This result is relevant to the discussion of the effects of parasitoid aggregation, in the section "Spatial Processes within Populations," below.)

It needs to be stressed that in some circumstances random spatial coupling can actually destabilize the equilibrium, and that the behavior of even this simple model is not altogether transparent. Nisbet et al. (1993) show that movement is never destabilizing if only the prey, or only the predator, migrates. But it can be destabilizing if both species move. A. Stewart-Oaten and colleagues (unpubl. results) showed instability is more likely if the predator moves preferentially to the patch with a lower prey rate of increase, but they were unable to find simple criteria for distinguishing conditions leading to instability.

Nisbet et al. (1993) looked at a "master-slave" two-patch system in which one of the patches has constant densities of prey and predators and its dynamics are not affected by immigration from the other patch, whereas the second patch is affected by immigration. We can think of this as representing a system with a large central refuge, or source area, and a single peripheral patch (Harrison 1991), with the large patch serving as the "environment" for the small patch. The conclusions from this model reinforce the importance of the relationship between immigration rate and density in the recipient patch: density-dependent per-head immigration arising from random movement is typically stabilizing.

As might be expected, instability can arise from a per-head immigration rate that increases with density in the patch, which happens when prey migrate to where prey density is higher. Predator aggregation to the area with higher prey density can be stabilizing, but again the outcome depends on the details.

Reeve (1988) showed that temporal environmental variation, leading to random differences between parameters in the different patches through time, can also reduce the amplitude of the fluctuations in Nicholson-Bailey subpopulations linked together by limited amounts of movement. These stochastic models were not analyzed for their local stability properties.

A central message from equation 10.1, reinforced by the other models, is that stability requires spatial heterogeneity. That is, at least one parameter must take a different value in the two subpopulations. Otherwise small-amplitude fluctuations in the two patches come into synchrony and we regain neutrally stable oscillations.

Results by Jansen (1995) reinforce the central role of spatial heterogeneity in metapopulation stability. He looked at equation 10.1 but set each parameter to the same value in each patch:

$$\frac{dH_1}{dt} = rH_1 - aH_1P_1 - m_hH_1 + m_hH_2$$

$$\frac{dP_1}{dt} = cH_1P_1 - dP_1 - m_pP_1 + m_pP_2$$

$$\frac{dH_2}{dt} = rH_2 - aH_2P_2 - m_hH_2 + m_hH_1$$ (10.2)

$$\frac{dP_2}{dt} = cH_2P_2 - dP_2 - m_pP_2 + m_pP_1.$$

Any initial differences in density between the patches die away through time because of the random movement between patches, and as expected we end up with a neutrally stable equilibrium, which is also exhibited by the non-spatial version (Jansen and de Roos 2000). However, Jansen (1995) established the striking result that the final cycle typically has a smaller amplitude than the original cycles.

Our explanation of Jansen's primary result is again a density-dependent per-head immigration rate (Fig. 10.1). The amplitude of a

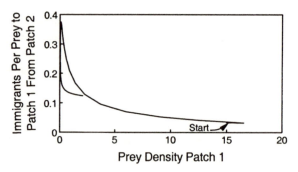

FIGURE 10.1. Density-dependent per-head prey immigration (number arriving per resident per unit time) to patch 1 from patch 2 over the first 15 time units of a simulation of equation 10.2. Parameter values are $r = 1, a = 0.1$, $m = 0.1, e = 1, d = 0.2$. y-axis is the number of prey immigrants to patch 1 per resident prey.

neutrally stable Lotka-Volterra cycle is determined by the initial conditions. But in these linked subpopulations, the "initial conditions" change continuously with immigration, until the two subpopulations come into synchrony. Because the changes take the form of density-dependent per-head immigration, the amplitude decreases. A simulation of the model shows how the initial asynchronous and asymmetrical fluctuations on the two patches induce density-dependent immigration to one of the patches (Fig. 10.1). *In spite of the decrease in cycle amplitude, however, the equilibrium does not become stable, as it can in equation 10.1, which has spatial heterogeneity.*

Jansen (1995) showed that reduction in cycle amplitude via spatial linkage is not restricted to the case of neutrally stable equilibria. In particular, he showed that a two-patch Rosenzweig-MacArthur model (equation 10.3, see below) with an unstable equilibrium exhibits exactly the same phenomenon. Although a range of dynamics may appear, the main result is that the amplitude of the fluctuations in density, averaged across the patches, is often less than in the non-spatial version of the model. Spatial separation, without spatial heterogeneity, can thus bound the fluctuations more narrowly.

Jansen and de Roos (2000) extend this result by looking at a collection of 90 linked patches, each patch having Rosenzweig-MacArthur dynamics (equation 10.3). They show that the amplitude of fluctuations in mean density across the patches is much less than in the non-spatial

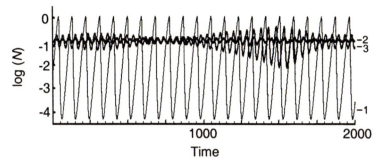

FIGURE 10.2. Prey densities over time from non-spatial and spatial versions of the Rosenzweig-MacArthur model. The large-amplitude cycles in curve 1 arise from the non-spatial version with no patches (equation 10.3) where there is an unstable equilibrium. Note that the y-axis is on a log scale, so these are large-amplitude cycles. Curves 2 and 3 were obtained from simulations in which the model retains the same parameters as for curve 1, but now there are 90 patches arranged in a linear chain. Prey are stationary, but predators move to adjacent patches at a specified rate. Curve 2 is prey density averaged over all patches, and given the log scale, the variation in density is many thousands of times less than in curve 1. Curve 3 is prey density in a single randomly chosen patch, showing that local as well as global fluctuations are decreased by a factor more than 1000. Redrawn with the permission of Cambridge University Press from Fig. 11.6b in Jansen and de Roos 2000.

model, and that the fluctuations in any particular patch are also much reduced (Fig. 10.2; notice the logarithmic y-axis).

The result in Fig. 10.2 is important and arises in another form in the next section. It tells us that mere spatial structure, and restricted movement, can impose narrow bounds on fluctuations in density. From a practical point of view, in a field population this may be indistinguishable from true stability.

ASYNCHRONOUS TRAJECTORIES IN SPATIALLY EXPLICIT INDIVIDUAL-BASED MODELS

We now look at restricted movement in a spatially explicit individual-based model (i.e., a lattice model) with a spatially homogeneous environment. We will see that restricted movement among areas with asynchronous population trajectories can again lead to reduced fluctuations in the total population. This has been called *statistical stability* (de Roos et al. 1991).

de Roos et al. (1991) explored a lattice model analog of the Rosenzweig-MacArthur model. The square lattice has 128×128 grid points. Prey growth is density dependent, since each grid point can contain only one individual. Excess prey offspring die if they cannot find an empty grid point within some specified distance from their parent (say the adjacent four points). Prey reproduction is stochastic: every prey individual has the same fixed probability of producing an offspring at each time step, but at each time step any particular prey individual either does or does not produce an offspring. Every predator that encounters a prey eats it with the same, fixed probability, and the predator reproduces with some probability that depends on its nutrient status. There is a handling time associated with consuming a prey: the predator cannot eat for some number of time steps after a meal. This gives an analog of a type 2 functional response in the standard Rosenzweig-MacArthur model. There is no density dependence in the predator since it never attains large enough populations.

de Roos et al. (1991) and McCauley et al. (1993) first establish a firm connection between a well-mixed version of the lattice model and the standard non-spatial Rosenzweig-MacArthur model (see chapter 3):

$$\frac{dH}{dt} = rH\left(1 - \frac{H}{K}\right) - \frac{aHP}{1 + aT_h H}$$

$$\frac{dP}{dt} = e\frac{aHP}{1 + aT_h H} - dP. \tag{10.3}$$

In one version of the lattice model only the predators move: prey offspring are placed on grid points adjacent to the adult, and otherwise prey are stationary. The authors show that when predators mix randomly across the entire lattice at each time step (i.e., unrestricted mobility), the dynamics of the lattice model correspond very closely to those of equation 10.3. In particular, the lattice model gives large-amplitude predator-prey oscillations when it has parameter values equivalent to those giving an unstable equilibrium in equation 10.3 (Fig. 10.3). The well-mixed spatial model is thus a good analog of the non-spatial model.

de Roos et al. (1991) then examine the effects of restricted mobility of both prey and predator. At each time step, predator individuals again move in a random direction, but movement is now restricted to the four adjacent grid points. Dynamics are now local. The consequences are

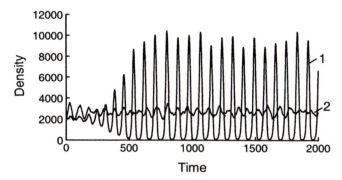

FIGURE 10.3. Dynamics of total prey density in the lattice model of de Roos et al. (1991; see their Fig. 2a). The model is a spatial version of equation 10.3. The large-amplitude cycles are obtained when predators have unrestricted movement across the grid (curve 1); parameter values correspond with those yielding an unstable equilibrium in equation 10.3. The almost constant densities (curve 2) are obtained with the same parameter values but restricted predator movement, and hence locally distinct dynamics. Data provided by A. de Roos.

dramatic. With the parameter values that gave large-amplitude cycles when mobility was unlimited (the cycles in Fig. 10.3), restricted mobility greatly compresses the fluctuations (curve 2 in Fig. 10.3) (de Roos et al. 1991).

The spatial dynamics for this case are illustrated by the simulation results in Fig. 10.4. Each plot in the figure is a snapshot of the grid at one point in time. Each dot is an individual prey (for clarity we omit the predators). At each grid point, an individual is either present or absent. On the left-hand panel predator individuals move a large distance (64 grid units) at each time step, so the predator population is well mixed. This gives Rosenzweig-MacArthur-type cycles: pictures taken at intervals show that population abundance on the grid increases and decreases roughly in synchrony throughout the grid. By contrast, the right-hand panel arises when movement is restricted to one grid point at each time step. Now the global population has about the same density at each time step (McCauley et al. 1993). Prey are distributed patchily, with sparse areas and dense areas, but the patches are ephemeral—the places where prey are abundant vary with time.

de Roos et al. (1991) call the restricted fluctuations of the global population statistical stability for reasons made clear in Fig. 10.5. In this

Time   Well-mixed   Restricted Mobility

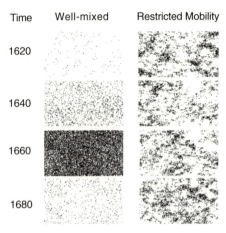

1620

1640

1660

1680

FIGURE 10.4. Snapshots of prey density in space at 4 time steps in the simulations shown in Fig. 10.3 (see Fig. 4 in McCauley et al. 1993). Each dot is a single prey individual at a grid point. In left-hand panels predators move 64 grid units at each times step, giving approximately synchronous changes in density across the grid and large-amplitude cycles in total abundance. In right-hand panels predators move 1 grid unit at each time step, synchrony is lost, and total density remains quite constant. Data provided by A. de Roos.

Well-mixed   Restricted Mobility

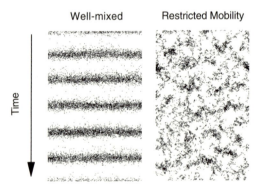

Time

FIGURE 10.5. Prey abundance along a single line in the lattice grid versus time. Data are from the same simulations as in Figs. 10.3 and 10.4. As in Fig. 10.4, each dot is a single prey individual at a grid point, but now each panel follows the individuals present on one line of the lattice. In left-hand panels predators move 64 grid units at each time step, giving approximately synchronous changes in density across the grid. In right-hand panels predators move 1 grid unit at each time step and synchrony is lost. Data provided by A. de Roos.

figure, each dot is a prey individual (we do not show the predators). Each panel follows the individuals present on one line of the lattice, over time. Each horizontal line is a time step in the model, and the picture contains almost 2000 time steps.

In the left-hand panel, both prey and predators again move 64 grid units at each time step. You can see cycles of high and low prey density that are almost synchronous over the entire line, corresponding with the cycles of Fig. 10.3. Thus, rapid movement gives long-range spatial correlation and synchronous cycles. In the right-hand panel, predators move only one grid point at each time step. This breaks up the spatial synchrony—changes in density are now not correlated over long distances, and the global population fluctuates very little.

de Roos et al. (1991) describe the outcome as statistical stability because it is the asynchrony of the population trajectories in different parts of the lattice that reduces the amplitude of fluctuations in total density: while density is increasing in some areas it is decreasing in others. A similar effect would occur if we averaged a series of sine waves started at different times, or a set of "white noise" runs.

de Roos et al. (1991) give insight into the role of local events by calculating a metric they term the *characteristic spatial scale*. This is the spatial scale below which the predator-prey dynamics resemble a well-mixed system. It is defined as the spatial scale at which the coefficient of variability of the system with limited mobility starts to differ from that of the well-mixed system. The characteristic scale effectively defines a local "patch," or subpopulation. Of course, unlike in the patch models discussed above, these patches are not fixed features of the landscape; they drift about in space, breaking up and coalescing on an intermediate time scale. The crucial point is that limited mobility imposes its own structure on a homogeneous environment, effectively creating drifting subpopulations.

COMPARISON BETWEEN PATCH MODELS AND SPATIALLY EXPLICIT INDIVIDUAL-BASED MODELS

We suspect there is a strong analogy between the results just presented and those of Jansen and de Roos (2000), discussed above. Both models are spatially homogeneous. In both, restricted movement alone produces a global population that fluctuates very little, even though unconnected local areas show large-amplitude fluctuations. There are, however, differences between the models, and more work is needed to determine their significance.

First, when Jansen and de Roos (2000) linked 90 patches, each with unstable dynamics when isolated, the amplitude of the fluctuations was greatly reduced *in each patch* (Fig. 10.2, curve 3 vs. curve 1) as well as in the global population. That is, the Jansen and de Roos result is not mere averaging of asynchronous fluctuations. In the de Roos et al. (1991) lattice model, however, we cannot make analogous measures of fluctuations in a small area, because there are no defined and fixed patches. The model is like a population of plankton affected only by diffusion. Although there may be patches with some temporal integrity, they drift on the lattice, albeit sometimes slowly.

The lattice model results do reinforce the conclusion we reached above, that while weakly linked subpopulations in homogeneous space can greatly reduce the amplitude of fluctuations, they cannot induce true stability. That still appears to require spatial heterogeneity, a subject not yet investigated in spatially explicit individual-based models.

Second, the lattice models contain demographic stochasticity. This could actually enhance fluctuations in each local collection of individuals. It may also be a hidden source of spatial heterogeneity. It is also clear that there are patches where prey are actually driven extinct, and Wilson (1998) argues that such local extinction, and reinvasion, of prey in small areas of the grid is a key to statistical stability.

Gurney et al. (1998) also studied a discrete-space version of the MacArthur-Rosenzweig model, but in this case there is a density of prey and/or predators in each quadrat, rather than just zero or 1 individual. As in earlier models with this feature (Hassell et al. 1991a; Rohani et al. 1996), spatial patterns such as spirals appear and are associated with less variable global temporal dynamics. When Gurney et al. (1998) placed realistic restrictions on local predator recovery from very low density with parameter values that give unstable equilibria in the nonspatial model, the predator population went globally extinct following overexploitation of the prey. The predator and prey could persist, however, provided there was some recolonization by predators in quadrats at very low predator density. These authors also showed that the large-scale spiral patterns persisted in the face of some spatial variation in prey carrying capacity.

Route 2: Spatial Processes Induce Stability by Altering Spatially Averaged Properties

In this section we present one example in detail, and another briefly, that illustrate how spatial variation in prey density, combined with a non-linear process, can stabilize a spatial Rosenzweig-MacArthur model. This route to stability is interesting because it links the two apparently disparate situations analyzed in this chapter: spatial processes that occur among subpopulations and those that occur within a single population. We then discuss an experiment that illustrates Route 2.

This example, again in a spatially homogeneous environment, is from the research group whose work we just discussed (McCauley et al. 1996b; de Roos et al. 1998). In this model, which is like the previous model but has 256 × 256 grid points, prey always have low mobility: they move only 1 grid unit at each time step and prey offspring are placed at a random site next to the parent. The model distinguishes juvenile (neonate) and adult predators. The authors first show that provided a newborn predator also settles close to the parent (i.e., 1 grid unit away), increasing adult predator mobility alone leads to larger and larger cycles in density that are increasingly synchronous across the lattice. The left-hand panel of Fig. 10.6 gives an example of a roughly spatially synchronous cycle arising when predators are moderately mobile (32 grid units per time step). Here the black dots are prey and the gray dots are predators on a one-dimensional line of the lattice through time. Close inspection shows predator peaks (intense gray bands) lagging prey peaks (intense black bands). In Fig. 10.6, an example of peak prey density is indicated by the numeral 1 and a peak of predator density by the numeral 2.

The interesting case is where the adult predators are mobile, as above, but now, at each time step, the neonate predators are dispersed at random across the grid instead of adjacent to the adult predator. The right-hand panel of Fig. 10.6 shows that the cycles disappear. Prey become very patchily distributed: most local areas have no or very few prey, whereas a few regions have prey in almost every cell. Furthermore, areas of high and low prey abundance are rather fixed through time, so local as well as global prey density changes little over long time periods. Notice that the predators in the right-hand panel of Fig. 10.6

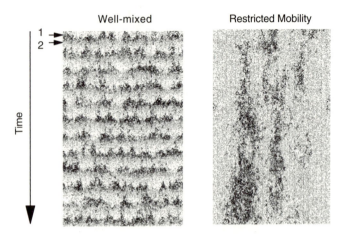

FIGURE 10.6. Densities of prey (black dots) and predators (gray dots) on a single line of a lattice over time. The model is a spatially explicit analog of equation 10.3 (McCauley et al. 1996b; de Roos et al. 1998). Left-hand panel: predators move 32 grid units, prey 1 grid unit, at each time step. Newborn prey and predators are placed adjacent to parent. Roughly spatially synchronous cycles in abundance are seen. Numeral 1 indicates a band of high-density prey, which is followed by a band of high-density predators, indicated by numeral 2. Right-hand panel: neonate predators are distributed randomly across the gird. Global and local abundance is quite constant: predators are evenly distributed across the grid, and areas of high and low prey abundance are rather fixed through time. Data provided by A. de Roos.

are evenly spread across the lattice as a result of thorough mixing of neonates.

We conjecture that the mechanism producing the extreme uneven-ness in prey distribution is as follows. Because of limited mobility and stochastic events, prey density soon varies over space as in earlier simulations. This variation is then greatly enhanced by the predator's type 2 functional response, as follows. Since predators are evenly dis-tributed, the fraction of prey killed in any small area is determined by the non-linear functional response, and the response induces inversely density-dependent mortality on each local patch of prey: the per-head attack rate is low in patches with many prey and high in patches with few prey (chapter 3). As a result, prey at locally low density are driven toward extinction, whereas locally dense prey escape from predator control and increase toward the carrying capacity.

The stabilizing effect can be best understood by comparing the lattice model and the stability criterion of its analytic analog (Box 10.1) (de Roos et al. 1998). The crucial effect of extreme spatial differences in prey density plus the type 2 functional response is to reduce the effective rate at which the predator population converts prey to new predators. This reduction in predator efficiency is stabilizing (chapter 3).

The reduction in predator efficiency arises in the lattice model as follows. Predators are evenly distributed—they are at the same abundance in areas with no or few prey as in areas with many prey. As a consequence, because predator intake rate saturates with local prey density, the average per-predator attack rate is lower than it would be if prey were everywhere at the mean prey density. This is an example of Jensen's Inequality: the mean of a concave down function is less than the value of the function at the mean, as illustrated in Fig. 10.7b.

de Roos et al. (1998) show nicely how the simulation results connect with existing analytic consumer-resource theory. They modify the original Rosenzweig-MacArthur model so that the populations are distributed among a set of N patches, with standard dynamics in each patch, i. Now, however, the prey do not disperse, or they move among patches only at a low rate, whereas the predators are evenly distributed across patches (Box 10.1).

An example of what happens is shown in Fig. 10.7a. In this case there are five patches, and the parameter values are such that if there were only one large patch, the system would be unstable. With five patches the equilibrium is stabilized: prey in three of the patches reach the same positive equilibrium abundance, whereas prey in two patches are driven extinct.

The reason for stability is that the predators spend equal time in patches with and without prey (see predator distribution in the right-hand panel of Fig. 10.6). This reduces their effective rate of converting prey to new predators; i.e., their realized conversion rate is lower than the nominal parameter value. The reduction is directly proportional to the fraction of empty patches (Box 10.1).

The stabilizing mechanism in this example illustrates a much broader class of population processes, one stressed in a different context by Chesson (1990). The result hinges on a combination of non-linearity and, in this case, spatial variability. The non-linearity is in the type 2 functional response and operates in this case at two places.

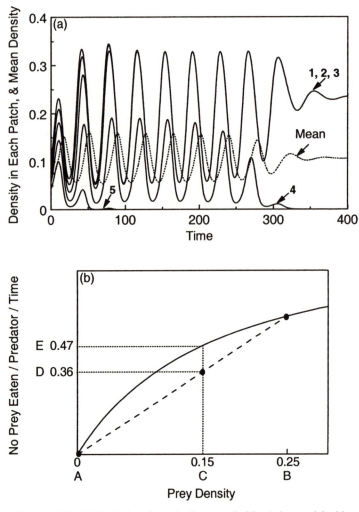

FIGURE 10.7. (a) Simulations from the Rosenzweig-MacArthur model with five patches (equation B10.1) (de Roos et al. 1998). Each solid curve is the density of prey in a patch. Dotted curve is global prey abundance. Densities in the three initially high-density patches (patches 1, 2, and 3) converge to the same constant density (about 0.2); prey go extinct in the two initially low-density patches (patches 4 and 5). Parameter values are $r = 0.5$, $K = 0.5$, $a = 6$, $T_H = 1$, $m_f = 0$ (no prey movement), $e = 0.5$, $d = 0.175$. Data provided by A. de Roos. de Roos et al. (1998) show that the conclusions hold when m_f is greater than zero but small. (b) Illustration of Jensen's Inequality applied to the de Roos et al. functional response (solid line) with parameter

First, the type 2 response increases the spatial variance by accentuating the initial differences in prey density. This can be seen by tracking the fate of the five patches in Fig. 10.7a. The two patches in which prey go extinct are those with fewest prey initially (patches 4 and 5).

Second, this spatial variance stabilizes the equilibrium by reducing predator efficiency, as follows. In the 60% of patches that are occupied, prey density is more than twice as high as in the non-spatial case; the final equilibrium prey density in the patches with prey is about 0.25, whereas in the non-spatial model the (unstable) equilibrium prey density with the same parameters is approximately 0.09. But the type 2 response causes the number of prey eaten per predator per unit time in these patches to be much less than twice as high as in the non-spatial case. The spatially averaged rate of converting prey, per predator per unit time, is therefore lower in the spatial case (Fig. 10.7b). As indicated by the stability criterion (equation B10.3), this stabilizes the interaction. This mechanism thus illustrates the suppression-stability trade-off mentioned throughout this book.

The connection drawn by de Roos et al. (1998) between individual-based simulations and analytical models, discussed above, has recently been explored more thoroughly using individual-based models and partial differential equations (PDEs) in which the organisms are distributed in only one dimension. These models reinforce the basic message of de Roos et al. (1998). Gurney and Veitch (2000) used a one-dimensional PDE model in which predators diffused homogeneously but prey were immobile. At equilibrium the prey occurred in high-density segments separated by empty segments. Predator density, by contrast, varied little in space because of homogeneous predator movement. This pattern was stabilizing and could be maintained against a low rate of prey invasion of empty patches, provided predator density was high enough. In addition, Gurney and Veitch show that spatial variation in prey density induced

---

values as in (a). In the de Roos example, 40% of predators are in patches with prey density equal to zero (point A), 60% in patches with prey density equal to 0.25 (point B). The weighted mean of the prey density experienced by predators is thus 0.15 (point C). Average predator intake per unit time under these conditions is 0.36 (point D). If instead all predators experienced the weighted mean prey density (point C), intake rate per predator per unit time would be the value of the functional response at C, namely 0.47 (point E).

BOX 10.1.

SPATIAL ROSENZWEIG-MACARTHUR MODEL IN WHICH PREY MOVE
SLOWLY FROM PATCH TO PATCH AND PREDATORS ARE EVENLY
DISTRIBUTED ACROSS PATCHES

The spatial Rosenzweig-MacArthur model (based on de Roos et al. 1998), with the type 2 functional response in the form of the Holling disc equation (Holling 1959), is

$$\frac{dH_i}{dt} = rH_i\left[1 - \frac{H_i}{K}\right] - \frac{aH_iP}{1 + aT_hH_i} - m_f\left[H_i - \frac{1}{N}\sum_{j=1}^{N}H_j\right]$$

$$\text{for } i = 1 \text{ to } N \qquad\qquad\qquad\qquad (\text{B}10.1)$$

$$\frac{dP}{dt} = \frac{1}{N}\sum_{j=1}^{N}e\frac{aH_jP}{1 + aT_hH_j} - dP$$

where r is per-capita prey growth rate, a is the attack rate per unit search time, T_h is handling time per prey item, m_f is the fraction of prey that leave the patch per unit time, e is the conversion efficiency of the predator, and d is its death rate. H_i is the prey density in patch i, but the predators have the same density, P, in all N patches.

de Roos et al. (1998) analyze a case with five patches. They choose parameter values for the non-spatial model (i.e., there is only 1 patch) that give an unstable equilibrium. They then show that, with these parameter values, a stable equilibrium can occur in the spatial version (equation B10.1) if prey movement is zero or low (small m_f). The stable equilibrium occurs because each patch reaches one of two states: zero prey or a single constant density of prey. Fig. 10.7a shows an example in which the number of patches without prey, N_0, equals 2.

The equilibrium is stable if

$$\left[1 - \frac{N_0}{N}\right]\frac{aT_hK - 1}{aT_hK + 1} < \frac{d}{e} \qquad\qquad (\text{B}10.2)$$

and global predator density, P is greater than r/a. The second condition ensures that predator density is high enough to keep patches with no prey empty.

(Box 10.1. continued)

We can rewrite the stability criterion for the non-spatial model (see Table 3.2) as

$$\frac{aT_h K - 1}{aT_h K + 1} < \frac{d}{e}. \tag{B10.3}$$

So the criterion for the spatial model is simply that for the non-spatial model, modified by the fraction of patches occupied by prey, namely $(1 - N_0/N)$. Thus the smaller the fraction of occupied patches, the more easily is the criterion met. In the example in Fig. 10.7a, this fraction is 0.6.

From equations B10.2 and B10.3 we can see that space increases stability by effectively decreasing the predator's conversion efficiency, e; i.e., efficiency is reduced by the factor $(1 - N_0/N)$ (this can be seen by dividing both sides of B10.2 by $(1 - N_0/N)$). This effect in turn arises because the predators, being evenly spread over the five patches regardless of local prey density, spend just as much time in empty patches as in patches with prey.

by environmental variation is likely to be reinforced by the processes in the model. Hosseini (2001) further showed that the PDE and a one-dimensional individual-based model with analogous assumptions gave virtually identical distributions of prey in segments separated by empty segments.

Of course, in real life, predators may be more likely to be where their prey is, and this route to stability will break down the more effective the predators are at concentrating near prey. This conclusion is echoed later in this chapter when we look at predator aggregation dependent on local prey density within a single prey population. Nevertheless, predators are never perfect at distributing their time in space, and this mechanism may contribute to stability in real systems.

Nisbet et al. (1998) show another way in which spatial patchiness can develop in a homogeneous environment and can, in connection with non-linearity, induce stability. The model and its implications are summarized in Box 10.2. Briefly, space is subdivided into tiny patches,

BOX 10.2.

CONSUMER-RESOURCE MODEL WITH SMALL-SCALE RESOURCE
HETEROGENEITY

Nisbet et al. (1998) developed a consumer-resource model in which individual consumers subsist on a continuum of resource distributed over a large number of small "bite-sized" patches. The patches are so small that all resource in a patch is consumed immediately whenever a consumer visits. This leads to a heterogeneous distribution of resource among the patches. Consumers are assumed to move randomly among patches, but there is no explicit representation of space.

Mathematically, the model is expressed using formalism traditionally used to describe age- and size-structured populations (e.g., Gurney and Nisbet 1998, chapter 8). The "age" of a patch is the time that has elapsed since the most recent visit by a consumer. Resource density on a patch is the analog of "size." An attack on a patch by a consumer involves the "death" of that patch and the "birth" of a new patch aged zero. At each visit by a consumer, the density of resource on the patch drops immediately to zero and the patch's age and size are reset to zero.

If τ represents the age of a patch; $n(\tau, t)d\tau$ is the fraction of patches aged between τ and $\tau + d\tau$ at time t; and $h(\tau, t)$ is the resource density on a patch aged τ at time t, then the average resource density is

$$H(t) = \int_0^\infty h(\tau, t)n(\tau, t)d\tau.$$

Consumers have a type 2 functional response (see chapter 3), so that the per-patch death rate $\mu(t)$ is

$$\mu(t) = \frac{aP(t)}{1 + aT_h H(t)},$$

where $P(t)$ is the (global) consumer density.

The effect of the births and deaths on the patch age distribution can then be described by a partial differential equation and an associated boundary condition:

$$\frac{\partial n}{\partial t} = -\frac{\partial n}{\partial \tau} - \mu(t)n \quad \text{with } n(0, t) = \mu(t) \text{ for all values of } t.$$

(Box 10.2. continued)

A second partial differential equation together with its boundary condition describes the changes in the resource density over time. If $g(h, t)$ is the resource growth rate on a patch aged τ at time τ, then

$$\frac{\partial h}{\partial t} = -\frac{\partial h}{\partial \tau} + g(h, \tau) \quad \text{with } h(0, t) = 0 \text{ for all values of } t.$$

Finally, it is assumed that there is some random mixing of resource among patches (rate ε), and that resource growth in a patch is logistic, i.e.,

$$g(h, \tau) = rh\left(1 - \frac{h}{K}\right) - \varepsilon h + \varepsilon H.$$

Because there is no explicit representation of space, or of consumer (st)age structure, the consumer dynamics are given by the same ordinary differential equation as in the Rosenzweig-MacArthur model (equation 3.15). Consequently the resource equilibrium does not increase with enrichment, a prediction that distinguishes this model from models (such as that in Box 10.1) that invoke changes in consumer efficiency and/or indirect consumer density dependence as a stabilizing mechanism. Here, local stability analysis reveals that, as discussed in the text, the resource heterogeneity has a stabilizing effect by reducing the rate of primary production and changing the relationship between resource growth and resource density.

the food in each representing a single "bite" for a grazing consumer. The authors had in mind systems for which the spatial scales for prey and predator are very different—e.g., benthic microalga and a grazing aquatic insect such as a mayfly.

The stabilizing mechanism in this model again involves the fact that local prey density varies greatly among patches, and in any single patch is typically either much above or much below average prey density, H. Since the graph of total rate of prey growth (dH/dt) versus H is humped, Jensen's Inequality applies once again: total rate of prey growth is lower than it would be if all patches were held at the average prey density. The density dependence of total prey growth also changes as a result of the spatial heterogeneity. Consequently, stability is more likely, and Nisbet et al. showed that all else being equal, the transition from stability to instability occurred at higher values of the prey carrying capacity than in the Rosenzweig-MacArthur model. Thus, once again, spatial variation in prey density combines with a non-linear process (here prey rate of increase as a function of prey density) to induce stability. The combination of variation and non-linearity in competitive systems is of course the basis for many results in stochastic models of competition by Chesson (see chapter 8).

These two sets of models show how spatial processes can stabilize the interaction by altering some spatially averaged property of the system. S. Ellner, E. McCauley, and co-workers (McCauley et al. 2000; Ellner et al. 2001) analyze an experiment that illustrates one mechanism. Janssen et al. (1997) compared the dynamics of prey and predatory mite populations on 80 small bean plants. The experiment was begun by releasing prey mites on focal plants, then releasing predatory mites 7 days later on the same plants. In one treatment all 80 plants were placed together on a uniform "super-island." In the structured treatment there were 8 islands, each with 10 plants, the islands being separated by a partial barrier to dispersal. Plants were replaced when they were 7 weeks old, or when they had been destroyed by the prey mite, whichever came first.

The qualitative dynamics on any given plant were independent of treatment (McCauley et al. 2000). The plant was discovered by prey. Then, either the prey increased, destroyed the plant, and the population collapsed (with individuals dispersing in the meantime to other plants), or the prey increased, predators found the plant, and drove the prey to extinction. The predator population then collapsed, but predators in the meantime had dispersed. Thus, local extinction on individual plants was the rule.

On the super-island, the prey population spread rapidly and reached a peak density. The predators also spread across the plants, destroyed all the prey, and the system went extinct in 160 days. In the structured systems, however, although populations on individual plants and islands went extinct, the entire system persisted until the experiment was stopped on day 450. Ellner et al. (2001) show that persistence arose because spatial structure reduced predator dispersal rate and hence predator efficiency (Route 2). Persistence was not attributable to asynchronous population trajectories on different islands (Route 1).

Connections with Simpler Theory

The connections forged above between simple and more complex theory illustrate beautifully the role of simpler models in understanding those that are more complex. The connection is most clear-cut with regard to Route 2. de Roos et al. (1998) and Nisbet et al. (1997a; 1998) show how spatially explicit individual-based models can be effectively reduced to implicit-space analytical models in which stability properties can be understood in terms of changes in mean properties.

The connection may be more problematic with respect to the "statistical stability" seen in lattice models when both prey and predators have restricted movement. The Jansen and de Roos (2000) model, with many linked subpopulations in homogeneous space, may be a good analog, in which case asynchronous trajectories are the key. But Wilson (1998) suggests that any analytic analogs of these individual-based models need themselves to be explicitly spatial. He shows that simple deterministic reaction-diffusion models cannot yield these dynamical phenomena and suggests that stochastic, discretized, reaction-diffusion models are needed to capture the dynamics.

Pascual et al. (2001), using a different approach, also show how some of the essential features of complex spatially explicit simulation models can be captured in a simple analytical model.

Summary of Among-Subpopulation Results

Two broad insights emerge from this section. First, an almost universal outcome of introducing spatial processes at the among-subpopulation scale is to reduce the amplitude of global fluctuations in predator-prey interactions. It seems likely that even in the absence of spatial heterogeneity, local as well as global fluctuations in density will be reduced. This result emerges from both models of connected subpopulations and spatially explicit individual-based models.

There are, however, circumstances in which spatial processes reinforce the oscillatory tendencies of consumer-resource interactions. So far, such destabilizing effects apparently occur only when both prey and predators migrate, but there is as yet no general rule for when these effects emerge. The result has been seen in models of connected subpopulations, but not as far as we know in spatially explicit individual-based models. Further work is needed here.

Second, spatial heterogeneity increases the Route 1 stabilizing effect of spatial structure (asynchronous trajectories), and is likely to create true stability rather than just statistical stability. All real environments are spatially heterogeneous, so this is an important result. However, almost all spatially explicit individual-based models to date have examined spatially homogeneous models, and we believe there will be much to be gained by introducing into such models, and exploring the effects of, spatial heterogeneity at different spatial scales.

SPATIAL PROCESSES WITHIN POPULATIONS: AGGREGATED ATTACKS AND OTHER SOURCES OF VARIATION IN RISK AMONG INDIVIDUALS

Entomological readers in particular will be familiar with the suggestion that aggregation by predators stabilizes predator-prey interactions (May 1978; Hassell and Pacala 1990). The familiarity of this idea is reflected in dozens of field studies that have tried to determine whether parasitism or predation rate is density dependent in space, and whether "CV^2 is greater than 1" (i.e., whether attacks on hosts or prey are highly clumped) (Lessels 1985; Stiling 1987; Walde and Murdoch 1988; Hassell and Pacala 1990).

Two types of predator aggregation have been portrayed in models: prey- or host-density-dependent aggregation (HDD), and prey- or host-density-independent aggregation (HDI) (Pacala et al. 1990). In the first, predators spend more of their search time in patches or areas containing higher densities of prey. This implies that a prey's risk of being attacked increases with local prey density if predator aggregation is sufficiently strong. In the second, the predator's spatially aggregated distribution is not correlated with local prey density. In this case a small fraction of prey are at high risk, but their risk is unrelated to how many other prey are nearby. In fact, we will see that such aggregated risk can arise also from non-spatial processes, and may be more appropriately termed skewed risk rather than predator aggregation.

Below, we first deal with density-dependent aggregation. We then turn to density-independent aggregation, but place it in the broader context of the distribution of relative risk among prey, which might be influenced by non-spatial factors. In the process, we explore theory that has been published to date and end by presenting some new theory that resolves outstanding questions. Hassell (2000, chapter 4) also deals with these issues in some detail and should be consulted for a somewhat different interpretation of some salient points.

It will help to bear in mind a few main points whose origin and significance will unfold below.

1. When skewed risk affects population dynamics, it does so by altering the efficiency of the average predator. In particular, it is potentially stabilizing if mean efficiency (a) decreases as mean predator density increases or (b) increases as mean prey density increases.

2. The dynamically important factor is variation in the relative probability of attack (i.e., relative risk) among prey. (As an aside, when predators aggregate to patches with more prey, the prey in dense patches will suffer higher risk only if predators spend more time searching there; risk will not be higher if predators are more common just because they spend more total time handling prey in such patches.) When our concern is aggregation independent of local prey density (HDI), the key variable is the distribution of relative risk among prey regardless of their spatial distribution.

3. With respect to skewed risk (HDI), we will discover that a key property is whether the relative risk run by an individual prey persists over its lifetime or is transient.

Predators Aggregate to Areas with Higher Prey Density (HDD)

We have known at least since Banks (1957) studied ladybugs that predators sometimes spend more of their time searching where prey are dense than where they are sparse. Mechanisms producing such patterns can include increased turning or residence time in response to higher rate of encounter with prey (Murdie and Hassell 1973), but also distant detection of local concentrations of prey or some cue associated with them. Provided the predators spend more search time in patches that have more prey, they should produce spatial density dependence in the prey death rate.

We first set aside one possible source of confusion. It is tempting, but incorrect, to assume that spatial density dependence in mortality (from any source) will generally lead to temporal density dependence in mortality (i.e., per-head mortality increases with total population density in different years, or generations, or times): spatially density-dependent mortality, per se, typically does not lead to temporal density-dependent mortality (Stewart-Oaten and Murdoch 1990).

HDD AGGREGATION IN LOTKA-VOLTERRA MODELS

The models in this section omit life history details, so it does not matter whether we think of the protagonists as parasitoids and hosts (the terms we will use) or predators and prey. We begin with continuous-time models because there the effects of aggregation are more transparent.

The first evidence that density-dependent aggregation might not be a source of stability was a model with explicit parasitoid behavior. Murdoch and Oaten (1975) simulated a parasitoid that stayed longer in more rewarding patches (the "giving-up time" increased with absolute reward rate). The parasitoid reduced the local host density as it fed, and was assumed to respond instantly to the ever-changing local reward rate. Murdoch and Oaten found such aggregation was not stabilizing unless there was a positive transit time between patches.

These authors did find that a transit time between patches was stabilizing, and the reason is instructive. At higher mean host density, the parasitoid encountered more high-density patches, stayed longer in the average patch, and therefore moved among patches less. As mean host density increased, the parasitoid therefore spent a decreasing fraction of its total time in transit among patches, and more time actually searching in patches. Parasitoid efficiency thus increased with prey density, which led to temporally density-dependent (hence stabilizing) mortality in the host population. We ignore transit time for the remainder of this discussion but note that it is a potential source of stability that is typically overlooked. We also ignore handling time and egg depletion, i.e., we assume a type 1 functional response (chapter 3).

Murdoch and Stewart-Oaten (1989) developed an analytical model of spatially density-dependent aggregation in the continuous-time Lotka-Volterra framework. The parasitoid was assumed to be perfect at aggregating continuously to local host density as host distribution changed in response to both reproduction and parasitism.

The authors began with the neutrally stable Lotka-Volterra model,

$$\frac{dh}{dt} = rh - ahp$$

$$\frac{dp}{dt} = chp - dp$$

(10.4)

where now h and p are the average numbers of hosts and parasitoids, respectively, per patch and $c = ea$ (parasitoid conversion efficiency times attack rate). We use H and P to signify the numbers of hosts and parasitoids in a randomly chosen patch, so $h = E(H)$ and $p = E(P)$, where E(.) means expectation. The number of hosts parasitized per unit time in a single patch is still aHP, and the average number parasitized per unit time is $aE(HP)$. In this basic spatial model with no aggregation, the number parasitized per unit time in the average patch is simply $aE(H, P) = ahp$, the number in the non-spatial Lotka-Volterra model.

Aggregation to local host density means that P increases with H. With aggregation, H and P are therefore positively correlated and the covariance of H and P, cov(H, P), will be positive. The number parasitized per unit time, averaged over all patches, is thus now

$$aE(HP) = a[hp + \text{cov}(H, P)].$$

(10.5)

The average number parasitized per parasitoid, i.e., the functional response, is the right-hand side of equation 10.5 divided by p. Thus we see immediately that aggregation to local host density increases parasitoid efficiency: each parasitoid now kills more hosts than in the Lotka-Volterra model (equation 10.4). The stronger aggregation is (the greater is cov(H, P)) the more efficient is the parasitoid and the lower will h^* be (see Table 10.1).

Now consider the consequences for stability. Since only the functional response changes in the model, the equilibrium will be stable if the new functional response accelerates as h (mean prey density) increases, and unstable if it decelerates as h increases. Specifically, the functional response is stabilizing if $f'(h^*)$ is greater than $f(h^*)/h^*$ (Box 3.2).

In the absence of aggregation, the functional response averaged over patches is simply ah. From equation 10.5 we see that the functional

TABLE 10.1. How parasitoid aggregation to local host density increases
host's death rate and parasitoid's functional response

Host and parasitoid distribution	H_1	H_2	P_1	P_2	$\Sigma H_i P_i$	$acov(H, P)$
Both even	5	5	5	5	50	0
Patchy hosts, no aggregation	1	9	5	5	50	0
Patchy hosts, aggregated parasitoids	1	9	1	9	82	32

Ten hosts and 10 parasitoids are distributed between 2 patches. The distribution of each
population is either even (5 in each patch) or aggregated (1 in patch 1 and 9 in patch 2).
We assume the attack rate, $a = 1$. Number of prey killed is $H_i P_i + acov(H, P)$.

response with aggregation will have an additional term, $acov(H, P) = a_p Q(h)$, so

$$f(h) = ah + aQ(h), \qquad (10.6)$$

where $Q(h)$ depends on the form of aggregation, i.e., on $\text{cov}(H, P)$.
Since we are adding a component to the linear part, ah, whether the
functional response is stabilizing or destabilizing now depends on the
shape of $Q(h)$: it is stabilizing if $Q'(h^*)$ is greater than $Q(h^*)/h^*$ (see
Box 3.2).

For illustration we will assume that P increases linearly with H as
in Table 10.1 (Murdoch and Stewart-Oaten 1989 explore other cases).
Then the proportion of the parasitoid population in a patch, $G(H)$,
increases linearly with the proportion of the host population in the patch,
H/h, so

$$G(H) = \alpha + \beta(H/h), \qquad (10.7)$$

and β measures the strength of aggregation. Murdoch and Oaten show
that the functional response is now

$$f(h) = ah + a\beta h V(H|h)/h, \qquad (10.8)$$

where $V(H|h)$ is the variance of H among patches—the spatial variance
of the hosts.

Hundreds of studies show that spatial variance increases with mean
density according to Taylor's power law:

$$V(H) = Ah^x. \qquad (10.9)$$

It is also known that x lies between 1 and 2 for most populations (Murdoch and Stewart-Oaten 1989), so the second term in equation 10.8 is a decelerating function, meaning that the functional response is type 2 and destabilizing.

This example illustrates two results that are general (Murdoch and Stewart-Oaten 1989). First, when the parasitoids continually respond to the changing host distribution, aggregation to local host density is typically destabilizing; only in unusual circumstances is it stabilizing. Second, the reason is that it is difficult for already-efficient parasitoids to become even more efficient as mean host density increases; i.e., it is difficult for them to induce temporally density-dependent mortality on the host population.

It is difficult for a parasitoid population to increase efficiency as mean host density increases for the following reason. For a given strength of aggregation, increases in efficiency come from increases in host spatial variance (greater "clumping"). But host variance would need to increase faster than h^2 (equation 10.9), and that rarely occurs in real systems. Instead, aggregation tends to cause a decrease in per-head host death rate as host density increases, because overall efficiency decreases as mean host density increases.

Finally, in this model the parasitoids interact, in true Lotka-Volterra fashion, only by reducing the instantaneous number of hosts available to other parasitoids. They do not interfere with each other. Parasitoid efficiency is thus uncorrelated with parasitoid density.

Now recall our point 3, above (under "Spatial Processes within Populations"), namely that stability depends on whether the host's relative risk is persistent. The relative risk run by an individual host in the above model is transient: it changes throughout its life, and this is associated with instability. A reasonable interpretation of what happens to individual hosts in the Murdoch and Stewart-Oaten model is as follows. Recruited hosts are distributed among patches. Assume a host stays in its birth patch. The parasitoids search mainly in patches with above-average numbers of hosts, reduce density there, and continuously redistribute themselves so they are always mainly in more abundant patches. Thus the parasitoid population is continuously moving among patches. A host in an initially dense patch suffers a high relative risk initially. But the relative risk of survivors in that patch declines as host density in the patch declines, and the parasitoids concentrate their attacks elsewhere.

A host in an initially low-density patch suffers an increasing relative risk as such a patch becomes one of the more dense. If the hosts also move among patches, then relative risk will likely be even more changeable throughout the lifetime. The Murdoch and Stewart-Oaten (1989) results thus reinforce those found by the detailed simulation model of Murdoch and Oaten (1975).

Godfray and Pacala (1992) suggested that an alternative continuous-time model gave contrary results to those summarized above (see also Bernstein 2000). However, Murdoch et al. (1992a) and Nisbet et al. (1993) showed that the Godfray-Pacala model was a version of equation 10.1 and actually described aggregation across subpopulations. That is, it is the Lotka-Volterra metapopulation model. They showed that within this context, weak aggregation tends to be stabilizing in such situations and strong aggregation tends to be destabilizing.

HDD AGGREGATION IN NICHOLSON-BAILEY MODELS

Now consider the Hassell and May (1973) model, in which spatially density-dependent aggregation is stabilizing in some circumstances. The basic model is the Nicholson-Bailey model, equation 4.12, in which the fraction surviving parasitism is

$$f = e^{-aP_t},$$

where a is the per-head attack rate of the parasitoid. Hassell and May (1973) modified this model by letting the hosts be distributed among n patches, with the proportion of hosts in the ith patch, α_i, being fixed and independent of mean host density. (This translates into the assumption that $x = 2$ in equation 10.9, above; Murdoch and Stewart-Oaten 1989). As above, the parasitoids respond to the relative local host density so that the proportion of parasitoids in patch i, β_i, increases with α_i. Now the fraction surviving in the population is obtained by summing survival over the n patches:

$$f = \sum_{i=1}^{n} \alpha_i e^{-a\beta_i P_t}. \tag{10.10}$$

Hassell and May assumed that predator aggregation was described by $\beta_i = (\alpha_i)^\mu$. The model thus assumes that host and parasitoid populations reassemble in the same initial distributions each generation. To facilitate analysis, Hassell and May assumed that all patches but one have the same low fraction of the hosts and that the remaining patch has all the other hosts. The model does not give very general results, but stability is most likely if the hosts are very patchy and μ is greater than 1, i.e., aggregation is greater than linear.

The details of the model are less important than is the general reason why it sometimes yields stability. The model is stable because the efficiency of the average parasitoid decreases with increasing parasitoid density (Fig. 10.8; see also Hassell 2000, Fig. 4.8). Parasitoid per-head attack rate decreases with parasitoid density—it is density dependent and hence stabilizing.

The key to this result is that the parasitoid cannot respond to changes in the host distribution within the generation, because the model does not keep track of these changes (Murdoch and Stewart-Oaten 1989; also see Box 4.1). A reasonable interpretation of the model is that the parasitoids aggregate in response to the distribution of hosts among patches at the *start* of the generation. But as the parasitoids kill hosts, and so alter that initial distribution, parasitoids do not rearrange themselves to spend more time in the patches that at any given time have relatively higher host densities. Somewhat paradoxically, therefore, such *initial* HDD does not actually give spatially host-density-dependent aggregation as the generation unfolds.

Consider how this lack of redistribution affects parasitoid efficiency. Initially, most parasitoids are concentrated in patches with more hosts. This is efficient. But host density decreases in these patches much faster than in those with lower host density. The fixed parasitoid distribution therefore becomes less and less efficient as time goes on. Indeed, if the initial parasitoid distribution is optimal, as soon as any host is attacked all subsequent distributions are suboptimal, and become increasingly so as the generation progresses. This effect is nicely illustrated in Fig. 4.8 of Hassell 2000. As a result, over the range of aggregation that is stabilizing, host equilibrium density increases with the strength of parasitoid aggregation to (initial) local host density (Fig. 10.8).

This form of "initial aggregation" is stabilizing because the inefficiency gets worse the more parasitoids there are. As parasitoid density increases, even more parasitoids are crowded into the initially high-density host patches, and so the host density there falls even faster during the generation relative to other patches—but the parasitoids do not leave. So increases in parasitoid density above equilibrium reduce the per-head

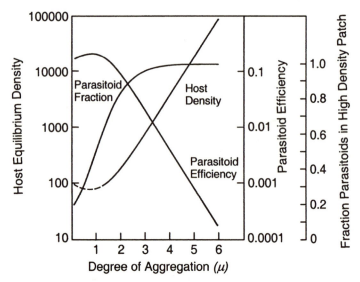

FIGURE 10.8. Trade-off between stability and parasitoid efficiency and host suppression in the model of Hassell and May (1973), in which parasitoids aggregate to the patch with most hosts. Half the hosts are in a high-density patch, and the remainder are divided equally among five low-density patches. Aggregation—the fraction of parasitoids in the high-density patch—increases as the parameter μ increases. Dashed section of the host density curve indicates unstable equilibria, solid section indicates stable equilibria. $\mu = 0$ is the basic Nicholson-Bailey case. Details are in Hassell and May 1973 and Murdoch and Stewart-Oaten 1989. Parasitoid efficiency is $\ln[H^*/(H^* -^* P^*)]/P^*$. Note logarithmic scale for host equilibrium density and parasitoid efficiency. Redrawn, with permission from University of Chicago Press, from Fig. 1 in Murdoch and Stewart-Oaten 1989.

parasitoid attack rate and so induce temporal density dependence in the parasitoid per-head recruitment rate, causing what Free et al. (1977) called *pseudo-interference*.

Now consider the persistence of an individual host's relative risk. If the host is in a patch that has many hosts at the start of the generation, it is subject to attack by many parasitoids and its risk is high. As the population develops within the vulnerable stage, and the density of hosts in the patch decreases, the same fraction of parasitoids stay in that patch, and the host's relative risk stays the same. Indeed, all hosts keep the same degree of risk throughout the generation. In contrast with the

Murdoch and Stewart-Oaten (1989) model, risk is persistent and the model can be stable.

The above interpretation is borne out by the results of Rohani et al. (1994), who examined a model with several patches in which dynamics are described by the standard Nicholson-Bailey model, but there is continuous reaggregation by the parasitoids to patches with more hosts during each generation. Hosts are distributed unevenly among patches at the start of the generation, but now, at each time during the vulnerable stage, parasitoids preferentially visit patches that currently have more hosts, rather than those that had more hosts at the start of the generation. They thus respond to the results of their own depredations. As a result, the relative risk of a host surviving in any particular patch changes throughout the vulnerable period. When the parasitoids respond quickly to the changes in relative density in patches, the stabilizing effect of parasitoid aggregation found by Hassell and May (1973) is lost.

As one would expect, the more rapidly the parasitoids reaggregate to the changing pattern of host density, the less stable is the system (Rohani et al. 1994). Increasing the rate of response effectively makes the model a closer approximation to the Murdoch and Stewart-Oaten (1989) model.

In terms of efficiency, reaggregation within the generation allows the parasitoids in the Rohani et al. model to remain efficient throughout the generation, and there is therefore no decrease in per-head efficiency as parasitoid density increases. This is why stability is lost. In terms of persistence of relative risk, hosts that survive attack in initially dense patches experience reduced risk as other patches attract parasitoids, so relative risk is a transient rather than a persistent property of each individual host.

We expect that if a consumer can aggregate to locally dense patches of its resource, it is likely to be able also to detect changes in resource density and to reaggregate to the changing distribution. That is, we suspect it is more likely to approximate the behavior in the Lotka-Volterra-based models, or the Nicholson-Bailey model with reaggregation, than that in the fixed aggregation models. Overall, we therefore suspect that aggregation by predators to local prey density, unless it involves transit times, has either no effect on stability or tends to be destabilizing. However, it is efficient, it should be selected for, and it should lead to lower mean prey densities than would otherwise be the case.

CONNECTION TO AGGREGATION INDEPENDENT OF
LOCAL HOST DENSITY (HDI)

In the next section we discuss variation in risk that does not depend on local host density, but here we alert the reader to the connection between the two types of variation in risk in Nicholson-Bailey models. Chesson and Murdoch (1986) show that aggregation to (initial) local host density tends to be stabilizing in Nicholson-Bailey models because relative risk has a clumped distribution, with hosts in a few patches being at much higher risk than hosts in the majority of patches. *It is this (lifelong) skewed risk among individuals, rather than the fact that parasitoids are aggregating in relation to local host density, that induces stability.* This explains how the Nicholson-Bailey equilibrium can also sometimes be stabilized if the parasitism rate decreases in patches with more hosts (Hassell 1984; 2000).

The above insight serves to emphasize a crucial difference between two types of models discussed in this section. When the parasitoids can rearrange themselves continually in response to changes in host distribution (as in the Lotka-Volterra models), stability hinges on whether the parasitoids can increase their efficiency as mean host density increases—stability, when it occurs, results from temporal density dependence in the host death rate. By contrast, when parasitoids do not rearrange themselves in response to changing host distribution, stability ensues because parasitoid efficiency decreases with parasitoid density, inducing temporal density dependence in the parasitoid attack rate.

Skewed Relative Risk Independent of Local Host Density (HDI)

In this section we look at the dynamical effects of variation in risk among prey, independent of their local density. The first and seminal paper on this subject assumed that individual hosts might be differentially vulnerable to parasitism for any one of a variety of reasons (Bailey et al. 1962). One possible reason suggested was spatial location: some sites might be more accessible to parasitoids than others.

We show below that there are two necessary conditions for stability in this class of models. First, the distribution of relative risk among host individuals needs to be "clumped," or skewed. That is, a small fraction of the hosts need to be at high relative risk, and most hosts must run a low relative risk (point 2 above, under "Spatial Processes within Populations"). Second, the relative risk run by an individual needs to persist over the vulnerable period (point 3 above). Also crucial, and resulting from these properties, is the relation between parasitoid efficiency and parasitoid density (point 1 above).

SKEWED RELATIVE RISK IN NICHOLSON-BAILEY MODELS (HDI)

This time we begin with the discrete-generation Nicholson-Bailey formulation and later compare the results with those from a continuous-time model. Bailey et al. (1962) imagined that a host individual "presents" to the searching parasitoid an "area of discovery," a, within which it is successfully attacked if a parasitoid enters the area; a thus measures the relative risk of that host being parasitized. (It can equally well represent the per-head attack rate of the parasitoid; these are just two interpretations of the same parameter.)

Different host individuals in the Bailey et al. model "present" different areas; for example, those in exposed places present a larger target, as it were, than those partially hidden. Alternatively, those in sites visited by more parasitoids present a larger area, the assumption in this case being that some aspect of the environment, independent of local host density, makes the site more or less attractive to parasitoids. There is therefore a distribution of relative risk, $r(a)$, across the host population. Bailey et al. assumed the distribution of a is continuous and might be defined, for example, by a gamma distribution whose shape is determined by its "clumping" parameter, k.

We then replace the usual fraction of hosts surviving attack by P parasitoids, namely $f = e^{-aP}$, with the fraction surviving over the heterogeneous population. The density of hosts surviving to the next generation, when the density at the start of the current generation is λH_t, is then

$$H_{t+1} = \lambda H_t \int_0^\infty e^{-aP} r(a) da, \qquad (10.11)$$

where P is the density of searching parasitoids, which is constant throughout the generation. The probability a host survives attack is thus now e^{-aP} weighted by its relative risk, which is defined by the distribution $r(a)$, and we obtain the fraction of the population surviving by integrating over the entire host population. Bailey et al. (1962) showed that with equation 10.11 as the host equation in the Nicholson-Bailey model, the equilibrium is stable if $r(a)$ is highly skewed, so that some hosts are extremely vulnerable but the majority have low risk. In the case of the gamma distribution, with coefficient of variation $\sqrt{1/k}$, stability requires k to be small.

This model once again attempts to describe in a discrete-time framework a process that is actually taking place continuously throughout the duration of the host vulnerable period (see Box 4.1). In this case, a reasonable interpretation is that hosts suffer a relative risk associated with the site in which they are born, do not move from the site, and that the relative risk remains the same throughout the generation (it is persistent). As in the previous section, this model does not allow the parasitoid population to redistribute its effort during the generation as its depredations alter the relative abundance of hosts across sites. Individual hosts are assigned a relative risk at the start of the generation and retain this value thereafter. We emphasize these statements because they contrast with an alternative interpretation (Rohani et al. 1994; Hassell 2000) discussed below.

Once again, inefficiency is the key to stability. When $r(a)$ is very skewed, as required for stability, most parasitoids are essentially continually reattacking the same few individuals. This induces inefficiency in the parasitoid, which causes the host equilibrium to be higher than it would be otherwise (May 1978; Murdoch 1992). The inefficiency increases with parasitoid density, as it does in the HDD Nicholson-Bailey model (equation 10.10), and it is this temporal parasitoid density dependence that is stabilizing.

In the next few paragraphs we give an outline of developments in this modeling framework. May (1978) assumed that parasitoids are gamma distributed among patches and showed that if they attack hosts at random within a patch, the overall distribution of attacks among hosts is negative binomial. The number of hosts and parasitoids in generation $t + 1$, averaged over patches, is therefore

$$H_{t+1} = \lambda H_t \left[1 + \frac{aP_t}{k} \right]^{-k}$$

$$P_{t+1} = H_t \left\{ 1 - \left[1 + \frac{aP_t}{k} \right]^{-k} \right\},$$

(10.12)

where k is now the clumping parameter of the negative binomial distribution and the term in parentheses is the zero term of the negative binomial distribution. This model has the same properties as the Bailey et al. model (Chesson and Murdoch 1986).

Chesson and Murdoch (1986) showed that the Bailey et al. and May models effectively assume that relative risk is independent of local host density (HDI) and cannot be derived from the assumption that parasitoids aggregate to local host density (HDD) without severe constraints that are not likely to be met in real populations. They also showed that aggregation independent of local host density is a more general stabilizing force than aggregation to local host density, reinforcing the conclusions drawn in the previous section.

Hassell et al. (1991b) use and extend this approach to establish that in several Nicholson-Bailey–type models, the equilibrium will be stable if the (squared) coefficient of variation of relative risk among hosts is greater than unity—the "$CV^2 > 1$" rule—regardless of the source of heterogeneity of risk. (The result requires some constraints on host and parasitoid distributions.) This rule is associated with the earlier result (May 1978) that stability requires k to be less than 1 if parasitoid density is distributed among patches according to a gamma distribution because, as noted, in that case the coefficient of variation is $\sqrt{1/k}$. They also show that CV^2 can be partitioned between the component contributed by risk associated with local host density and the component that is independent of local host density.

Pacala and Hassell (1991) then develop procedures for estimating these different contributions in field data. They analyze more than 26 field data sets and show that less than a quarter have sufficiently high values to suggest the clumped distribution of relative risk is stabilizing. They also show that in these cases the great majority of variance arises from aggregation (or some other process) that operates independently of local host density, thus confirming the theoretical results in Chesson and Murdoch (1986). These models and the empirical data are also discussed by Hassell (2000) and Bernstein (2000).

Finally, to establish that aggregation independent of local host density is broadly stabilizing, Rohani et al. (1994) studied a patchy HDI Nicholson-Bailey model. The salient features are as follows. (1) Parasitoids were distributed unevenly across patches at the start of the season, but independently of how many hosts were present in the patch.

(2) The number of hosts in a patch changed within the season as hosts were killed. In particular, at the start of the season, hosts were distributed uniformly across five patches and were implicitly assumed not to move thereafter. (3) Parasitoids were allowed to move among patches; *however, the number of searching parasitoids visiting any particular patch remained constant throughout the generation* (the italics are our emphasis, for the reason explained below).

This model retains the same stability properties as the HDI model in equation 10.11; its stability properties are unaffected by parasitoid movement. The equilibrium is stable if the parasitoid distribution, and hence the distribution of host relative vulnerabilities, is sufficiently skewed.

We now draw together the results from the two Nicholson-Bailey-type models studied by Rohani et al. (1994). In the version where parasitoids continually reaggregated to local host density within the generation (HDD), stability was lost and the model approximated the Murdoch and Stewart-Oaten (1989) model. But when parasitoids aggregated independently of host density (HDI) and moved among patches within the generation, stability through aggregation was retained.

Rohani et al. (1994) and Hassell (2000) concluded, from this contrast between HDD and HDI models, that stability is not lost when parasitoids reaggregate under HDI. We believe, however, that such a conclusion is incorrect, for the following reason. Although individual parasitoids move among patches in the HDI model by Rohani et al., a parasitoid that leaves a patch is simply replaced by another: hosts in any given patch are visited by a fixed fraction (and in this case, number) of parasitoids throughout the vulnerable period. Parasitoid movement therefore has no effect on parasitoid distribution or on the interaction, and effectively does not change the model. Thus, stability is retained in the Rohani et al. HDI model because the relative risk of any given host is fixed over the vulnerable period, as in equation 10.11. We present a model below ("An HDI Model Separating the Effects of Skew and Persistence of Risk") that establishes this conclusion directly.

CLUMPED RELATIVE RISK IN LOTKA-VOLTERRA MODELS

In this section we ask what happens if (1) relative risk among hosts is always skewed but (2) the parasitoids continually rearrange themselves through the generation and, as a result, the overall parasitoid distribution continually changes. Because of (2), the relative risk run by each host individual changes through the generation, and changes faster if the parasitoid distribution changes faster. We might expect that we would lose the stability seen in the Rohani et al. model. This turns out to be the case.

We first make the point with a model at one end of the continuum of constant movement. Murdoch and Stewart-Oaten (1989) noted that the dynamics of the basic neutrally stable Lotka-Volterra model are unaffected if the host and parasitoid are independently distributed, regardless of how aggregated the parasitoid distribution is.

If H and P are again the numbers on a randomly chosen patch and predators search randomly within patches, then on that patch

$$\frac{dH}{dt} = rH - aHP$$
$$\frac{dP}{dt} = cHP - dP. \tag{10.13}$$

If we average over all patches we get

$$\frac{dh}{dt} = rh - a\mathrm{E}(HP)$$
$$\frac{dp}{dt} = c\mathrm{E}(HP) - dp, \tag{10.14}$$

where the average numbers of hosts and parasitoids per patch are, respectively, $h = \mathrm{E}(H)$ and $p = \mathrm{E}(P)$, and the expected number of encounters between hosts and parasitoids per unit time is proportional to the expectation of HP, namely $\mathrm{E}(HP)$. The actual distribution of parasitoids across patches can be as aggregated as we like; we need only be concerned with the expected number in a randomly chosen patch, $\mathrm{E}(P)$. If parasitoids and hosts are distributed independently of each other, $\mathrm{E}(HP) = \mathrm{E}(H)\mathrm{E}(P) = hp$, and we get

$$\frac{dh}{dt} = rh - ahp$$
$$\frac{dp}{dt} = chp - dp, \tag{10.15}$$

which is just the original Lotka-Volterra model (equation 10.4). Thus this type of aggregation has no effect on stability.

In this model, the distribution of relative risk among hosts is embodied in the joint distribution of H and P. The model assumes that whatever the parasitoid and host spatial distributions are, they are constant through time. This can be achieved, as hosts and parasitoids on patches die and recruit, only if both species move around continuously (Godfray and Pacala 1992). It is not clear how fast this movement has to be for the model to be a reasonable approximation in the face of continuous births and deaths and continuously changing densities. Presumably it needs to be fast relative to the time scale set by the vital rates: the location (and hence relative risk) of any particular host must change quickly relative to, for example, the expected host life span. The relative risk run by a given host individual is thus a transient property and has no effect on dynamics.

In this model, aggregated attacks do not reduce efficiency, as they do in the Nicholson-Bailey framework, because parasitoid attack is not repeatedly aimed at the same small fraction of the population, even though it might be concentrated in some small fraction at any one point in time. Per-head efficiency thus does not decrease with parasitoid density, and aggregation is not stabilizing.

To summarize to this point: the Nicholson-Bailey and Lotka-Volterra models portray both extreme possibilities: persistent risk (i.e., no change in relative risk run by a given host individual over the entire vulnerable period) and transient risk (i.e., continuous change in relative risk).

Persistent Relative Risk in Continuous Time

The two previous sections might suggest that the crucial feature determining whether clumped relative risk is stabilizing is whether the models are discrete-generation or continuous-time models. This is not the case, as is nicely illustrated by a model by Reeve et al. (1994a). Theirs is a continuous-time model, but skewed risk is stabilizing. Again, their model suggests the key is whether an individual's relative risk is persistent or transient.

The model is based on the saltmarsh planthopper, *Prokelisia marginata*, which is common in saltmarshes along the Gulf Coast of Florida, and its egg parasitoid, *Anagrus delicatus*. The planthopper lays its eggs at different tidal levels, so it appears that some are more vulnerable to parasitism than others, though the adult parasitoid appears to be evenly distributed.

The model has the continuous-time stage-structured form of those in chapter 5. The host has a vulnerable egg stage and invulnerable nymphal and adult stages. Adult hosts lay eggs continuously and distribute them evenly among 10 patches. Parasitoids move around freely and are also evenly distributed: all patches are visited by the same number of adult parasitoids. However, eggs in different patches have different fixed relative risk because patches differ in how exposed they are to parasitism. Thus risk is determined entirely by patch identity, and since eggs do not move, each egg has a permanent relative risk, a, throughout the stage. Reeve et al. assumed that a is gamma distributed.

Unsurprisingly, given the discussions above, clumped relative risk (small values of the skew parameter, k) is strongly stabilizing. The model thus gives a result analogous to that of the Nicholson-Bailey models. (However, for the particular parameter values in the real system, clumped risk cannot alone stabilize the model's equilibrium [Reeve et al. 1994a]; the invulnerable adult host stage is needed for stability.)

An HDI Model Separating the Effects of Skew and
Persistence of Risk

The models with skewed relative risk that we have examined to this point have been at one extreme or the other. Each host has either effectively had a fixed relative risk through the vulnerable period, or its relative risk has changed rapidly but not in a quantified way. In this section we separate the effects of skewed risk from the degree of persistence of relative risk.

We assume a continuous-time framework with overlapping generations. Suppose hosts and parasitoids are distributed along a continuous one-dimensional spatial axis, x. Hosts are vulnerable to parasitism during the entire juvenile stage, where $J(x, a, t)$ is the density function of juvenile hosts at position x, of age a, at time t. $P(x, t)$ is the density function of adult parasitoids at position x at time t. Adult female hosts, $A(t)$, are assumed to distribute their eggs across the environment according to a distribution function, $F(x)$. The juvenile hosts do not move from the site at which they were oviposited. Mobile adult hosts emerge after a development time of T_J days. Parasitoids have an attack rate, α, and are distributed along the axis according to a clumped distribution, $G(x, t)$, which remains clumped, but may change position continuously. Juvenile parasitoids develop into adults after T_P days. d_A and d_P are the background death rates of adult hosts and parasitoids, respectively (the background death rate of juvenile hosts, d_J, is assumed to be zero, and the through-stage survival of immature parasitoids, S_I, is assumed to be 1).

The equations for this model are as follows (C. Briggs and W. Murdoch, unpubl. results):

$$\frac{\partial J}{\partial a} + \frac{\partial J}{\partial t} = -\alpha\, J(x, a, t) P(x, t)$$

$$\frac{dA(t)}{dt} = \int_{-\infty}^{\infty} J(x, T_J, t)dx - d_A A(t)$$

$$\frac{dP_{tot}(t)}{dt} = \alpha \int_{-\infty}^{\infty} \int_{0}^{T_J} J(x, a, t - T_P)P(x, t - T_P)da\, dx$$

$$- d_P P_{tot}(t) \qquad\qquad (10.16)$$

$$P_{tot}(t) = \int_{-\infty}^{\infty} P(x, t)dx$$

$$J(x, 0, t) = rF(x)A(t)$$

$$J(x, T_J, t) = J(x, 0, t - T_J) \exp\left\{-\alpha \int_{t-T_J}^{t} P(x, t')dt'\right\}$$

$$P(x, t) = P_{tot}(t)G(x, t).$$

The key feature of this model is the clumped, but potentially mobile, distribution of parasitoids. For a rough approximation, we can think of the parasitoid population consisting mainly of a swarm that moves around the environment independent of the local density of hosts. We show below that if the position of the swarm does not move, and is, for example, distributed according to a gamma distribution, then we will regain the features of the Reeve et al. (1994a) model, in which skewed risk, independent of host density but persistent through time,

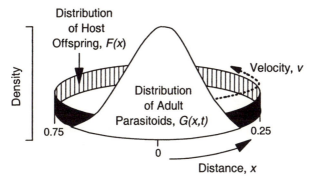

FIGURE 10.9. Diagram of one simple implementation of the HDI model separating the effects of skew and persistence of risk (equation 10.16). Host offspring are distributed on a one-dimensional spatial axis according to distribution $F(x)$; the axis has periodic boundaries and a circumference of 1. In this example, hosts are distributed uniformly, such that $F(x)$ is equal to 1. Parasitoids are distributed according to the distribution $G(x, t)$, which can be a clumped distribution whose position moves around space in one direction at velocity v. In our example, a beta distribution is used for clumped parasitoid distributions: $G(x, t) = (x^{\hat{a}-1}(1 - x)^{\hat{b}-1})/(B(\hat{a}, \hat{b}))$, where $B(\hat{a}, \hat{b})$ is the beta function and $\hat{a} > 1$ and $\hat{b} > 1$ are two parameters of the distribution.

is stabilizing. If instead the position of the swarm moves continually around the environment, then this stabilizing effect is lost.

As an example, one particular implementation of this model is shown in Fig. 10.9. In this simple example, space, x, spans from zero to 1, and for simplicity it is assumed that the two ends are connected, such that space forms a continuous ring. Adult hosts distribute their eggs through space according to the distribution $F(x)$. In this example, we assume that this distribution is uniform, such that $F(x)$ is equal to 1 (for x between zero and 1). Adult parasitoids are distributed through space according to $G(x, t)$. All other assumptions in the model are analogous to those in the basic stage-structured host-parasitoid model (see Box 5.1 and Fig. 5.6).

If the adult parasitoids are distributed through space in a uniform manner ($G(x, t) = 1$, for x between zero and 1), then all hosts have an equal risk of parasitism and the dynamics of this spatial model are identical to those in the basic host-parasitoid model (Fig. 5.6). The equilibrium is stable if the invulnerable adult host stage is relatively long and if the parasitoid development time is relatively short; otherwise we

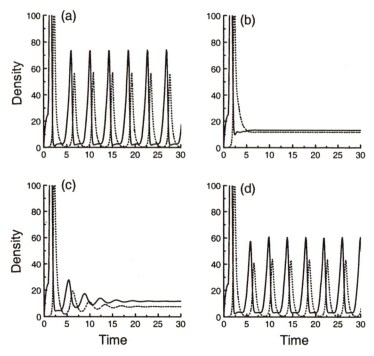

FIGURE 10.10. Simulations of the implementation of the HDI model separating the effects of skew and persistence of risk illustrated in Fig. 10.9. Total juvenile host density is shown in solid lines, adult parasitoid density in dotted lines. In all simulations, adult hosts distribute offspring uniformly through space, $F(x) = 1$. (a) Parasitoids are also distributed uniformly, so the model behaves exactly as the well-mixed system. (b) Parasitoids are distributed in a clumped but stationary distribution; $G(x, t)$ is a beta distribution with $\hat{a} = 2, \hat{b} = 2, v = 0$. (c) Parasitoids have a clumped distribution, but the location of the peak of the clump moves slowly through space; $G(x, t)$ is a beta distribution with $\hat{a} = 2, \hat{b} = 2, v = 0.1$. (d) As in (c), but parasitoid distribution moves through space at a higher velocity, $v = 0.5$. For all simulations, $T_J = 1, T_A = 1/d_A = 0.5*T_J, T_P = 0/4*T_J, r = 33/T_A, d_P = 8/T_J, \alpha = 1/T_J$.

get consumer-resource cycles (see chapter 5). For illustrative purposes, we will choose parameter values that give consumer-resource cycles when $G(x, t)$ is equal to 1 (Fig. 10.10a) and show what happens when we alter this assumption about the distribution of adult parasitoids.

First, if the distribution of parasitoids is clumped, but fixed in space, some hosts experience a persistent high risk of parasitism, whereas others experience a persistent low risk. This is highly stabilizing (Fig. 10.10b), as in the Reeve et al. (1994a) model.

Next, we assume that the clumped distribution of adult searching parasitoids moves through space. To implement movement in this simple example, we assume that the shape of the clumped distribution is fixed but that the distribution moves continuously around the ring of space in one direction at velocity, v, as illustrated in Fig. 10.9. Fig. 10.10c shows that if v is small, such that the "swarm" of parasitoids moves slowly through space, the equilibrium is still stable, but there are now damped cycles, demonstrating a reduction in the stabilizing effect. As v is increased, such that the distribution of parasitoids moves more rapidly through space, consumer-resource cycles return (Fig. 10.10d). As v becomes very large, the dynamics become indistinguishable from those of the basic host-parasitoid model and of the model with a uniform distribution of adult parasitoids (Fig. 10.10a).

These results illustrate that a persistent skewed distribution of risk of attack across the host population is the key to stability. As v is increased, the relative risk run by a given host individual becomes more transient and the stabilizing effect is lost, even though the risk at any time remains skewed. For real populations, the issue becomes how quickly parasitoids move around relative to other rates in the system.

Other Sources of Variation in Relative Risk

A SPATIAL REFUGE

The distinction between persistent and transient properties extends to the dynamic effects of a physical refuge. Physical refuges represent the most extreme case of spatial variation in individual risk. The contrast in Lotka-Volterra models between a constant fraction and a constant number in the refuge is illuminating.

When a fraction of the host population, s, is safe and a fraction $1 - s$ is vulnerable, we have

$$\frac{dH}{dt} = r[sH + (1-s)H] - a(1-s)HP$$

$$\frac{dP}{dt} = ea(1-s)HP - dP. \tag{10.17}$$

But this collapses to the basic neutrally stable Lotka-Volterra model with $a(1 - s)$ equal to z:

$$\frac{dH}{dt} = rH - zHP$$

$$\frac{dP}{dt} = ezHP - dP.$$

(10.18)

Thus stability is unaffected when a constant fraction is in the refuge. In such a Lotka-Volterra population, with continuously changing densities and continuously occurring births and deaths, a constant fraction can be maintained in the refuge only if individuals are continuously moving in and out of it; i.e., individual properties are transient.

By contrast, a constant number in the refuge does stabilize the equilibrium in a Lotka-Volterra model. Where H is the number of vulnerable hosts, R is the number of hosts in the refuge, and S is the total number of hosts, we have

$$\frac{dS}{dt} = rH - aPH + aR$$

$$\frac{dP}{dt} = eaPH - dP,$$

(10.19)

so the R individuals never take part in the interaction with the parasitoid, though they reproduce at rate r. We can think of the refuge as being populated by a distinct set of individuals that remain there for their lifetimes and leak offspring into the true parasitoid-host interaction (chapter 3).

FACTORS OTHER THAN SPACE

Bailey et al. (1962) recognized that space was only one possible cause of differences in risk. In fact, their first mechanism involved host individuals having a range of "intrinsic defensive properties" (though some intrinsic differences might be heritable and this would add a so far unexamined complication). Other differences in risk that could persist for most of the host's life include those arising from early accidents or early attacks by viruses. These could vary in severity among individuals and induce a variation in relative risk to attack. In true prey, another source might be initial random differences in, say, size, that become reinforced if initially larger prey can eat faster, and larger prey are less vulnerable.

Transient individual differences could also arise in a variety of ways. In true prey, parasite load, and hence risk of predation, might change continuously. An ever-changing concatenation of many small randomly acting processes could have the same effect. These could include how active the host is, how well its camouflage operates at any given time, how the light filtering through a forest canopy makes it more or less visible at any given time, etc. Such combinations of environmental and spatial variation should lead to rapidly changing vulnerabilities in a given individual.

Finally, Abrams and Walters (1996) explored models directly related to the degree of persistence of individual properties. Theirs have the basic structure of a Rosenzweig-MacArthur model, with logistic prey and type 2 functional response (chapter 3). The prey, however, are of two types: vulnerable and invulnerable (which in principle might or might not relate to spatial location), and prey move between the two classes at a constant rate. Their results are consistent with those in this chapter: equilibria are more likely to be stable when the transition rate between prey classes is low; i.e., when the differences between individuals are more persistent. As the transition rate increases, and individual differences are more transient, they recover the instability of the unstructured model.

Implications for Field Populations

A summary of conclusions so far is as follows.

1. Aggregation by parasitoids in response to local host density (HDD) is not likely to be an important source of stability in real systems, and may often be destabilizing. This result can be reversed if time spent in transit is important. As noted, HDD is of course efficient and should lead to greater suppression of host density.

2. The dynamical effect of aggregation by parasitoids, or any other factor affecting the distribution of relative risk, independent of local density (HDI), will depend on how persistent an individual's relative risk is. If it is persistent, a skewed distribution of relative risk is likely to be stabilizing.

The new modeling results derived above give insight into where skewed risk is likely to be dynamically important. For example, hosts will carry their risk levels for long periods if they settle in sites that differ intrinsically in their accessibility to parasitoids and if the parasitoids can do little to alter these site properties as hosts in the more accessible sites are depleted. Red scale attacked by the parasitoid *Aphytis* might provide an example. The crawlers settle on a substrate (leaf, stem, interior bark) and never move. It appears that *Aphytis* is attracted to some substrates more than others, so risk of parasitism seems to be rather permanently associated with substrate (Murdoch et al. 1989). We do not yet know, however, if relative risk also depends on local scale density, which would weaken the stabilizing effect.

At the opposite extreme, skewed risk probably is not a key to stability in most pelagic communities. For example, although zooplankton and their fish predators are certainly clumped in their distribution, it is likely that clumps of zooplankton are continually breaking up and reforming, so that an individual zooplankter's relative risk is continually varying.

How frequently is a clumped distribution of relative risk strong enough to account for stability? We expect insects to provide some of the best examples because the juvenile usually is both the vulnerable stage and is relatively immobile. However, a careful study by Pacala and Hassell (1991) shows how difficult it is to establish the role of clumped relative risk. These authors analyzed 26 cases in which the distribution of parasitism had been measured in the field under natural conditions. Only 6 of these 26 data sets (5 species) showed CV^2 greater than 1 (see Hassell 2000 and Bernstein 2000 for further discussion of these data).

Even in these six cases, the match between data and theory is weak. Estimates of CV^2 typically varied, over time, above and below 1, whereas the model assumes a fixed value (Hassell 2000) (chapter 4 this text). It was usually estimated at only one point in the vulnerable stage, but we would need to know whether skewed risk persists over the vulnerable period; and the entire population was not always sampled.

Clumped risk may well contribute to the stability of populations, without being alone responsible for stability. These results suggest, however, that if the data analyzed by Pacala and Hassell are representative of host-parasitoid systems, strongly clumped risk is not a prevalent feature of such interactions.

Perhaps the most convincing example is provided by Reeve et al. (1994b), which we referred to above. These authors studied parasitism of a Florida saltmarsh planthopper, *Prokelisia*, by its egg parasitoid, *Anagrus*, over a period of 30 weeks. The planthopper can pass through up to about 5 generations in 30 weeks; eggs take 2 weeks to develop. The planthopper lays its eggs in cordgrass leaves, where they are attacked by *Anagrus*. *Anagrus* moves freely about the habitat and appears to be randomly or close to randomly distributed among patches. Any clumping of parasitism among leaves is thus likely owing to variation in how vulnerable the eggs are.

Reeve et al. sampled on 29 occasions over the 30-week period and found that CV^2 was greater than 1 in all but a few dates. This alone does not establish that cumulative risk over the entire vulnerable period was strongly clumped. However, the eggs were probably laid in rather even-aged cohorts, and it seems likely that cohorts of eggs were at or nearing the end of their development period on a few of the sampled dates. So it is possible that this example illustrates the theory well.

Empirical support for stabilization via aggregated risk is even more tenuous than the above analyses suggest. Gross and Ives (1999) show that one cannot infer the distribution of risk of parasitism among hosts from observations of the distribution of the fraction parasitized among patches, which is how such inferences are made. The among-host variance in risk will be inflated when, as is usually the case, individual parasitoids can attack hosts that share a patch, increasing covariances among these hosts.

Summary of Within-Population Results

1. Predators may often spend time in transit between patches or areas containing prey. Stability can be enhanced if the predators stay longer in patches, and hence spend less time in transit, when overall prey density is higher.
2. Aggregation by predators in response to local prey density is best dealt with separately from aggregation independently of local prey density and other sources of skewed relative risk. The former typically is not stabilizing and may usually be destabilizing. It is stabilizing in Nicholson-Bailey models if there is

no reaggregation after the generation begins. But aggregation is efficient, and presumably of selective advantage, and it is difficult to imagine that predators could aggregate at the start of the vulnerable period but not thereafter as prey distribution changes.

3. Strongly skewed relative risk probably mostly results from some property of the prey population, for example its being distributed among sites that are differentially exposed to the predator. It is never destabilizing in the models explored so far. For stability, two processes need to operate: the distribution of relative risk must be clumped among individuals, and the relative risk run by an individual must change little during the vulnerable period. If the distribution of relative risk is skewed but an individual's relative risk changes rather quickly, there is a negligible stabilizing effect. Skewed risk is stabilizing only when an individual's relative risk is a more or less permanent property.

Equation 10.16 gives a quantitative framework for evaluating the likely stabilizing effect of skewed risk in different types of consumer-resource interactions. The key is the rate at which the distribution of relative risk among individual hosts or prey changes relative to the duration of the vulnerable stage.

CONNECTION BETWEEN PROCESSES WITHIN POPULATIONS AND AMONG SUBPOPULATIONS

Little or no attempt has been made to relate results from models of the sort discussed in the previous section to those discussed in the first section. Yet it seems likely that the two approaches will in some instances converge to give similar insights. Recent work by Hosseini (2001) illustrates the point.

Hosseini examined a spatially explicit individual-based model, with the usual rule that only one individual can occupy a grid point, and the model simulates the MacArthur-Rosenzweig model (equation 10.3). The predator can move freely across the space, but now there is a transit time. In addition, instead of being a stupid predator that spends the same amount of time searching each point, the predator has a fixed giving-up time if no prey is encountered. Hosseini (2001) shows first that optimizing the giving-up time tends to result in an unstable equilibrium, echoing the conclusion in this section that predator aggregation to patches with more prey is destabilizing. Second, the transit time tends to be stabilizing, just as in the early model by Murdoch and Oaten (1975). We suspect that the mechanism is the same in both types of models: per-predator attack rate can increase as mean prey density increases (i.e., it can be density dependent), because with a fixed giving-up time, total transit time declines as mean prey density increases.

The distinction between within-population and among-subpopulation dynamics is perhaps arbitrary for many sets of interacting species. For example, most California red scale individuals move very little: most newborn scale settle within a few centimeters of their immobile mothers (though a few may be blown for long distances in the wind). By contrast, the adult female of the parasitoid, *Aphytis*, probably covers many meters in her lifetime. Such differences in spatial scale between prey or host and predator or parasitoid are probably common. They imply that at intermediate spatial scales the prey is divided into locally interacting groups whereas the predator constitutes a well-mixed population.

This situation was described in an among-subpopulation context by both the spatially explicit individual-based model discussed in Route 2, and a more standard consumer-resource model (equation B10.1). The results suggest the dynamics are close to those of the within-population models discussed in the second part of the chapter, which are determined by a combination of spatial (or other) variance among hosts and non-linearity. Thus we may think of spatial structure as a continuum from pure among-subpopulation dynamics to pure within-population dynamics.

CHAPTER ELEVEN

Synthesis and Integration across Systems

In chapters 5–8 we presented stage- and state-structured theory, formulated with mainly parasitoid-host interactions in mind. In chapters 6 and 7, especially, we showed how that theory is both more realistic than its predecessors and yet achieves generality. In this chapter we broaden our scope by showing how this theory is an integral part of a larger and coherent body of theory for population dynamics, including theory for single-species populations. We use this last connection to argue that it is often appropriate to use single-species models to describe the dynamics of generalist consumers in species-rich food webs.

We begin by noting that theory for most consumer-resource interactions originates from a common source—the Lotka-Volterra model. We then discuss a deep connection between consumer-resource and single-species theory: consumer-resource models under some circumstances collapse to single-species models. Finally, we explore cycles in nature to ask whether this collapse also occurs in real systems.

SHARED THEORY FOR DIFFERENT KINDS OF CONSUMER-RESOURCE INTERACTIONS

Simple Models

Mathematical theory for almost all types of consumer-resource interactions shares basic assumptions with the Lotka-Volterra or Nicholson-Bailey consumer-resource models. Since the latter is formally a discrete-time version of the former (Box 4.1), it is clear that simple predator-prey and parasitoid-host models are joined at the hip. Here is a brief survey that extends this commonality, for both simple and stage-structured consumer-resource models.

Herbivore-plant models began directly in the Lotka-Volterra framework (Crawley 1983). They include models examined in chapter 3 and variants on them. van de Koppel et al. (1996) used modifications of the Rosenzweig-MacArthur model (chapter 3 and see below) to investigate plant-herbivore dynamics along a productivity gradient. Grover and Holt (1998) extended the standard framework to include a dynamic nutrient resource, competition between two plant species, and a herbivore attacking both plant species.

In the past decade or so plant-herbivore models have incorporated, as fundamental properties, plant quality and its interaction with herbivores (Edelstein-Keshet 1986; Lewis 1994). This work is nevertheless linked to other consumer-resource models by basing the formulation of the distribution of plant quality on the classic von Foerster equation, which is also the basis of the various approaches dealing with age and stage distribution in consumer-resource models (including those in chapters 5–7).

True parasite- and disease-host theory also shares many assumptions with the Lotka-Volterra family of models. Although disease models have their roots in human epidemiology (e.g., Kermack and McKendrick 1927), it has often been noted that childhood diseases share fundamental properties with other consumer-resource systems. If we think of susceptible individuals as the prey and infected individuals as the predator, the basic interaction between host and disease is a predator-prey interaction (Edelstein-Keshet 1988; Anderson and May 1991; Grenfell and Harwood 1997). We give some examples here.

In the simplest models, with no time lags, it is often assumed that the adult human population is constant, so the birth rate, and hence the recruitment rate of susceptible individuals, is constant, and we need not track the adult population. If recovery is possible, with subsequent lifelong immunity as in measles, the recovered individuals disappear into the unspecified adult population. The simplest model for such a disease, with recovery but no reinfection (a SIR model), is therefore just a Lotka-Volterra model, with the crucial modification that recruitment to the susceptible population occurs at a constant rate, B (Box 11.1, Model 1). Such constant susceptible recruitment from "somewhere else" is the analog of external prey recruitment in the predator-prey equation 3.18 and is of course stabilizing; any feasible equilibrium of this model is always stable (Box 11.1). As we stress in the next section, such constant recruitment partially decouples the consumer-resource interaction and prevents true consumer-resource cycles from appearing.

In other childhood diseases, such as whooping cough, individuals who have been infected and have recovered can be infected again (Box 11.1, Model 2, modified from Anderson and May 1981). Such a SIRS model then has the new feature, unique to disease models, that the "predator" (infected population) contributes recruits to the "prey" (susceptible population). This process is strongly stabilizing (Anderson and May 1981; Briggs and Godfray 1995a), and this model also has a stable equilibrium.

Although this new feature appears to be very unlike other consumer-resource theory, it actually nicely illustrates a process we have emphasized as central to the dynamics of stage- and spatially structured consumer-resource models. Stability in Model 2 is caused by what we have called indirect density dependence (Briggs and Godfray 1995a). The number of recruits to the susceptible population from the infected population is not correlated with the current density of susceptibles. Recruitment per head from this source therefore declines with the density of susceptibles; it is density dependent. This model therefore illustrates again the potential stabilizing effect of asynchronous fluctuations in different components of the interaction.

If all hosts are susceptible, and if infected individuals cannot recover, we lose the constant adult host population and therefore the constant stabilizing recruitment of Model 1. However, the "predator" can still contribute to the host population if some infected individuals continue to reproduce. Once again, recruitment of hosts from the infected population is strongly stabilizing, for the reason just given, and any feasible equilibrium is stable (Anderson and May 1981; Briggs and Godfray 1995a) (Box 11.1, Model 3).

Finally, simple models of virus-insect interactions also fit the general Lotka-Volterra framework. Anderson and May (1981) and Briggs and Godfray (1995a) discuss an extension of the Lotka-Volterra model that adds an intervening free-living stage of the disease. This adds

BOX 11.1.
SIMPLE DISEASE-HOST MODELS AS PREDATOR-PREY MODELS

Model 1. H (usually S) and P (usually I) are the number of susceptible and infected individuals, respectively, B is the constant recruitment rate, a is the transmission rate, and d combines the death rate of infected individuals from infection and their recovery rate and disappearance into the unspecified adult population. As in many predator-prey models, we ignore background death rates of susceptible and infected individuals, assume there are no time lags, and get:

$$\frac{dH}{dt} = B - aHP$$

$$\frac{dP}{dt} = aHP - dP.$$

This is a simplification of equation 3.18. The equilibrium is always stable.

Model 2. In a modification of Model 1, infection does not necessarily confer immunity. Infected hosts (P) may recover at rate z and rejoin the susceptible population (H). We then have a SIRS model with a fixed adult host population. Whooping cough is an example:

$$\frac{dH}{dt} = B - aHP + zP$$

$$\frac{dP}{dt} = aHP - dP - zP.$$

Again, the equilibrium is always stable.

Model 3. Some infected individuals can reproduce new susceptibles. Now r is the per-head birth rate of susceptible hosts, z is the reproductive rate of infected hosts, and H is the total uninfected host population. We have:

$$\frac{dH}{dt} = rH - aHP + zP$$

$$\frac{dP}{dt} = aHP - dP.$$

Any feasible equilibrium is stable.

(Box 11.1. continued)

Model 4. This is a model for a virus and its invertebrate host population. Here susceptible individuals, H, become infected when they contact the free-living stage, which now is therefore the analog of the predator population, P. We can think of the infected population as (non-feeding) juvenile predators, J, with infection-induced death rate, d. They produce the free-living stage at per-head rate m, which we can think of as an analog of maturation. The free-living stage has a constant death rate, z, that greatly exceeds losses through infection (which are ignored). A simple model for this type of interaction, in which the susceptibles have a constant intrinsic rate of increase, r, is then:

$$\frac{dH}{dt} = rH - aHP$$

$$\frac{dJ}{dt} = aHP - dJ$$

$$\frac{dP}{dt} = mJ - zP.$$

This model always has an unstable equilibrium and produces unstable, diverging oscillations. This is because of the delay (in this case a distributed delay with an exponential distribution) between infection of the host and the production of free-living, infectious pathogens. The model includes no stabilizing process to bound the diverging oscillations. Anderson and May (1981) showed that the addition of either recovery of infected individuals or reproduction by infected individuals could contribute a stabilizing effect. These processes both add an input from the infected class into the susceptible class. The resulting stabilizing effect can convert the diverging oscillations into either long-period consumer-resource cycles or a stable equilibrium.

stage structure to the consumer, which in the absence of any stabilizing process destabilizes the equilibrium (Box 11.1, Model 4). Insects typically die from infection by a virus, and infected individuals typically do not reproduce. As we would expect, however, if either of these processes does occur, they are potentially stabilizing for the reason discussed above.

Stage-Structured Models

The broad domain of consumer-resource theory extends beyond simple models to our stage-structured framework. First, as stressed throughout this book, that theory is an extension of the Lotka-Volterra-like models. For example, chapter 5 shows that our Basic parasitoid-host model has the same dynamics as a Lotka-Volterra model when the adult host stage has infinitesimally short duration and the parasitoid has no developmental time lag: neutrally stable Lotka-Volterra cycles, with the standard Lotka-Volterra period, occur at the origin in Fig. 5.6b.

Second, our stage-structured framework applies to parasitoid-host, predator-prey, and virus-insect interactions. In fact, our Basic host-parasitoid model, modified to have vulnerable adults and invulnerable immatures (equation 5.32), was previously proposed by Hastings (1984) as a structured predator-prey model. Below we discuss a stage-structured model for the interaction between *Daphnia* and algae that improves the fit to predators by taking into account changes in predator properties as they grow. That model also retains the basic Lotka-Volterra-like properties—in particular, the properties of the Rosenzweig-MacArthur model.

The extension of the parasitoid-host framework to virus-insect interactions is even more straightforward. In most species, the virus attacks only the larval insect, so we need stage structure in the insect host and a properly stage-structured model turns out to have almost exactly the same form as our Basic model in chapter 5 (Briggs and Godfray 1995a, appendix D). The invulnerable adult stage can then supply a stabilizing effect if it is sufficiently long-lived (see the Basic model of chapter 5).

CONNECTION BETWEEN CONSUMER-RESOURCE DYNAMICS AND SINGLE-SPECIES DYNAMICS IN THEORY AND NATURE

In chapters 4 and 6 we gave examples of how some process that induces weak coupling (chapter 2) can collapse consumer-resource dynamics to single-species-like dynamics. This is a potentially powerful result. First, it establishes a deep connection between theory for single-species and multispecies systems. Second, it leads us to ask whether multispecies systems in nature can act like single-species systems.

Cycles play a key role in establishing the connection between consumer-resource and single-species dynamics and in defining the difference between them. This is because cycles, unlike most equilibrium dynamics, give insight into the mechanisms that produce them (Kendall et al. 1999). In chapters 4–6 we used the period of the cycle as the main "probe" for detecting mechanism. By probe we mean a property of real cycles that we can compare with that of the cycles predicted from mechanistic models.

To recapitulate, we have distinguished three major types of cycles. The first two are seen in single-species consumer populations (feeding on implicit resources), or their period is determined by consumer dynamics in a consumer-resource interaction. Single-generation cycles have the shortest period (1–2 times the consumer developmental delay), and delayed-feedback cycles have short to intermediate periods (2–4 times the consumer developmental delay). Consumer-resource cycles have longer periods: typically at least 6 time units when consumer and resource each take 1 time unit to develop (chapters 4–6). (A fourth type, with which we are not concerned here, is single-generation cycles whose period is determined by the resource [host] population; chapter 5, "Single-Generation Cycles in Parasitoid-Host Models.")

The obviously striking fact is that although single-generation and delayed-feedback cycles are characteristic of single-species models, they are also seen in consumer-resource models under appropriate circumstances. Two circumstances are required. (1) A stabilizing process suppresses consumer-resource instability and partially uncouples the dynamics of one species from the other; this causes single-species-like dynamics. (2) The cycles are induced by age- or stage-dependent density dependence that is either direct (giving single-generation cycles) or delayed (giving delayed-feedback cycles).

Below we summarize and expand on our theoretical results for single-generation and delayed-feedback cycles. Then we investigate the implications of these results for real populations.

Single-Generation Cycles

The connection between single-species dynamics and partially un-coupled consumer-resource dynamics is easily seen and demonstrated via single-generation cycles. We first summarize results from chapters 5–6 and then explore a model for plankton populations.

SINGLE-GENERATION CYCLES IN SINGLE-SPECIES SYSTEMS

Single-generation cycles are more common in laboratory experiments than are delayed-feedback cycles. They are manifested as more or less discrete generations, even though reproduction is continuous (chapter 5). When the adult is short-lived, the period is close to T, the immature developmental delay. Gurney and Nisbet (1985) analyze two stage-structured single-species models, with competition affecting juvenile survival or maturation time, and show that the period ranges from $1T$ to $2T$.

The key to single-generation cycles is density dependence that either takes effect immediately or, if not, within the same generation. Discrete generations appear, regardless of the details, because a dominant cohort is able to suppress other cohorts in the generation. Suppression may affect the production of new immature cohorts (through reduced adult fecundity or adult survival) or the survival of other immature cohorts. The laboratory populations of the Indian meal moth (*Plodia inter-punctella*) and the associated model illustrate the point well (chapter 5). The larvae compete for constant amounts of food. Adult life is brief, so cycles are caused by bursts of adults that produce bursts of eggs. In the simplest *Plodia* model, the cycles appear if there is uniform competition among all the larvae. But a more plausible explanation, incorporated in a more complex model, is that older larvae have a greater effect on the food than do younger larvae, and also cannibalize eggs (Briggs et al. 2000). Larvae from a burst of eggs reach a critical size and thereafter suppress the survival of younger and older larvae. Only when they have developed into adults is a replacement cohort produced that is again large enough to suppress other cohorts.

Single-generation cycles are also seen in laboratory populations of the flour beetle *Tribolium*, which has long-lived adults. In this case, there is an almost constant stream of eggs; in fact, Hastings and Costantino (1987) caricature the situation with a simplified model (which behaves like a full model of the system) in which it is assumed there is a constant supply of eggs from a purely implicit constant adult population. The density dependence is again immediate. The older larvae are cannibals and have high survival. A large cohort entering the older larval age class cannibalizes all successive younger immature cohorts. Only when that dominant cohort has metamorphosed into adults can a new wave of young larvae survive to become old cannibals and then repeat the process. Even though the recruitment rate of young immatures is constant, it is mainly only those in the dominant cohort that give rise to older larvae.

A key feature of these and other single-species laboratory populations is that the food supply is either held constant (*Plodia*) or turns over very rapidly since it is resupplied at a constant rate. Experimentally, these are equivalent to semi-chemostat conditions, but the resource is only implicit in the single-species model.

SINGLE-GENERATION CYCLES IN GAIN MODELS

As noted in chapter 6, the gain mechanism gives rise to single-generation cycles when the predator development time exceeds that of the prey (Fig. 6.10). The period is approximately the predator development time, T_p.

The gain model is especially instructive. The stabilizing, decoupling mechanism is the long-lived invulnerable prey stage. Because the invulnerable adult is long-lived, adult density varies little, and so rather like a refuge, the adult population "leaks" an approximately constant stream of juvenile prey. This is akin to semi-chemostat dynamics in the laboratory, in which a constant rate of food is supplied from a reservoir. This process was explicitly modeled in a simplified model with a constant stream of juvenile prey, which gave identical results to the original model. Although the total prey population oscillates in response to cyclic pressure of predation, there is little variation in density of the youngest prey stage (Fig. 6.9b). It is this semi-chemostat process that turns consumer-resource dynamics into single-species dynamics.

The cycles arise because a dominant cohort of adult predators produce a dominant cohort of juvenile predators and, in the process, reduce all age classes of juvenile prey. Only the older juvenile prey class produces predator offspring. The density of these older prey remains low for some time because of the prey's developmental lag. The dominant predator cohort thus effectively suppresses new predator cohorts until the abundance of older juvenile prey recovers. Since the predator takes longer to develop than the prey, abundant older juvenile prey are available when the new dominant predator cohort appears, which again produces a burst of offspring. The density dependence thus acts within each predator generation.

SINGLE-GENERATION CYCLES IN A ZOOPLANKTON-ALGAL MODEL AND IN THE REAL SYSTEM

This section illustrates the idea with a predator-prey model for a system where single-generation cycles have been seen in the field. *Daphnia* is the major zooplankter in many northern temperate lakes where, in the absence of heavy predation, it is able to suppress the abundance of edible algae far below the limit set by algal nutrients. This interaction is one of the most thoroughly studied at all levels from individual physiology to population dynamics in the lab, microcosms, and field (Gurney et al. 1990). We discussed it briefly in chapter 3, under "Suppression-Stability Trade-Off."

Daphnia is an especially useful case because laboratory populations show single-species-like single-generation cycles both when *Daphnia* is raised "alone" (with dead algae) and in the presence of dynamic prey, i.e., live algae (McCauley and Murdoch 1987; McCauley et al. 1996a). *Daphnia* also exhibits such cycles in microcosms with dynamic algae, and in lakes (Murdoch and McCauley 1985; McCauley and Murdoch 1987). These cycles are caused by dominant cohorts of juvenile *Daphnia* that suppress adult reproduction.

Nisbet et al. (1989) and McCauley et al. (1996a) developed a consumer-resource model in which the (female only) *Daphnia* population is divided into stages: juveniles, young adults that cannot produce offspring, and "adults" that can (Fig. 11.1). The general form of this model is familiar to the reader, so we give only a summary explanation here.

As in all such models, all individuals in a given stage have the same properties, some of which are indicated in Fig. 11.1 as per-head rates.

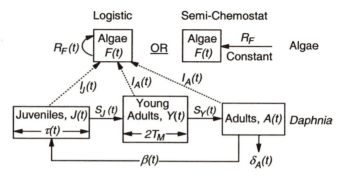

FIGURE 11.1. Diagram of stage-structured model of the interaction between *Daphnia* and edible algae. It illustrates both the strongly coupled interaction, in which algal recruitment ($R_F(t)$) is logistic, and a weakly coupled interaction, in which there is constant algal recruitment from an external source (R_F), giving effectively semi-chemostat dynamics. Young adults and adults have the same ingestion rage, $I_A(t)$.

Unlike our parasitoid-host models, the consumer, but not the resource, is stage structured. All consumer stages feed. Individual properties, such as maintenance requirement, intrinsic death rate, sensitivity of the death rate to food stress, and maximum feeding rate, change with stage, though young and mature adults ingest food at the same rate. Feeding rate in a given stage varies with time because it depends on algal density (via a type 2 functional response).

A major novelty, compared with parasitoid-host models, is that a juvenile *Daphnia* develops at a rate determined by the amount by which its food intake exceeds its maintenance requirement. A juvenile changes into a young adult when it reaches a threshold size. So the time spent in the immature stage varies. Mortality of each stage depends on feeding rate, again in relation to maintenance requirement. In addition, adult death rate increases with age. The algal population is assumed in the first instance to recruit at a rate defined by the logistic model, and dies off as a consequence of feeding by all three stages of the *Daphnia* population.

The model is well parameterized on the basis of extensive physiological studies and testing of models of individual performance under different food regimes (Gurney et al. 1990; McCauley et al. 1990). It is able to predict the dynamics observed in laboratory populations (McCauley et al. 1996a; also see chapter 3).

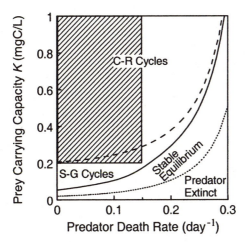

FIGURE 11.2. Dynamical properties, as a function of lake nutrient status and *Daphnia* death rate, of the *Daphnia*-algal model diagramed in Fig. 11.1, with algal dynamics following the logistic model. Range of realistic parameter values in nature is indicated by the shaded area. Modified, with permission from Blackwell Publishers, from Fig. 8a in Murdoch et al. 1999.

As we would expect, since the model is a stage-structured extension of the Rosenzweig-MacArthur model, the stability properties of the two models are similar, i.e., Fig. 11.2 is similar to Fig. 3.6. Also as expected, the equilibrium is unstable over most of the biologically reasonable parameter space. Fig. 11.3a shows an example of the large-amplitude long-period consumer-resource cycles in the unstable region. Unlike in the real single-generation cycles described above, adult and immature *Daphnia* cycle in phase with each other and dominant cohorts do not appear (McCauley et al. 1999). (There is a small region in parameter space, at low nutrient levels next to the stable region, where single-generation cycles can appear [Fig. 11.2] because the predator-prey interaction is weak there.)

The consumer-resource interaction collapses to a single-species-like system under the now familiar semi-chemostat conditions. We partially decouple the *Daphnia*-algal interaction, and make it more stable, by making algal recruitment constant rather than logistic. Where R_F is the recruitment rate of algae, we replace logistic with constant recruitment:

$$R_F = rK. \tag{11.1}$$

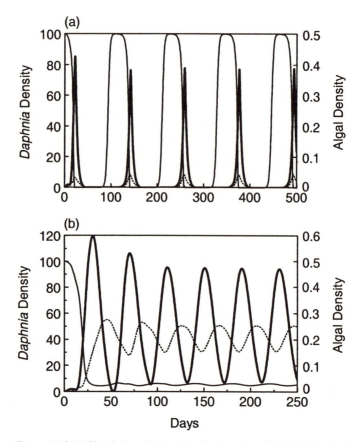

FIGURE 11.3. (a) Simulation of the logistic-food version of the *Daphnia*-algal model from the unstable region of Fig. 11.2, with $r = K = 0.5$, showing consumer-resource cycles. (b) Simulation of the *Daphnia*-algal model with constant algal recruitment showing single-generation cycles in *Daphnia*. Algal density (thin solid curve) falls rapidly to a low, relatively constant density; immature stages (thick solid curve) show larger amplitude cycles than adult population (dotted curve).

This might occur, for example, if the algae have a refuge from which a roughly constant flow of algae leaks into the interaction (chapter 3). Persson et al. (1998) were the first to use this approach, in a stage-structured model to account for single-generation cycles in fish.

Because constant recruitment of algae is strongly stabilizing, consumer-resource cycles disappear, even at the highest levels of enrichment (K). Single-generation cycles can now occur over the entire range of K, depending on other parameter values. An example, for K equal to 0.5, is shown in Fig. 11.3b.

There are two striking features in this simulation. First, the *Daphnia* and algal dynamics are indeed partially decoupled. The algal population goes quickly to a low equilibrium set by *Daphnia* predation and then changes very little, just as seen in some stock tank populations (Murdoch and McCauley 1985). (The tiny fluctuations in algal density are matched exactly by tiny reciprocal [not lagged] fluctuations in total *Daphnia* "feeding pressure," i.e., the density in each stage weighted by its feeding rate.)

Second, although total *Daphnia* feeding pressure varies little, the two developmental stages execute substantial single-generation cycles. The period reflects the fact that the population is on the brink of starvation and development is stretched to the maximum. Pulses of immatures suppress the adult population and fecundity and do not allow a new pulse through until the dominant cohort has declined in density and its remnants have matured. In contrast with the consumer-resource cycles, the immature and adult stages are out of phase, as we expect (McCauley et al. 1999).

Single-generation cycles occur because of direct density-dependent competition for algae between adult and juvenile *Daphnia*. The competitive ability of a stage can be defined by the density of food at which an individual can just meet its maintenance needs (Persson et al. 1998); the lower this critical food density, the better is the stage at competing. Competitive ability changes with stage because feeding rate and maintenance requirement change with stage. Given *Daphnia*'s physiology, juveniles are better competitors than are adults. If we reduce the maintenance requirement of adult *Daphnia* by about 30%, the two stages become equal competitors, and the cycles disappear and are replaced by stable equilibria and constant stage distribution.

SINGLE-GENERATION CYCLES IN REAL AND MODEL FISH

Persson et al. (1998) obtain equivalent results in a model of fish feeding on zooplankton in Swedish lakes. The fish breed once per year but take 7 years to mature. In one version of the model, juveniles can again survive on a lower zooplankton density than can adults. When there is a constant recruitment rate to the zooplankton from a notional refuge, single-generation cycles appear. A large cohort of juveniles suppresses adult fecundity and increases adult mortality, so a new cohort cannot be produced until the dominant cohort matures. It was this model that suggested asymmetric competition in the *Daphnia* model above.

A real example of the mechanism in the Persson et al. (1998) model may occur in yellow perch (*Perca flavescens*) in Crystal Lake in Wisconsin. Sanderson et al. (1999) found cycles with a period of 5 years, which is the development time of this fish population. The peak in each cycle is made up mainly of a pair of exceptionally large young-of-the-year classes, each produced by an unusually abundant class of recently matured adults. The mechanisms of inter-cohort competition may include cannibalism or competition for food.

Delayed-Feedback Cycles

DELAYED-FEEDBACK CYCLES IN SINGLE-SPECIES SYSTEMS

Delayed-feedback cycles in single-species models result, as the name implies, from delayed density dependence and, in particular, from density dependence that is delayed across generations. The archetypal delayed-feedback cycle is found in the discrete-time single-species Ricker model (chapter 4). The cycles are induced by the inherent delay in the discrete-time structure: competition between adults that creates density-dependent fecundity at time t does not have its dynamical effect until a new generation of adults is recruited at time $t + 1$. The fundamental period is $2T_P$ where the development time is T_P time units. The period increases if some adults survive to breed in more than one generation (Higgins et al. 1997b). In that case, the period ranges between $2T_P$ and $4T_P$. As noted in chapter 4, Higgins et al. show that this range of periods typically holds in the Ricker model even away from the stability boundary, i.e., when non-linear effects occur.

The same result arises in stage-structured single-species continuous-time models. Nicholson's (1957) adult blowflies competed for a fixed amount of food whereas larvae were given unlimited amounts of food (chapter 5). Both the real population (Fig. 5.1a) and the blowfly model (Figs. 5.3a and b) showed delayed-feedback cycles. Again, the requirement for such cycles is that the dynamical effect of density dependence is delayed to the next generation. This happened in the blowflies because fecundity declined with adult density, but survival through the immature period was high and invariant, so adult recruitment on day t depended on adult density at time $t - T_P$. Gurney and Nisbet (1985) and Jones et al. (1988) show that, on the stability boundary, the minimum period of such cycles is $2T_p$, and as in the Ricker model, the period increases with adult longevity and typically has a maximum of $4T_p$. (Nisbet and Bence [1989], however, show that in an extreme model [of kelp] with long adult longevity and large-amplitude cycles, longer periods are possible away from the stability boundary.)

DELAYED-FEEDBACK CYCLES IN CONSUMER-RESOURCE GAIN MODELS

In chapter 6 we saw how delayed-feedback cycles occur in our stage-structured parasitoid-host models when the gain to the future female parasitoid population increases with the age of the host encountered, a general feature of real parasitoid-host interactions. The cycles arise only when the parasitoid develops faster than the host, which is usually the case in real systems. The cycles are in the parasitoid population, whose development time determines the period.

We showed in chapter 6 that the properties of the cycles are retained if we literally uncouple host recruitment from host density by allowing a constant number of newborn hosts to recruit per unit time. This is an analogy of a semi-chemostat system. The detailed properties of the cycles are virtually unchanged, including the increase in period at approximately $2T_P$ (Figs. 6.10a and b).

The delayed density dependence in the parasitoid, which causes the cycles (Figs. 6.2a and 6.9), arises from the gain mechanism as explained in chapter 6. A higher-than-equilibrium cohort of adult female parasitoids imposes a higher-than-equilibrium attack rate on young immature hosts. This immediately reduces the recruitment rate to the old immature host stage, hence leading immediately to a lower-than-equilibrium per-head recruitment rate to the juvenile parasitoid stage. This in turn yields a lower-than-equilibrium per-head recruitment rate to the adult parasitoid stage T_p days later. The basic time delay is therefore the parasitoid development time, T_p.

Just as in the delayed-feedback cycles of a single-species model, the period of the cycles in this model increases at approximately $2T_P$. As discussed in chapter 6, however, other processes also affect the period. They effectively add a constant to period length.

CYCLES IN REAL SYSTEMS: SINGLE-SPECIES MODELS FOR MANY-SPECIES SYSTEMS

We made two broad claims in the above discussion. First, there are three types of cycles that are general to structured consumer-resource models. Second, the first two types of cycle (single generation and delayed feedback) occur in the consumer population when it acts effectively like a single-species population. Two conditions appear to be essential for such single-species-like cycles to appear. First, there is a strong stabilizing process in the resource population, such as constant recruitment, which partially decouples the consumer-resource interaction and suppresses the consumer-resource cycles. Second, there is competition between consumer age or stage classes that causes the cycles by inducing either direct or delayed density dependence.

Weakly Coupled Interactions: A Conjecture about Generalist Consumers

In this section we ask where in real communities we are most likely to find the decoupling between consumer and resource that allows single-species dynamics to emerge in consumer populations. We conjecture that such decoupling is probably most prevalent in generalist consumers that feed on several to many different resource species; i.e., in many-species food webs. The idea is that a consumer that is connected to many species cannot be tightly connected to any one. Multiple connections can arise because the consumer feeds on multiple resource species at any one time and/or changes resources as it develops (ontogenetic shift), or may even feed across many spatially separated resource populations. Effectively, the total recruitment rate of the resource populations should be largely independent of consumer density. As a corollary, and not surprisingly, we expect consumer-resource cycles to be most common in specialist consumers tightly coupled to a single resource population (Murdoch et al. 2002).

To test this conjecture, we need to distinguish between a set of generalist and specialist consumers, and then we need to show that the former show mainly single-species-like dynamics. We expect the latter to show mainly tightly coupled consumer-resource interactions. We look to cyclic populations to distinguish between these two types of dynamics, because as we illustrate below, we can use properties such as their period to provide insight into the mechanism that produces them. To this end, in the next subsection we show how single-species-like cycles can be distinguished by their period from true consumer-resource cycles.

Classification of Cycles Based on Cycle Period

As noted earlier, we think of the population in a single-species model as a consumer (of an implicit resource). We also saw earlier that single-generation cycles in single-species models have a period between $1T_c$ and $2T_c$, where T_c is the development time of the population (the subscript "c" indicating "consumer"). Delayed-feedback cycles in single-species models have a period between $2T_c$ and $4T_c$ (chapter 4). No population in nature is alone, so we define single-species-like cycles as those having periods between $1T_c$ and $4T_c$.

It turns out we can distinguish single-species-like cycles from true consumer-resource cycles because the latter have, under most realistic parameter values, a minimum period of $4T_c + 2T_r$, where T_r is the development time of the resource population (Lauwerier and Metz 1986; Murdoch et al. 2002; R. Nisbet, C. Briggs, and W. Murdoch, unpubl. results). The periods of single-species cycles and consumer-resource cycles thus do not overlap. We saw specific cases of the minimum consumer-resource period when we discussed the Nicholson-Bailey model in chapter 4 and the Basic model in chapter 5. Since these two models cover discrete- and continuous-time approaches, the rule is likely to be quite general.

When both consumer and resource species take 1 time unit (say 1 year) to develop, the minimum period of consumer-resource cycles is 6 (= $4T_c + 2T_r$). Murdoch et al. (2002) and R. Nisbet, C. Briggs, and W. Murdoch (unpubl. results) show that density dependence in the resource population can reduce the period below 6. For example, it appears that in the Nicholson-Bailey model with resource density dependence, the period can decrease to about 5.7 when R, the resource multiplication rate in the absence of the consumer and density dependent mortality, is at the upper end of the realistic range ($R = 10$).

Real consumer-resource periods are likely to be at least as long as the minimum. The period increases as R decreases. In both the Nicholson-Bailey and Basic models, the minimum period increases when adults survive and reproduce for an extended period. In discrete-time models, this means some consumers live to reproduce in more than one generation. In the Basic model it simply means that the consumer adult stage is longer lived. Finally, the minimum period holds along the stability boundary, and the period will almost always be greater than the minimum in the unstable region away from the boundary. For practical purposes we can thus expect $4T_c + 2T_r$ to be the minimum period of real consumer-resource cycles.

One class of cycles does not fit this neat distinction between single-species-like and consumer-resource dynamics. These are long-period cycles generated in single-species models in which some density-dependent process acts with a delay of two generations. These can be single-species analogs of consumer-resource interaction, or they can describe, for example, a process (maternal effect) such as the following. Current high density suppresses the size of offspring produced by a female, these offspring are consequently less fecund when they become adult, and this effect is expressed in the number of adults recruiting to the following generation (Ginzburg and Taneyhill 1994). We assume this type of process does not drive the observed long-period cycles we discuss. If it does drive some, it adds to the prevalence of single-species-like behavior.

ANALYSIS OF REAL POPULATION CYCLES

We analyzed more than 100 cyclic population time series comprising 40 species, each at least 25 years long and recorded annually (Kendall et al. 1998). The data are from the Global Population Dynamics Database (NERC Centre for Population Biology, Imperial College 1999), plus a few found while searching the literature for relevant natural history information. To conform to the discrete-time theory, we included only species that take at least 1 year to mature and that breed annually.

We classified each species into one of two trophic roles, based mainly on the literature. *Generalists* (e.g., predatory marine or freshwater fish) feed on many resource species. Generalist consumers—usually top predators with broad diets—typically could be categorized easily and unambiguously. *Specialists* (e.g., lynx) feed largely on one species, or are prey of specialist predators (e.g., snowshoe hares). Lepidopteran forest pests were classified as specialist prey (of parasitoids) (Berryman 1996). The remaining specialists are the main protagonists in mammalian cycles in north temperate regions.

We then estimated cycle periods and calculated scaled periods (details of the time series analyses are in Murdoch et al. 2002). Scaled period, τ, was calculated by dividing the period, in years, by time to maturity in years, T_c. For cases in which τ was greater than 4, we then asked if the cycle period in years was equal to or greater than $4T_c + 2T_r$.

We are mainly concerned with generalists. The specialists in the database are well known to have long-period cycles, so cannot be used to test the conjecture, though they allow us to test the prediction about minimum consumer-resource cycle length. The great majority of cyclic series from specialists are true consumer-resource cycles, and as predicted these periods all exceed $4T_c + 2T_r$ (Fig. 11.4b).

The data give strong support to our conjecture. First, specialists and generalists do not separate cleanly when we examine period measured in years: this is mainly because a substantial fraction of generalist periods are longer than 4, or even 6, years (Fig. 11.4a). Scaled period, however, separates the two classes almost perfectly (Fig. 11.4b). In contrast with the specialists, all but 2 of the 66 generalist consumer series show single-species-like cycles (SGC and DFC in Fig. 11.4b), including those with periods greater than 4 years. There are no cycles with intermediate scaled period, i.e., between $4T_c$ and $4T_c + 2T_r$, even though many periods lie between 4 and 6 years (Fig. 11.4a). The same patterns occur if we consider species rather than populations (legend, Fig. 11.4b).

This analysis suggests that the strong connection between single-species dynamics and partially coupled consumer-resource dynamics found in models also occurs in nature.

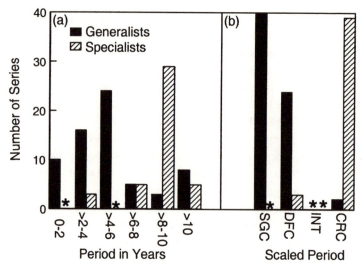

FIGURE 11.4. Cycles from natural populations classified by period; * indicates there were no cycles in this class. (a) Number of cyclic populations with various periods, in years. (b) Distribution of cycles among classes defined by period scaled by development time. SGC: single-generation cycles ($\tau = 1$); DFC: delayed-feedback cycles ($2 \leq \tau \leq 4$); CRC: consumer-resource cycles (period in years $\leq 4T_C + 2T_R$). No cycles fall in the intermediate class (INT) between single-species and consumer-resource cycles. The 66 generalist series came from 25 species: 11 marine and freshwater fish, 5 birds, 5 mammals, and 4 invertebrates, 2 of which (carabid beetles) had CRC periods. The 42 specialist series came from 9 forest Lepidoptera and 6 high-latitude mammals; 14 species showed CRCs, 2 (1 of which also had a CRC series) showed DFCs. Copyright 2002, MacMillan Magazines Limited, reprinted with permission from Murdoch et al. 2002.

GENERAL CONCLUSIONS/CONSIDERATIONS

In this chapter we have emphasized connections among different types of theory for population dynamics. Our first major message relates to the stage-structured parasitoid-host models that constitute chapters 5–8: although their structure appears quite different from more familiar, simpler, consumer-resource models, they are seamlessly connected to the rest of population dynamics theory. This connection extends to single-species models, to consumer-resource models in continuous-time (Lotka-Volterra-like models), and to single-species and consumer-resource models in discrete time. The broader message is that consumer-resource theory is thus quite unified and internally coherent.

The second major message concerns the link between consumer-resource and single-species dynamics. This connection is especially useful in giving insight into the cause and nature of different types of cycles. Shorter-period cycles in the consumer can be usefully thought of as arising because the consumer acts as if it were a single-species population supplied with food at a rate unrelated to the system's internal dynamics. Decoupling of consumer-resource dynamics is key, and paradoxically, it appears that species diversity can be the cause of such dynamical and conceptual simplification.

We know that most populations exist in complex food webs. Most populations interact with many others, as resources, competitors, or consumers. Yet for almost a century theoretical ecologists have effectively ignored this complexity. They have instead written models for population dynamics that include one, two, sometimes three, but rarely more species. Our theoretical and empirical results give support to this traditional simplifying approach. Cyclic populations appear to fall into those for which a single-species model may be adequate, and those for which a two-species consumer-resource model is needed.

Murdoch and Walde (1989) and Murdoch (1993) made the case that weak coupling might be widespread in nature, not only in cyclic populations. They argued that if this is the case, few-species models can be good representations in general of population dynamics in many-species communities. Murdoch (1993) discussed how weak coupling can happen via a separation of time scales. We here suggest that multiple prey species provide another mechanism for weak coupling. These are encouraging results for those who hope for relatively simple theory.

CHAPTER TWELVE
Concluding Remarks

Here we summarize insights developed in earlier chapters. We also comment on some promising areas of future research.

Biological Realism: New Phenomena and New Explanations

We began developing stage-structured models more than a decade ago because we were interested in how individual properties affect population dynamics. The models have uncovered new dynamic phenomena and hence provide new predictions, and explanations, for observed patterns not otherwise explicable.

Among the phenomena emerging from stage-structured models are novel stabilizing mechanisms and three new types of cycles in consumer-resource systems: single-generation cycles whose periods are determined mainly by the resource development lag, and single-generation and delayed-feedback cycles, both of whose periods are determined mainly by the consumer development lag. As noted below, these cycles also provide novel connections to single-species dynamics. Stage-structured models are also more reliable guides than unstructured models to the expected period of true consumer-resource cycles. Such models in this book have also revealed new ways for species to win in competition, new ways to coexist in competition, and new factors to consider when choosing natural enemies for pest control.

Although we have found real examples that seem to illustrate some of these new theoretical phenomena, research is now needed to discover which occur in the real world, and under what circumstances. There has been relatively little population dynamics done on appropriate systems, for example parasitoid-host interactions in climates inducing more or less continuous reproduction. Such systems are the norm in the tropics and subtropics, yet the current literature on insect field populations is dominated by annual (univoltine) species from the north temperate zone.

Biological Realism: Scope and Generality

Although at first sight the most apparent strength of stage-structured models is their ability to describe particular species, they have turned out to have wide scope. The structured parasitoid-host models of chapters 5–7 are able to incorporate most of the observed variation in life history, natural history, and parasitoid "decisions" seen in real systems. They are broad in scope. Because many different hosts and parasitoids share what turn out to be fundamental properties, we were able to arrive in chapter 6 at a single generic gain model, with invulnerable adults and arbitrary "gain" function, that serves to define the dynamical outcomes of their interactions.

Immediately, however, we need to recognize a fundamental limitation of ecological theory. It is true that the generic gain model is general: it fits many apparently disparate life and natural histories. The model then tells us that if these are indeed the salient features of the interaction, the dynamics will be as predicted. But it does not tell us that parasitoid-host systems with these features will behave as predicted. That is because other aspects of natural history such as spatial processes, predators, hyperparasitoids, and so on may be more important. This is of course a feature of all ecological theory—and of ecological systems.

The focal mechanisms in these models are the existence of one or more vulnerable stages/ages in the host's life history, and the fact that the gain to the parasitoid population changes (typically increases) with the age of the attacked host. These features are of course not peculiar to parasitoid-host interactions, and research extending them to true predator-prey systems should be fruitful. In fact, although we had parasitoids and hosts in mind when developing the models, we showed in chapter 11 that the models can be adapted to a wide range of consumer-resource interactions.

Biological Realism: Generality and Testability

Creating models that are both general and testable is a fundamental difficulty in ecology. Models lose generality if they are too particular;

but a model needs enough specificity to be tested against a particular system. Lacking adequate specificity, we are often forced to make do with models that offer plausible explanations rather than directly testable predictions; many confrontations between model and nature in ecology are arguably of this sort. Recent advances in statistical methodology, notably in non-linear time series analysis, are likely to increase the opportunities for true tests of theory, but the lack of sufficient specific information will always remain a problem.

We believe that developing theory in the form of a hierarchy of models is a broadly useful response to this dilemma (Murdoch and Nisbet 1996), and we briefly summarize the approach using studies of red scale and *Aphytis* as illustration. The problem to be explained is that *Aphytis* suppresses red scale to extremely low densities relative to the limit set by red scale resources (a citrus tree) but that, contrary to theoretical expectations, the populations are stable (chapter 2).

Preliminary analyses placed this field system well into the unstable region of parameter space of our Basic stage-structured model (Fig. 5.6b) but in the locally stable region of the sex-allocation model (Fig. 6.3c). This suggests that the invulnerable stage plus the gain mechanism can account for stability. The problem with accepting this explanation (in addition to the presence of multiple equilibria), however, is that these models, and even the more realistic models of chapters 6 and 7, still lack features of the real system that may be significant. In particular, they do not limit the rate of *Aphytis* egg production, yet *Aphytis* cannot mature more than about 8 new eggs per day; they also lack some host-stage-specific details such as probability of host-feeding versus parasitism and number of eggs per host meal.

We therefore developed a day-by-day simulation that is stage and state structured, just as are the models of chapters 6 and 7, but that contains the missing details (C. Briggs, W. Murdoch, and S. Swarbrick, unpubl. results). To test the model, we asked it to predict the outcome of a field experiment in which citrus trees were caged and scale outbreaks were created. The abundance of scale, the response of the *Aphytis* population in the cage, and various *Aphytis* behaviors were tracked over the duration of this experiment, which was ending as this book was being written. Preliminary data analysis shows, astonishingly, that *Aphytis* suppressed the experimental scale outbreaks within roughly a single scale generation. A preliminary version of the model predicts this remarkable result. Thus there is evidence that the model is directly testable in the field.

The approach has two salient features. First, the model makes detailed predictions (e.g., scale stage distributions, fractions parasitized, and immature *Aphytis* distributions on each date) that can be tested against the data. Second, the model, though it corresponds to this particular system, is clearly linked, in structure and output, to the more general models of chapters 6 and 7. In particular, output from the simulation model, in spite of its many details, is similar to output from an analytical model that is a small extension of those in chapters 6 and 7 (the "Parasitoid state- and host size-dependent sex-allocation and host-feeding model" in Murdoch et al. 1997).

The hierarchy in this example is laid out in Fig. 12.1. The general model in the top level of Fig. 12.1 covers many life- and natural-history patterns, and probably hundreds of thousands of species (chapter 6). Each model in the second level describes a single mechanism, but they are all manifestations of the general gain phenomenon. Any particular parasitoid-host interaction, in the third level, is some combination of these mechanisms, and with the addition of detail can be made testable. Thus, although no single model is both general and directly testable, the hierarchy of models is. Chesson (1994) has similarly developed a hierarchy for interspecific competition theory.

Coherent Theory for Population Dynamics

There are important differences among the different types of organisms for which consumer-resource theory has been developed, and models in the major subareas of the field reflect these differences. Parasitoids and hosts differ in important ways from true predators and prey, from herbivores and plants, and from disease organisms and their hosts. Consumer-resource theory is also concerned with a range of different processes. Age or stage structure, as we have seen, affects dynamics in a variety of ways, as do different types of spatial structure and spatial processes. There may be several or many competing resource or consumer species. One strength of consumer-resource theory is that it has evolved to handle these variations as they have caught our attention.

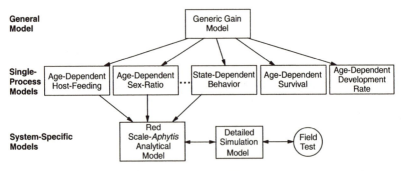

FIGURE 12.1. Hierarchy of models for parasitoid-host interactions. The collection of models in the hierarchy extends from a non-specific general model to a model testable in the field.

We hope, however, that this book has made it clear that consumer-resource theory is not simply a disparate collection of models. The previous section, for example, reinforced the message from chapters 5–7 that our stage-structured parasitoid-host models collectively form an internally coherent set of theory. We saw in chapters 3, 4, and 11 that the variants of consumer-resource theory can be seen as branching off from the common origin of the Lotka-Volterra model. Perhaps surprisingly, two major variants that cut across kinds of organisms—age structure and spatial structure—take the same form in their simplest manifestations, and hence share a fundamental stabilizing mechanism (chapter 3, also see chapters 5 and 11).

A wide range of apparently diverse consumer-resource models have common properties. These include the relationship between resource parameters and consumer equilibrium, a tendency to cycle, shared stabilizing and destabilizing features (though the contrast between aggregation in continuous time and discrete time shows there are exceptions), and the trade-off between suppression of the resource population and stability. When cycles appear in various forms of consumer-resource models, they appear to fall into the same small number of types—true consumer-resource cycles, single-generation cycles, and delayed-feedback cycles.

Finally, from our running theme of population cycles, we saw that consumer-resource theory merges seamlessly with single-species theory. In particular, the former collapses to dynamics very much like the latter under conditions that we can define quite rigorously (chapter 11).

The Importance of Simple Models

In the first section of this chapter we emphasized that new dynamic phenomena appear when we add realism, and hence complexity, to consumer-resource models. We emphatically do not want to leave an implication, however, that simple models are not useful. To the contrary, we believe simple models play an essential role in ecology.

As an aside, occasionally we can use a simple model directly to interpret field data, as we did in the *Daphnia*-algal model (chapter 3). This was possible because the parameters in the model corresponded quite directly with real vital rates under field conditions. Commonly, comparison of field results and simple models is an exercise in plausibility, and may mislead if the parameters do not correspond to real biological entities. This is why in earlier chapters we took care to formulate models in a way that incorporated specific mechanisms, a good example being the derivation of the Nicholson-Bailey model in chapter 4. However, simple models derive much of their importance from a less direct use.

Simple models of population dynamics typically simplify reality in three ways. First, they ignore many differences among individuals. Much of this book has explored the consequences of relaxing this simplification. Second, spatial processes are usually simplified or ignored. Finally, most models deal with only a small number of interacting species.

Consider the first simplification—ignoring differences among individuals. We noted above that differences among individuals give rise to new dynamics that cannot be understood without dealing explicitly with such differences. But we also noted that many properties of simple models survive in more complex models, including factors affecting equilibrium values and stability. It is this combination that leads to the major usefulness of simple models: they are essential tools in understanding more complex models and hence, we hope, real systems.

The persisting similarities between simple and complex models allow us to see processes that are operating in complex models, processes that would otherwise be obscured by the complexity. A single example will suffice. We saw in chapter 6 that gain increasing with the size of the encountered host can both stabilize and destabilize the equilibrium and leads to two new kinds of cycles. It is very difficult to understand how these effects arise when they are first encountered. They can be understood only when we factor out host dynamics and think about direct or delayed density dependence in the parasitoid population. This comes naturally because we are used to thinking of such density dependences in simple single-species models.

Simple models are also useful because they serve as a benchmark for evaluating differences between them and complex models. There are many examples in the book. Indeed, the most dramatic is within simple models themselves. We can use the Lotka-Volterra model, with its neutrally stable equilibrium, to detect and understand each of many stabilizing or destabilizing processes. And this carries over to how they work in more complex models. The new phenomenon of single-generation cycles (chapter 5) appears when we add density dependence to the parasitoid because this damps the consumer-resource cycles and exposes the single-generation period inherent in the Basic model. Chapter 10 showed how the simple Rosenzweig-MacArthur model, in both a spatial and non-spatial form, helped de Roos et al. (1998) uncover the essential mechanisms of their spatially explicit individual-based models.

Indeed, we would make a stronger argument. Simple models are not only useful, they are essential in understanding complex models. Unless we can explain the processes and outcomes of complex models in terms of, or as extensions and modifications of, simpler models, we do not understand the complex model. This is one reason we feel that hierarchies of models are central in theory for population dynamics.

The second simplification was space, and here we think the case for simple models is particularly strong. Much simulation work has been done on complicated spatially explicit models, and there has also been substantial growth in complex semi-analytical models. We have not surveyed this vast field. We believe it is clear, however, that two major processes that tend to damp global dynamics in spatial consumer-resource models are best understood as variants on processes seen in simple models. The first is coupled asynchronous population trajectories in different parts of space. The second is the combination of spatial variance and non-linear processes that serve to move a system toward a stable region of parameter space (chapter 10).

We must not ignore the possibility, however, that we may need a different kind of (relatively) simple model for some spatial processes.

In an important book, Dieckmann et al. (2000) argue that models of spatial processes that give generic insight (and that move on from spatially explicit individual-based models) may include diffusion approximations using partial differential equations, or new methods such as pair approximations and correlation dynamics. The latter types of "simple" models are in their infancy.

Finally, models of population dynamics rarely have more than three species and usually have one or two. But we know that each population is typically part of a complex food web and interacts with many other species. This reductionism among theorists (and indeed, usually among empiricists) has been a constant source of unease among ecologists. Can we safely extract for analysis these smaller components without losing crucial effects induced by the missing interactions?

The final answer is of course not yet in. But we are encouraged by our analyses of types of cycles. First, a consumer-resource cycle is itself, a strong signal that dynamics are determined mainly by a strongly-coupled two-species consumer-resource interaction. P. Turchin (pers. comm.) provides a real example. Some 90% of the variability in larch budmoth populations (Fig. 2.2a), and the parasitism rate of their caterpillars, is explained by a simple parasitoid-host model. Adding the effect of food (pine needle) quality explains only a few percent more.

Second, the models predict that single-species-like cycles may occur when the consumer and resource populations are only weakly coupled. We argued that was likely to occur in real communities when the consumer attacked many independently fluctuating prey, and so could not be strongly coupled with any one of them. Thus, paradoxically, we might expect single-species models to serve us well when the consumer is in a species-rich web. Real cyclic populations gave evidence that this indeed is the case (chapter 11). Thus, there is support for continuing to use few-species models for many-species systems.

Connecting Theory to Nature

We have concentrated mostly on theory, not on testing it in the field. But of course theory is meant to explain reality, and hence needs to be tested against real systems. Making this connection is surely the hardest part of ecology, and also the least explored.

We hope that theorists, experimentalists, and field biologists reading this book will be encouraged to forge this connection. By this we do not mean to suggest that the theory is complete. Indeed, we suspect it is still missing major parts of the explanation for the regulation of many real populations and interacting systems. We also believe that a more complete theory will most easily arise from analysis of real systems done with theory in mind.

If field ecologists are to play their full role, however, they need to have some confidence that theory can connect with nature. In our view, models that do this best are likely to strike a balance between realism and simplicity, and we hope that many of the models in this book do so.

Literature Cited

Abdelrahman, I. 1974. Studies in ovipositional behaviour and control of sex in *Aphytis melinus* DeBach, a parasite of California red scale, *Aonidiella aurantii* (Mask.). *Australian Journal of Zoology* 22:231–247.

Abrams, P. A., and L. R. Ginzburg. 2000. The nature of predation: prey dependent, ratio dependent or neither. *Trends in Ecology and Evolution* 15:337–341.

Abrams, P. A., and C. J. Walters. 1996. Invulnerable prey and the paradox of enrichment. *Ecology* 77:1125–1133.

Adams, E. S., and W. R. Tschinkel. 2001. Mechanisms of population regulation in the fire ant *Solenopsis invicta:* an experimental study. *Journal of Animal Ecology* 70:355–369.

Adler, F. R. 1998. *Modeling the Dynamics of Life: Calculus and Probability for Life Scientists.* Pacific Grove, Calif.: Brooks/Cole.

Alroy, J. 2000. New methods for quantifying macroevolutionary patterns and processes. *Paleobiology* 26:707–733.

Amarasekare, P. 1998. Allee effects in metapopulation models. *American Naturalist* 152:298–302.

———. 2000a. Coexistence of competing parasitoids on a patchily distributed host: local vs. spatial mechanisms. *Ecology* 81:1286–1296.

———. 2000b. Spatial dynamics in a host-multiparasitoid community. *Journal of Animal Ecology* 69:201–213.

Amarasekare, P., and R. M. Nisbet. 2001. Spatial heterogeneity, source-sink dynamics, and the local coexistence of competing species. *American Naturalist* 158:572–584.

Anderson, R. M., and R. M. May. 1981. The population dynamics of microparasites and their invertebrate hosts. *Philosophical Transactions of the Royal Society of London*, B 291:451–524.

———. 1991. *Infectious Diseases of Humans: Dynamics and Control.* Oxford: Oxford University Press.

Andrewartha, H. G., and L. C. Birch. 1954. *The Distribution and Abundance of Animals.* Chicago: University of Chicago Press.

Argov, Y., and Y. Rossler. 1993. Biological control of the Mediterranean black scale *Saissetia oleae* Hom. Coccidae in Israel. *Entomophaga* 38:98–100.

Arino, A., and S. L. Pimm. 1995. On the nature of population extremes. *Evolutionary Ecology* 9:429–443.

Armstrong, R. A., and R. McGehee. 1980. Competitive exclusion. *American Naturalist* 115:151–170.

Atkinson, W. D., and B. Shorrocks. 1981. Competition on a divided and ephemeral resource: a simulation model. *Journal of Animal Ecology* 50:461–471.

———. 1984. Aggregation of larval Diptera over discrete and ephemeral breeding sites: the implications for coexistence. *American Naturalist* 124:336–351.

Auslander, D. M., G. F. Oster, and C. B. Huffaker. 1974. Dynamics of interacting populations. *Journal of the Franklin Institute* 297:345–367.

Bailey, V. A., A. J. Nicholson, and E. Williams. 1962. Interaction between hosts and parasites when some host individuals are more difficult to find than others. *Journal of Theoretical Biology* 3:1–18.

Banks, C. J. 1957. The behavior of individual coccinellid larvae on plants. *Animal Behaviour* 5:12–24.

Bartlett, B. R. 1964. Patterns in the host feeding habit of adult parasitic Hymenoptera. *Annals of the Entomological Society of America* 57:344–350.

Bartlett, M. S. 1957. On theoretical models for competitive and predatory biological systems. *Biometrika* 44:27–42.

Bauer, E., T. Trenczek, and S. Dorn. 1998. Instar-dependent hemocyte changes in *Pieris brassicae* after parasitization by *Cotesia glomerata*. *Entomologia Experimentalis et Applicata* 88:49–58.

Beddington, J. R. 1975. Mutual interference between parasites or predators and its effects on searching efficiency. *Journal of Animal Ecology* 44:331–340.

Beddington, J. R., C. A. Free, and J. H. Lawton. 1975. Dynamic complexity in predator-prey models framed as difference equations. *Nature* 255:58–60.

Bernstein, C. 2000. Host-parasitoid models: the story of a successful failure. Pp. 41–57 *in* M. E. Hochberg and A. R. Ives, eds., *Parasitoid Population Biology*, Princeton, N.J.: Princeton University Press.

Berryman, A. A. 1996. What causes population cycles of forest Lepidoptera? *Trends in Ecology and Evolution* 11:28–32.

Bertschy, C., T. C. J. Turlings, A. Bellotti, and S. Dorn. 2000. Host stage preference and sex allocation in *Aenasius vexans*, an encyrtid parasitoid of the cassava mealybug. *Entomologia Experimentalis et Applicata* 95:283–291.

Boavida, C., M. Ahounou, M. Vos, P. Neuenschwander, and J. J. M. van Alphen. 1995. Host stage selection and sex allocation by *Gyranusoidea tebygi* (Hymenoptera: Encyrtidae), a parasitoid of the mango mealybug, *Rastrococcus invadens* (Homoptera: Pseudococcidae). *Biological Control* 5:487–496.

Bokonon-Ganta, A. H., P. Neuenschwander, J. J. M. van Alphen, and M. Vos. 1995. Host stage selection and sex allocation by *Anagyrus mangicola*

(Hymenoptera: Encyrtidae), a parasitoid of the mango mealybug, *Rastrococcus invadens* (Homoptera: Pseudococcidae). *Biological Control* 5:479–486.

Bonsall, M. B., and M. P. Hassell. 1995. Identifying density-dependent processes: a comment on the regulation of winter moth. *Journal of Animal Ecology* 64:781–784.

Brauer, F. 1979. Boundedness of solutions of predator-prey systems. *Theoretical Population Biology* 15:268–273.

Briggs, C. J. 1993a. Competition among parasitoid species on a stage-structured host and its effect on host suppression. *American Naturalist* 141:372–397.

———. 1993b. The effect of multiple parasitoid species on the gall-forming midge, *Rhopalomyia californica*. Ph.D. diss., University of California, Santa Barbara, Calif.

Briggs, C. J., and H. C. J. Godfray. 1995a. The dynamics of insect-pathogen interactions in stage-structured populations. *American Naturalist* 145:855–887.

———. 1995b. Models of intermediate complexity in insect-pathogen interactions: population dynamics of the microsporidian pathogen, *Nosema pyrausta*, of the European corn borer, *Ostrinia nubilalis*. *Parasitology* 111:S71–S89.

———. 1996. The dynamics of insect-pathogen interactions in seasonal environments. *Theoretical Population Biology* 50:149–177.

Briggs, C. J., and J. Latto. 2000. The effect of dispersal on the population dynamics of a gall-forming midge and its parasitoids. *Journal of Animal Ecology* 69:96–105.

Briggs, C. J., W. W. Murdoch, and R. M. Nisbet. 1999a. Recent developments in theory for biological control of insect pests by parasitoids. Pp. 22–42 *in* B. A. Hawkins and H. V. Cornell, eds., *Theoretical Approaches to Biological Control*, Cambridge: Cambridge University Press.

———. 1993. Coexistence of competing parasitoid species on a host with a variable life cycle. *Theoretical Population Biology* 44:341–373.

———. 1999b. Host age-specific parasitoid gain, delayed-feedback, and multiple attractors in a host-parasitoid model. *Journal of Mathematical Biology* 38:317–345.

Briggs, C. J., R. M. Nisbet, W. W. Murdoch, T. R. Collier, and J. A. J. Metz. 1995. Dynamical effects of host-feeding in parasitoids. *Journal of Animal Ecology* 64:403–416.

Briggs, C. J., S. M. Sait, M. Begon, D. J. Thompson, and H. C. J. Godfray. 2000. What causes generation cycles in populations of stored-product moths? *Journal of Animal Ecology* 69:352–366.

Brillinger, D. R., J. Guckenheimer, P. Guttorp, and G. Oster. 1980. Empirical modelling of population time series data: the case of age and density dependent vital rates. *Lectures on Mathematics in the Life Sciences* 13:65–90.

Burnham, K. P., and D. R. Anderson. 1998. *Model Selection and Inference: A Practical Information-Theoretic Approach.* New York: Springer-Verlag.

Casas, J. 1989. Foraging behaviour of a leafminer parasitoid in the field. *Ecological Entomology* 14:257–265.

Case, T. J. 2000. *An Illustrated Guide to Theoretical Ecology.* New York: Oxford University Press.

Caswell, H. 2001. *Matrix Population Models: Construction, Analysis, and Interpretation.* Sunderland, Mass.: Sinauer Associates.

Chan, M. S., and H. C. J. Godfray. 1993. Host-feeding strategies of parasitoid wasps. *Evolutionary Ecology* 7:593–604.

Chang, G. C., and P. Kareiva. 1999. The case for indigenous generalists in biological control. Pp. 103–115 *in* B. A. Hawkins and H. V. Cornell, eds., *Theoretical Approaches to Biological Control,* Cambridge: Cambridge University Press.

Charnov, E. L. 1982. *The Theory of Sex-Allocation.* Princeton, N.J.: Princeton University Press.

Chatfield, C. 1996. *The Analysis of Time Series, An Introduction.* London: Chapman and Hall.

Chau, A., and M. Mackauer. 2000. Host-instar selection in the aphid parasitoid *Monoctonus paulensis* (Hymenoptera: Braconidae, Aphidiinae): a preference for small pea aphids. *European Journal of Entomology* 97:347–353.

Chesson, P. L. 1978. Predator-prey theory and variability. *Annual Review of Ecology and Systematics* 9:288–325.

———. 1981. Models for spatially distributed populations: the effect of within-patch variability. *Theoretical Population Biology* 19:288–323.

———. 1982. The stabilizing effect of a random environment. *Journal of Mathematical Biology* 15:1–36.

———. 1986. Environmental variation and the coexistence of species. Pp. 240–256 *in* J. Diamond and T. J. Case, eds., *Community Ecology,* New York: Harper & Row.

———. 1990. Geometry, heterogeneity and competition in variable environments. *Philosophical Transactions of the Royal Society of London,* B 330:165–173.

———. 1994. Multispecies competition in variable environments. *Theoretical Population Biology* 45:227–276.

———. 1996. Matters of scale in the dynamics of populations and communities. Pp. 353–368 *in* R. B. Floyd and A. W. Sheppard, eds., *Frontiers of Population Ecology,* Melbourne, Australia: Commonwealth Scientific & Industrial Research Organization.

————. 2000. General theory of competitive coexistence in spatially-varying environments. *Theoretical Population Biology* 58:211–237.

Chesson, P. L., and N. Huntly. 1988. Community consequences of life-history traits in a variable environment. *Annales Zoologici Fennici* 25:5–16.

————. 1997. The roles of harsh and fluctuating conditions in the dynamics of ecological communities. *American Naturalist* 150:519–553.

Chesson, P. L., and W. W. Murdoch. 1986. Aggregation of risk: relationships among host-parasitoid models. *American Naturalist* 127:696–715.

Clark, C. W., and M. Mangel. 2000. *Dynamic State Variable Models in Ecology: Methods and Applications*. New York: Oxford University Press.

Clausen, C. P. 1978. *Introduced Parasites and Predators of Arthropod Pests and Weeds: A World Review*. Agricultural handbook no. 480. Washington, D.C.: Agricultural Research Service, United States Department of Agriculture.

Collier, T. R. 1995a. Adding physiological realism to dynamic state variable models of parasitoid host feeding. *Evolutionary Ecology* 9:217–235.

————. 1995b. Host feeding, egg maturation, resorption, and longevity in the parasitoid *Aphytis melinus* (Hymenoptera: Aphelinidae). *Annals of the Entomological Society of America* 88:206–214.

Collier, T. R., W. W. Murdoch, and R. M. Nisbet. 1994. Egg load and the decision to host-feed in the parasitoid, *Aphytis melinus*. *Journal of Animal Ecology* 63:299–306.

Comins, H. N., and M. P. Hassell. 1996. Persistence of multispecies host-parasitoid interactions in spatially distributed models with local dispersal. *Journal of Theoretical Biology* 183:19–28.

Comins, H. N., M. P. Hassell, and R. M. May. 1992. The spatial dynamics of host-parasitoid systems. *Journal of Animal Ecology* 61:735–748.

Crawley, M. J. 1983. *Herbivory: The Dynamics of Animal-Plant Interactions*. Berkeley, Calif.: University of California Press.

Croft, P., and M. J. W. Copland. 1995. The effect of host instar on the size and sex ratio of the endoparasitoid *Dacnusa sibirica*. *Entomologia Experimentalis et Applicata* 74:121–124.

Crowley, P. H. 1981. Dispersal and the stability of predator-prey interactions. *American Naturalist* 118:673–701.

DeBach, P. 1965. Some biological and ecological phenomena associated with colonizing entomophagous insects. Pp. 287–306 *in* H. G. Baker and G. Ledyard Stebbins, eds., *The Genetics of Colonizing Insects*, New York: Academic Press.

DeBach, P., E. J. Dietrick, C. A. Fleschner, and T. W. Fisher. 1950. Periodic colonization of *Aphytis* for control of the California red scale. Preliminary tests, 1949. *Journal of Economic Entomology* 48:783–806.

DeBach, P., D. Rosen, and C. E. Kennett. 1971. Biological control of coccids by introduced natural enemies. Pp. 165–194 *in* C. B. Huffaker, ed., *Biological Control*, New York: Plenum.

DeBach, P., and P. Sisojevic. 1960. Some effects of temperature and competition on the distribution and relative abundance of *Aphytis lingnanensis* and *Aphytis chrysomphali* (Hymenoptera: Aphelinidae). *Ecology* 41:153–160.

DeBach, P., and R. A. Sundby. 1963. Competitive displacement between ecological homologues. *Hilgardia* 34:105–166.

de Roos, A. M. 1997. A gentle introduction to physiologically structured population models. Pp. 119–204 *in* S. Tuljapurkar and H. Caswell, eds., *Structured-Population Models in Marine, Terrestrial, and Freshwater Systems*, New York: Chapman and Hall.

de Roos, A. M., E. McCauley, and W. G. Wilson. 1991. Mobility versus density-limited predator-prey dynamics on different spatial scales. *Proceedings of the Royal Society of London*, Series B 246:117–122.

———. 1998. Pattern formation and the spatial scale of interaction between predators and their prey. *Theoretical Population Biology* 53:108–130.

de Roos, A. M., J. A. J. Metz, E. Evers, and A. Leipoldt. 1990. A size-dependent predator-prey interaction: who pursues whom? *Journal of Mathematical Biology* 28:609–643.

Dieckmann, U., R. Law, and J. A. J. Metz, eds. 2000. *The Geometry of Ecological Interactions: Simplifying Spatial Complexity*. Cambridge: Cambridge University Press.

Diekmann, O., S. A. Van Gils, S. M. V. Lunel, and H. O. Walther. 1995. *Delay Equations: Functional-,Complex-, and Nonlinear Analysis*. New York: Springer-Verlag.

Dijkstra, L. J. 1986. Optimal selection and exploitation of hosts in the parasitic wasp *Colpoclypeus florus* (Hym: Eulophidae). *Netherlands Journal of Zoology* 36:177–301.

Djemai, I., R. Meyhöfer, and J. Casas. 2000. Geometrical games between a host and a parasitoid. *American Naturalist* 156:257–265.

Durrett, R., and S. A. Levin. 1994. Stochastic spatial models: a user's guide to ecological applications. *Philosophical Transactions of the Royal Society of London*, B 343:329–350.

———. 2000. Lessons on pattern formation from planet WATOR. *Journal of Theoretical Biology* 205:201–214.

Edelstein-Keshet, L. 1986. Mathematical theory for plant-herbivore systems. *Journal of Mathematical Biology* 24:25–58.

———. 1988. *Mathematical Models in Biology*. New York: Random House.

Ehler, L. E. 1977. Natural enemies of cabbage looper on cotton in the San Joaquin Valley. *Hilgardia* 45:73–106.

———. 1985. Species-dependent mortality in a parasite guild and its relevance to biological control. *Environmental Entomology* 14:1–6.

———. 1992. Guild analysis in biological control. *Environmental Entomology* 21:26–40.

———. 1995. Biological control of obscure scale (Homoptera: Diaspididae) in California: an experimental approach. *Environmental Entomology* 24:779–795.

Ehler, L. E., and R. W. Hall. 1982. Evidence for competitive exclusion of introduced natural enemies in biological control. *Environmental Entomology* 11:1–4.

Eisenberg, R. M. 1966. The regulation of density in a natural population of the pond snail, *Lymnea elodes*. *Ecology* 47:889–906.

Ellers, J., J. G. Sevenster, and G. Driessen. 2000. Egg load evolution in parasitoids. *American Naturalist* 156:650–665.

Ellers, J., and J. J. M. van Alphen. 1997. Life history evolution in *Asobara tabida*: plasticity in allocation of fat reserves to survival and reproduction. *Journal of Evolutionary Biology* 10:771–785.

Ellers, J., J. J. M. van Alphen, and J. C. Sevenster. 1998. A field study of size-fitness relationships in the parasitoid, *Asobara tabida*. *Journal of Animal Ecology* 67:318–324.

Ellner, S. 1987. Alternate plant life-history strategies and coexistence in randomly varying environments. *Vegetatio* 69:199–208.

———. 1991. Detecting low-dimensional chaos in population dynamics data: a critical review. Pp. 69–90 *in* J. A. Logan and F. P. Hain, eds., *Chaos and Insect Ecology*. Blacksburg, Va.: Virginia Agricultural Experiment Station and Virginia Polytechnic Institute and State University.

Ellner, S., and P. Turchin. 1995. Chaos in a noisy world: new methods and evidence from time-series analysis. *American Naturalist* 145:343–375.

Ellner, S. P., B. A. Bailey, G. V. Bobashev, A. R. Gallant, B. T. Grenfell, and D. W. Nychka. 1998. Noise and nonlinearity in measles epidemics: combining mechanistic and statistical approaches to population modeling. *American Naturalist* 15:425–440.

Ellner, S., E. McCauley, B. Kendall, C. J. Briggs, P. R. Hosseini, S. Wood, A. Janssen, M. W. Sabelis, P. Turchin, R. M. Nisbet, and W. W. Murdoch. 2001. Habitat structure and population persistence in an experimental community. *Nature* 412:538–543.

Fletcher, J. P., J. P. Hughes, and I. F. Harvey. 1994. Life expectancy and egg load affect oviposition decisions of a solitary parasitoid. *Proceedings of the Royal Society of London*, Series B 258:163–167.

Force, D. C. 1970. Competition among four hymenopterous parasites of an endemic insect host. *Annals of the Entomological Society of America* 63:1675–1688.

———. 1974. Ecology of insect-parasitoid communities. *Science* 184:624–632.

Free, C. A., J. R. Beddington, and J. H. Lawton. 1977. On the inadequacy of simple models of mutual interference for parasitism and predation. *Journal of Animal Ecology* 46:543–554.

Gilpin, M., and I. Hanski, eds. 1991. *Metapopulation Dynamics: Empirical and Theoretical Investigations*. London: Academic Press.

Ginzburg, L. R., and H. R. Akcakaya. 1992. Consequences of ratio-dependent predation for steady-state properties of ecosystems. *Ecology* 73:1536–1543.

Ginzburg, L. R., and D. E. Taneyhill. 1994. Population cycles of forest Lepidoptera: a maternal effect hypothesis. *Journal of Animal Ecology* 63:79–92.

Godfray, H. C. J. 1994. *Parasitoids: Behavioral and Evolutionary Ecology*. Princeton, N.J.: Princeton University Press.

Godfray, H. C. J., and M. P. Hassell. 1987. Natural enemies can cause discrete generations in tropical environments. *Nature* 327:144–147.

———. 1989. Discrete and continuous insect populations in tropical environments. *Journal of Animal Ecology* 58:153–174.

Godfray, H. C. J., and S. W. Pacala. 1992. Aggregation and the population dynamics of parasitoids and predators. *American Naturalist* 140:30–40.

Godfray, H. C. J., and J. K. Waage. 1991. Predictive modelling in biological control: the mango mealy bug (*Rastrococcus invadens*) and its parasitoids. *Journal of Applied Ecology* 28:434–453.

Gordon, D. M., R. M. Nisbet, A. de Roos, W. S. C. Gurney, and R. K. Stewart. 1991. Discrete generations in host-parasitoid models with contrasting life cycles. *Journal of Animal Ecology* 60:295–308.

Grenfell, B., and J. Harwood. 1997. (Meta)population dynamics of infectious diseases. *Trends in Ecology and Evolution* 12:395–399.

Griffiths, K. J. 1969. The importance of coincidence in the functional and numerical responses of two parasites of the European pine sawfly, *Neodiprion sertifer*. *Canadian Entomologist* 101:673–713.

Gross, K., and A. R. Ives. 1999. Inferring host-parasitoid stability from patterns of parasitism among patches. *American Naturalist* 154:489–496.

Grover, J. P. 1988. Dynamics of competition in a variable environment: experiments with two diatom species. *Ecology* 69:408–417.

———. 1990. Resource competition in a variable environment: phytoplankton growing according to Monod's model. *American Naturalist* 136:771–789.

———. 1991. Dynamics of competition among microalgae in variable environments: experimental test of alternative models. *Oikos* 62:231–243.

Grover, J. P., and R. D. Holt. 1998. Disentangling resource and apparent competition: realistic models for plant-herbivore communities. *Journal of Theoretical Biology* 191:353–376.

Gurney, W. S. C., S. P. Blythe, and R. M. Nisbet. 1980. Nicholson's blowflies revisited. *Nature* 287:17–21.

Gurney, W. S. C., E. McCauley, R. M. Nisbet, and W. W. Murdoch. 1990. The physiological ecology of *Daphnia*: a dynamic model of growth and reproduction. *Ecology* 71:716–732.

Gurney, W. S. C., and R. M. Nisbet. 1985. Fluctuation periodicity, generation separation, and the expression of larval competition. *Theoretical Population Biology* 28:150–180.

———. 1998. *Ecological Dynamics*. Oxford: Oxford University Press.

Gurney, W. S. C., R. M. Nisbet, and J. H. Lawton. 1983. The systematic formulation of tractable single-species population models incorporating age structure. *Journal of Animal Ecology* 52:479–495.

Gurney, W. S. C., and A. R. Veitch. 2000. Self-organization, scale and stability in a spatial predator-prey interaction. *Bulletin of Mathematical Biology* 62:61–86.

Gurney, W. S. C., A. R. Weitch, I. Cruickshank, and G. McGeachin. 1998. Circles and spirals: population persistence in a spatially explicit predator-prey model. *Ecology* 79:2516–2530.

Gutierrez, A. P., P. Neuenschwander, and J. J. M. van Alphen. 1993. Factors affecting biological control of cassava mealybug by exotic parasitoids: a ratio-dependent supply-demand driven model. *Journal of Applied Ecology* 30:706–721.

Gyori, I., and G. Ladas. 1991. *Oscillation Theory of Delay Differential Equations: With Applications*. Oxford: Oxford University Press.

Haigh, J., and J. Maynard Smith. 1972. Can there be more predators than prey? *Theoretical Population Biology* 3:290–299.

Hanski, I. 1981. Coexistence of competitors in patchy environments with and without predation. *Oikos* 37:306–312.

Hanski, I., and M. Gilpin, eds. 1997. *Metapopulation Biology: Ecology, Genetics, and Evolution*. San Diego: Academic Press.

Hanski, I., A. Moilanen, and M. Gyllenberg. 1996. Minimum viable metapopulation size. *American Naturalist* 147:527–541.

Hanski, I., T. Pakkala, M. Kuussaari, and G. Lei. 1995. Metapopulation persistence of an endangered butterfly in a fragmented landscape. *Oikos* 72:21–28.

Hanski, I., and C. D. Thomas. 1994. Metapopulation dynamics and conservation: a spatially explicit model applied to butterflies. *Biological Conservation* 68:167–180.

Hardin, G. 1960. The competitive exclusion principle. *Science* 131:1292–1297.

Hardy, I. C. W., N. T. Griffiths, and H. C. J. Godfray. 1992. Clutch size in a parasitoid wasp: a manipulation experiment. *Journal of Animal Ecology* 61:121–129.

Harris, P. 1989. The use of Tephritidae for the biological control of weeds. *Biocontrol News Information* 10:7–16.

Harrison, S. 1991. Local extinction in a metapopulation context: an empirical evaluation. Pp. 73–88 *in* M. E. Gilpin and I. Hanski, eds., *Metapopulation Dynamics: Empirical and Theoretical Investigations*, London: Academic Press.

Harrison, S., and A. D. Taylor. 1997. Empirical evidence for metapopulation dynamics. Pp. 27–42 *in* I. Hanski and M. E. Gilpin, eds., *Metapopulation Biology: Ecology, Genetics, and Evolution*, San Diego: Academic Press.

Harvey, J. A., M. A. Jervis, R. Gols, N. Jiang, and L. E. M. Vet. 1999. Development of the parasitoid, *Cotesia rubecula* (Hymenoptera: Braconidae) in *Pieris rapae* and *Pieris brassicae* (Lepidoptera: Pieridae): evidence for host regulation. *Journal of Insect Physiology* 45:172–182.

Harvey, J. A., and L. E. M. Vet. 1997. *Venturia canescens* parasitizing *Galleria mellonella* and *Anagasta kuehniella*: differing suitability of two hosts with highly variable growth potential. *Entomologia Experimentalis et Applicata* 84:93–100.

Hassell, M. P. 1978. *The Dynamics of Arthropod Predator-Prey Systems*. Princeton, N.J.: Princeton University Press.

———. 1984. Parasitism in patchy environments: inverse density dependence can be stabilizing. *IMA Journal of Mathematics Applied in Medicine and Biology* 1:123–133.

———. 2000. *The Spatial and Temporal Dynamics of Host-Parasitoid Interactions*. Oxford: Oxford University Press.

Hassell, M. P., and H. N. Comins. 1978. Sigmoid functional responses and population stability. *Theoretical Population Biology* 14:62–66.

Hassell, M. P., H. N. Comins, and R. M. May. 1991a. Spatial structure and chaos in insect population dynamics. *Nature* 353:255–258.

———. 1994. Species coexistence and self-organizing spatial dynamics. *Nature* (London) 370:290–292.

Hassell, M. P., J. H. Lawton, and R. M. May. 1976. Patterns of dynamical behavior in single species populations. *Journal of Animal Ecology* 45:471–486.

Hassell, M. P., and R. M. May. 1973. Stability in insect host-parasite models. *Journal of Animal Ecology* 42:693–736.

Hassell, M. P., R. M. May, S. W. Pacala, and P. L. Chesson. 1991b. The persistence of host-parasitoid associations in patchy environments: I. A general criterion. *American Naturalist* 138:568–583.

Hassell, M. P., and S. W. Pacala. 1990. Heterogeneity and the dynamics of host-parasitoid interactions. *Philosophical Transactions of the Royal Society of London*, B 330:203–220.

Hastings, A. 1980. Disturbance, coexistence, history, and competition for space. *Theoretical Population Biology* 18:363–373.

———. 1984. Delays in recruitment at different trophic levels: effects on stability. *Journal of Mathematical Biology* 21:35–44.

———. 1987. Cycles in cannibalistic egg-larval interactions. *Journal of Mathematical Biology* 24:651–666.

———. 1991. Structured models of metapopulation dynamics. *Biological Journal of the Linnean Society* 42:57–72.

————. 1995. A metapopulation model with population jumps of varying sizes. *Mathematical Biosciences* 128:285–298.

————. 1997. *Population Biology: Concepts and Models.* New York: Springer-Verlag.

Hastings, A., and R. F. Costantino. 1987. Cannibalistic egg-larva interactions in *Tribolium*: an explanation for the oscillations in population numbers. *American Naturalist* 130:36–52.

Hastings, A., and C. L. Wolin. 1989. Within-patch dynamics in a metapopulation. *Ecology* 70:1261–1266.

Hawkins, B. A., and H. V. Cornell, eds. 1999. *Theoretical Approaches to Biological Control.* Cambridge: Cambridge University Press.

Heimpel, G. E., M. Mangel, and J. A. Rosenheim. 1998. Effects of time limitation and egg limitation on lifetime reproductive success of a parasitoid in the field. *American Naturalist* 152:273–289.

Heimpel, G. E., and J. A. Rosenheim. 1995. Dynamic host feeding by the parasitoid *Aphytis melinus*: the balance between current and future reproduction. *Journal of Animal Ecology* 64:153–167.

————. 1998. Egg limitation in parasitoids: a review of the evidence and a case study. *Biological Control* 11:160-168.

Heimpel, G. E., J. A. Rosenheim, and D. Kattari. 1997. Adult feeding and lifetime reproductive success in the parasitoid *Aphytis melinus*. *Entomologia Experimentalis et Applicata* 83:305–315.

Heimpel, G. E., J. A. Rosenheim, and M. Mangel. 1996. Egg limitation, host quality, and dynamic behavior by a parasitoid in the field. *Ecology* 77:2410–2420.

Higgins, K., A. Hastings, and L. W. Botsford. 1997a. Density dependence and age structure: nonlinear dynamics and population behavior. *American Naturalist* 149:247–269.

Higgins, K., A. Hastings, J. N. Sarvela, and L. W. Botsford. 1997b. Stochastic dynamics and deterministic skeletons: population behavior of Dungeness crab. *Science* (Washington, D.C.) 276:1431–1435.

Hochberg, M. E. 1996. Consequences for host population levels of increasing natural enemy species richness in classical biological control. *American Naturalist* 147:307–318.

Hochberg, M. E., and R. D. Holt. 1999. The uniformity and density of pest exploitation as guides to success in biological control. Pp. 71–88 *in* B. A. Hawkins and H. V. Cornell, eds., *Theoretical Approaches to Biological Control*, Cambridge: Cambridge University Press.

Hogarth, W. L., and P. Diamond. 1984. Interspecific competition in larvae between entomophagous parasitoids. *American Naturalist* 124: 552–560.

Holling, C. S. 1959. Some characteristics of simple types of predation and parasitism. *Canadian Entomologist* 91:385–398.

Holt, R. D., and J. H. Lawton. 1993. Apparent competition and enemy-free space in insect host-parasitoid communities. *American Naturalist* 142:623–645.

Holt, R. D., and G. A. Polis. 1997. A theoretical framework for intraguild predation. *American Naturalist* 149:745–764.

Honda, J. Y., and R. F. Luck. 2000. Age and suitability of *Amorbia cuneana* (Lepidoptera: Tortricidae) and *Sabulodes aegrotata* (Lepidoptera: Geometridae) eggs for *Trichogramma platneri* (Hymenoptera: Trichogrammatidae). *Biological Control* 18:79–85.

Hosseini, P. R. 2001. The effects of localized interactions, movement rules, and space on predator-prey dynamics. Ph.D. diss., University of California, Santa Barbara.

Hu, J. S., and S. B. Vinson. 2000. Interaction between the larval endo-parasitoid *Campoletis sonorensis* (Hymenoptera: Ichneumonidae) and its host the tobacco budworm (Lepidoptera: Noctuidae). *Annals of the Entomological Society of America* 93:220–224.

Huffaker, C. B., ed. 1971. *Biological Control*. New York: Plenum.

Huffaker, C. B., and C. E. Kennett. 1966. Studies of two parasites of olive scale, *Parlatoria oleae* (Colvee). IV. Biological control of *Parlatoria oleae* (Colvee) through the compensatory action of two introduced parasites. *Hilgardia* 37:283–355.

Huffaker, C. B., and P. S. Messenger. 1976. *Theory and Practice of Biological Control*. New York: Academic Press.

Hughes, J. P., I. F. Harvey, and S. F. Hubbard. 1994. Host-searching behavior of *Venturia canescens* (Grav.) (Hymenoptera: Ichneumonidae): superparasitism. *Journal of Insect Behavior* 7:455–464.

Hulot, F. D., G. Lacroix, F. Lescher-Moutoue, and M. Loreau. 2000. Functional diversity governs ecosystem response to nutrient enrichment. *Nature* 405:340–344.

Islam, W. 1994. Effect of host age on rate of development of *Dinarmus basalis* (Rond.) (Hym., Pteromalidae). *Journal of Applied Entomology* 118:392–398.

Ives, A. R. 1991. Aggregation and coexistence in a carrion fly community. *Ecological Monographs* 61:75–94.

Ives, A. R., and W. H. Settle. 1997. Metapopulation dynamics and pest control in agricultural systems. *American Naturalist* 149:220–246.

Iwasa, Y., Y. Suzuki, and H. Matsuda. 1984. Theory of oviposition strategy of parasitoids: 1. Effect of mortality and limited egg number. *Theoretical Population Biology* 26:205–227.

Jansen, V. A. A. 1995. Regulation of predator-prey systems through spatial interactions: a possible solution to the paradox of enrichment. *Oikos* 74:384–390.

Jansen, V. A. A., and A. M. de Roos. 2000. The role of space in reducing predator-prey cycles. Pp. 183–201 *in* U. Dieckmann, R. Law, and

J. A. J. Metz, eds., *The Geometry of Ecological Interactions: Simplifying Spatial Complexity*, Cambridge: Cambridge University Press.

Jansen, V. A. A., R. M. Nisbet, and W. S. C. Gurney. 1990. Generation cycles in stage structured populations. *Bulletin of Mathematical Biology* 52:375–396.

Janssen, A., E. Van Gool, R. Lingeman, J. Jacas, and G. Van De Klashorst. 1997. Metapopulation dynamics of a persisting predator-prey system in the laboratory: time series analysis. *Experimental and Applied Acarology* 21:415–430.

Jensen, A. L., and D. H. Miller. 2001. Age structured matrix predation model for the dynamics of wolf and deer populations. *Ecological Modelling* 141:299–305.

Jervis, M. A., G. E. Heimpel, P. N. Ferns, J. A. Harvey, and N. A. C. Kidd. 2001. Life-history strategies in parasitoid wasps: a comparative analysis of "ovigeny". *Journal of Animal Ecology* 70:442–458.

Jones, A. E., R. M. Nisbet, W. S. C. Gurney, and S. P. Blythe. 1988. Period to delay ratio near stability boundaries for systems with delayed feedback. *Mathematical Analysis and Applications* 135:354–368.

Jones, W. A., and S. M. Greenberg. 1999. Host instar suitability of *Bemisia argentifolii* (Homoptera: Aleyrodidae) for the parasitoid *Encarsia pergandiella* (Hymenoptera: Aphelinidae). *Journal of Agricultural and Urban Entomology* 16:49–57.

Kakehashi, M., Y. Suzuki, and Y. Iwasa. 1984. Niche overlap of parasitoids in host-parasitoid systems: its consequences to single versus multiple introduction controversy in biological control. *Journal of Applied Ecology* 21:115–131.

Karamaouna, F., and M. J. W. Copland. 2000. Host suitability, quality and host size preference of *Leptomastix epona* and *Pseudaphycus flavidulus*, two endoparasitoids of the mealybug *Pseudococcus viburni*, and host size effect on parasitoid sex ratio and clutch size. *Entomologia Experimentalis et Applicata* 96:149–158.

Keeling, M. J., H. B. Wilson, and S. W. Pacala. 2002. Deterministic limits to stochastic spatial models of natural enemies. *American Naturalist* 159:57–80.

Kendall, B. E., C. J. Briggs, W. W. Murdoch, P. Turchin, S. P. Ellner, E. McCauley, R. M. Nisbet, and S. N. Wood. 1999. Why do populations cycle? A synthesis of statistical and mechanistic modeling approaches. *Ecology* 80:1789–1805.

Kendall, B. E., J. Prendergast, and O. N. Bjornstad. 1998. The macroecology of population dynamics: taxonomic and biogeographic patterns in population cycles. *Ecology Letters* 1:160–164.

Kermack, W. O., and A. G. McKendrick. 1927. Contributions to the mathematical theory of epidemics. 1. *Proceedings of the Royal Society* 115A:700–721.

Kfir, R., and R. F. Luck. 1979. Effects of constant and variable temperature extremes on sex ratio and progeny production by *Aphytis melinus* and *Aphytis lingnanensis*. *Ecological Entomology* 4:335–344.

Kidd, N. A. C., and M. A. Jervis. 1989. The effects of host-feeding behavior on the dynamics of parasitoid-host interactions, and the implications for biological control. *Researches on Population Ecology* 31:235–274.

———. 1991a. Host-feeding and oviposition by parasitoids in relation to host stage: consequences for parasitoid-host population dynamics. *Researches on Population Ecology* 33:87–100.

———. 1991b. Host-feeding and oviposition strategies of parasitoids in relation to host stage. *Researches on Population Ecology* 33:13–28.

King, B. H. 1990. Sex ratio manipulation by the parasitoid wasp *Spalangia cameroni* in response to host age—a test of the host-size model. *Evolutionary Ecology* 4:149–156.

———. 1994. Effects of host size experience on sex ratios in the parasitoid wasp *Spalangia cameroni*. *Animal Behaviour* 47:815–820.

Kittlein, M. J. 1997. Assessing the impact of owl predation on the growth rate of a rodent prey population. *Ecological Modelling* 103:123–134.

Klinkhamer, P. G. L., T. J. de Jong, and J. A. J. Metz. 1983. An explanation for low dispersal rates: a simulation experiment. *Netherlands Journal of Zoology* 33:532–541.

Kot, M. 2001. *Elements of Mathematical Ecology*. Cambridge: Cambridge University Press.

Kuang, Y. 1993. *Delay Differential Equations: With Applications in Population Dynamics*. Boston: Academic Press.

Lauwerier, H. A., and J. A. J. Metz. 1986. Hopf bifurcation in host-parasitoid models. *IMA Journal of Mathematics Applied in Medicine and Biology* 3:191–210.

Lei, G., and I. Hanski. 1998. Spatial dynamics of two competing specialist parasitoids in a host metapopulation. *Journal of Animal Ecology* 67:422–433.

Lessels, C. M. 1985. Parasitoid foraging: should parasitism be density dependent? *Journal of Animal Ecology* 54:27–41.

Levin, S. A. 1974. Dispersion and population interactions. *American Naturalist* 108:207–228.

Levins, R. 1969. Some demographic and genetic consequences of environmental heterogeneity for biological control. *Bulletin of the Entomological Society of America* 15:237–240.

Lewis, M. A. 1994. Spatial coupling of plant and herbivore dynamics: the contribution of herbivore dispersal to transient and persistent "waves" of damage. *Theoretical Population Biology* 45:277–312.

Luck, R. F., and H. Podoler. 1985. Competitive exclusion of *Aphytis lingnanensis* by *Aphytis melinus*: potential role of host size. *Ecology* 66:904–913.

Luck, R. F., H. Podoler, and R. Kfir. 1982. Host selection and egg allocation behavior by *Aphytis melinus* and *Aphytis lingnanensis*—comparison of two facultatively gregarious parasitoids. *Ecological Entomology* 7:397–408.

MacArthur, R. H., and R. Levins. 1967. The limiting similarity, convergence, and divergence of coexisting species. *American Naturalist* 101:377–385.

MacDonald, N. 1989. *Biological Delay Systems: Linear Stability Theory*. Cambridge: Cambridge University Press.

Mangel, M. 1989. Evolution of host selection in parasitoids: does the state of the parasitoid matter? *American Naturalist* 133:688–705.

Mangel, M., and C. W. Clark. 1988. *Dynamic Modeling in Behavioral Ecology*. Princeton, N.J.: Princeton University Press.

Mangel, M. and G. E. Heimpel. 1998. Reproductive senescence and dynamic oviposition behaviour in insects. *Evolutionary Ecology* 12:871–879.

May, R. M. 1972. Limit cycles in predator-prey communities. *Science* 177:900–902.

———. 1973a. *Stability and Complexity in Model Ecosystems*. Princeton, N.J.: Princeton University Press.

———. 1973b. Stability in randomly fluctuating versus deterministic environments. *American Naturalist* 107:621–650.

———. 1974. Biological populations with nonoverlapping generations: stable points, stable cycles and chaos. *Science* 186:645–647.

———. 1976a. Models for single populations. Pp. 5–29 *in* R. M. May, ed., *Theoretical Ecology: Principles and Applications*, Oxford: Blackwell Scientific Publications.

———. 1976b. Models for two interacting populations. Pp. 49–70 *in* R. M. May, ed., *Theoretical Ecology: Principles and Applications*, Oxford: Blackwell Scientific Publications.

———. 1976c. Simple mathematical models with very complicated dynamics. *Nature* 261:459–467.

———. 1978. Host-parasitoid systems in a patchy environment: a phenomenological model. *Journal of Animal Ecology* 47:833–844.

May, R. M., and M. P. Hassell. 1981. The dynamics of multiparasitoid-host interactions. *American Naturalist* 117:234–261.

May, R. M., M. P. Hassell, R. M. Anderson, and D. W. Tonkyn. 1981. Density dependence in host-parasitoid models. *Journal of Animal Ecology* 50:855–865.

May, R. M., and G. F. Oster. 1976. Bifurcations and dynamic complexity in simple ecological models. *American Naturalist* 110:573–599.

Mayhew, P. J. 1998. Offspring size-number strategy in the bethylid parasitoid *Laelius pedatus*. *Behavioral Ecology* 9:54–59.

McCauley, E., B. E. Kendall, A. Janssen, S. Wood, W. W. Murdoch, P. Hosseini, C. J. Briggs, S. P. Ellner, R. M. Nisbet, M. W. Sabelis, and P. Turchin. 2000. Inferring colonization processes from population dynamics in spatially structured predator-prey systems. *Ecology* (Washington, D.C.) 81:3350–3361.

McCauley, E., and W. W. Murdoch. 1987. Cyclic and stable populations: plankton as paradigm. *American Naturalist* 129:97–121.

———. 1990. Predator-prey dynamics in environments rich and poor in nutrients. *Nature* (London) 343:455–457.

McCauley, E., W. W. Murdoch, R. M. Nisbet, and W. S. C. Gurney. 1990. The physiological ecology of *Daphnia*: development of a model of growth and reproduction. *Ecology* 71:703–715.

McCauley, E., R. M. Nisbet, A. M. de Roos, W. W. Murdoch, and W. S. C. Gurney. 1996a. Structured population models of herbivorous zooplankton. *Ecological Monographs* 66:479–501.

———. 1999. Large-amplitude cycles of *Daphnia* and its algal prey in enriched environments. *Nature* (London) 402:653–656.

McCauley, E., W. G. Wilson, and A. M. de Roos. 1993. Dynamics of age-structured and spatially structured predator-prey interactions: individual-based models and population-level formulations. *American Naturalist* 142:412–442.

———. 1996b. Dynamics of age-structured predator-prey populations in space: asymmetrical effects of mobility in juvenile and adult predators. *Oikos* 76:485–497.

McLaughlin, J. F., and J. Roughgarden. 1991. Pattern and stability in predator-prey communities: how diffusion in spatially variable environments affects the Lotka-Volterra model. *Theoretical Population Biology* 40:148–172.

———. 1992. Predation across spatial scales in heterogeneous environments. *Theoretical Population Biology* 41:277–299.

McNair, J. N. 1987. A reconciliation of simple and complex models of age-dependent predation. *Theoretical Population Biology* 32:383–392.

Metcalfe, J. R. 1971. Observations on the ecology of *Saccharosydne saccharivora* (Westw.) (Hom., Delphacidae) in Jamaican sugar-cane fields. *Bulletin of Entomological Research* 60:565–597.

Micheli, F. 1999. Eutrophication, fisheries, and consumer-resource dynamics in marine pelagic ecosystems. *Science* 285:1396–1398.

Mills, N. J. 1994. The structure and complexity of parasitoid communities in relation to biological control. Pp. 397–417 *in* B. A. Hawkins and W. Sheehan, eds., *Parasitoid Community Ecology*, Oxford: Oxford University Press.

Morris, R. F. 1959. Single-factor analysis in population dynamics. *Ecology* 40:580–588.

Mueller, L. D., and A. Joshi. 2000. *Stability in Model Populations*. Princeton, N.J.: Princeton University Press.

Murdie, G., and M. P. Hassell. 1973. Food distribution, searching success and predator-prey models. Pp. 87–101 *in* R. W. Hiorns, ed., *The Mathematical*

Theory of the Dynamics of Biological Populations, London: Academic Press.

Murdoch, W. W. 1975. Diversity, complexity, stability and pest control. *Journal of Applied Ecology* 12:795–807.

———. 1990. The relevance of pest-enemy models to biological control. Pp. 1–24 *in* M. Mackauer, L. E. Ehler, and J. Roland, eds., *Critical Issues in Biological Control*, Andover, U.K.: Intercept, Ltd.

———. 1992. Ecological theory and biological control. Pp. 197–221 *in* S. K. Jain and L. W. Botsford, eds., *Monographiae Biologicae*, vol. 67, *Applied Population Biology*, Dordrecht, Netherlands: Kluwer Academic Publishers.

———. 1993. Individual-based models for predicting effects of global change. Pp. 147–162 *in* P. M. Kareiva, J. G. Kingsolver, and R. B. Hvey, eds., *Biotic Interactions and Global Change; Workshop, San Juan Island, Washington, USA, September 20–23, 1991*, Sunderland, Mass.: Sinauer Associates.

———. 1994. Population regulation in theory and practice. *Ecology* 75:271–287.

Murdoch, W. W., and C. J. Briggs. 1996. Theory for biological control: recent developments. *Ecology* 77:2001–2013.

Murdoch, W. W., C. J. Briggs, and T. R. Collier. 1998a. Biological control of insects: implications for theory in population ecology. Pp. 167–186 *in* J. Dempster and I. McLean, eds., *Insect Populations in Theory and Practice*, Dordrecht, Netherlands: Kluwer Academic Publishers.

Murdoch, W. W., C. J. Briggs, and R. M. Nisbet. 1996a. Competitive displacement and biological control in parasitoids: a model. *American Naturalist* 148:807–826.

———. 1997. Dynamical effects of host size- and parasitoid state-dependent attacks by parasitoids. *Journal of Animal Ecology* 66:542–556.

———. 1999. Dynamics of consumer-resource interactions: importance of individual attributes. Pp. 521–550 *in* H. Olff, V. K. Brown, and R. H. Drent, eds., *Herbivores: Between Plants and Predators*, Oxford: Blackwell Science Ltd.

Murdoch, W. W., C. J. Briggs, R. M. Nisbet, W. S. C. Gurney, and A. Stewart-Oaten. 1992a. Aggregation and stability in metapopulation models. *American Naturalist* 140:41–58.

Murdoch, W. W., J. Chesson, and P. L. Chesson. 1985. Biological control in theory and practice. *American Naturalist* 125:344–366.

Murdoch, W. W., B. E. Kendall, R. M. Nisbet, C. J. Briggs, E. McCauley, and R. Bolser. 2002. Single-species models for many-species food webs. *Nature* 417: 541–543.

Murdoch, W. W., R. F. Luck, S. L. Swarbrick, S. Walde, D. S. Yu, and J. D. Reeve. 1995. Regulation of an insect population under biological control. *Ecology* 76:206–217.

Murdoch, W. W., R. F. Luck, S. J. Walde, J. D. Reeve, and D. S. Yu. 1989. A refuge for red scale under control by *Aphytis*: structural aspects. *Ecology* 70:1707–1714.

Murdoch, W. W., and E. McCauley. 1985. Three distinct types of dynamic behavior shown by a single planktonic system. *Nature* 316:628–630.

Murdoch, W. W., and R. M. Nisbet. 1996. Frontiers of population ecology. Pp. 31–43 *in* R. B. Floyd and A. W. Sheppard, eds., *Frontiers of Population Ecology*, Melbourne, Australia: Commonwealth Scientific & Industrial Research Organization.

Murdoch, W. W., R. M. Nisbet, W. S. C. Gurney, and J. D. Reeve. 1987. An invulnerable age class and stability in delay-differential parasitoid-host models. *American Naturalist* 129:263–282.

Murdoch, W. W., R. M. Nisbet, R. F. Luck, H. C. J. Godfray, and W. S. C. Gurney. 1992b. Size-selective sex-allocation and host feeding in a parasitoid-host model. *Journal of Animal Ecology* 61:533–541.

Murdoch, W. W., R. M. Nisbet, E. McCauley, A. M. de Roos, and W. S. C. Gurney. 1998b. Plankton abundance and dynamics across nutrient levels: tests of hypotheses. *Ecology* 79:1339–1356.

Murdoch, W. W., and A. Oaten. 1975. Predation and population stability. *Advances in Ecological Research* 9:1–131.

Murdoch, W. W., and A. Stewart-Oaten. 1989. Aggregation by parasitoids and predators: effects on equilibrium and stability. *American Naturalist* 134:288–310.

Murdoch, W. W., S. L. Swarbrick, R. F. Luck, S. Walde, and D. S. Yu. 1996b. Refuge dynamics and metapopulation dynamics: an experimental test. *American Naturalist* 147:424–444.

Murdoch, W. W., and S. J. Walde. 1989. Analysis of insect population dynamics. Pp. 113–140 *in* P. J. Grubb and J. B. Whittaker, eds., *Toward a More Exact Ecology*, Oxford: Blackwell Scientific Publications.

Murray, J. D. 1993. *Mathematical Biology*. Berlin: Springer-Verlag.

Mylius, S. D., K. Klumpers, A. M. de Roos, and L. Persson. 2001. Impact of intraguild predation and stage structure on simple communities along a productivity gradient. *American Naturalist* 158:259–276.

Nee, S., and R. M. May. 1992. Dynamics of metapopulations: habitat destruction and competitive coexistence. *Journal of Animal Ecology* 61:37–40.

NERC Centre for Population Biology, Imperial College. 1999. The global population dynamics database; http://www.sw.ic.ac.uk/cpb/cpb/gpdd.html.

Neubert, M. G., and M. Kot. 1992. The subcritical collapse of predator populations in discrete-time predator-prey models. *Mathematical Biosciences* 110:45–66.

Neveu, N., L. Krespi, N. Kacem, and J.-P. Nenon. 2000. Host-stage selection by *Trybliographa rapae*, a parasitoid of the cabbage root fly *Delia radicum*. *Entomologia Experimentalis et Applicata* 96:231–237.

Nicholson, A. J. 1933. The balance of animal populations. *Journal of Animal Ecology* 2:131–178.

———. 1957. The self-adjustment of populations to change. *Cold Spring Harbor Symposia on Quantitative Biology* 22:153–173.

Nicholson, A. J., and V. A. Bailey. 1935. The balance of animal populations. Part I. *Proceedings of the Zoological Society of London* 3:551–598.

Nisbet, R. M. 1997. Delay-differential equations for structured populations. Pp. 89–118 *in* S. Tuljapurkar and H. Caswell, eds., *Structured-Population Models in Marine, Terrestrial, and Freshwater Systems*, New York: Chapman and Hall.

Nisbet, R. M., and J. R. Bence. 1989. Alternative dynamic regimes for canopy-forming kelp: a variant on density-vague population regulation. *American Naturalist* 134:377–408.

Nisbet, R. M., C. J. Briggs, W. S. C. Gurney, W. W. Murdoch, and A. Stewart-Oaten. 1993. Two-patch metapopulation dynamics. Pp. 125–135 *in* J. Steele, S. A. Levin, and T. Powell, eds., *Patch Dynamics in Freshwater, Terrestrial and Marine Ecosystems*, Berlin: Springer-Verlag.

Nisbet, R. M., A. M. de Roos, W. G. Wilson, and R. E. Snyder. 1998. Discrete consumers, small scale resource heterogeneity and population stability. *Ecology Letters* 1:34–37.

Nisbet, R. M., S. Diehl, W. G. Wilson, S. D. Cooper, D. D. Donalson, and K. Kratz. 1997a. Primary-productivity gradients and short-term population dynamics in open systems. *Ecological Monographs* 67:535–553.

Nisbet, R. M., and W. S. C. Gurney. 1982. *Modelling Fluctuating Populations*. New York: John Wiley & Sons.

———. 1983. The systematic formulation of population models for insects with dynamically varying instar duration. *Theoretical Population Biology* 23:114–135.

Nisbet, R. M., W. S. C. Gurney, W. W. Murdoch, and E. McCauley. 1989. Structured population models a tool for linking effects at individual and population level. Pp. 79–100 *in* P. Calow and B. R. J, eds., *Evolution, Ecology and Environmental Stress; Symposium, London, England, UK, June 1988*, London: Academic Press.

Nisbet, R. M., E. McCauley, A. M. de Roos, W. W. Murdoch, and W. S. C. Gurney. 1991. Population dynamics and element recycling in an aquatic plant-herbivore system. *Theoretical Population Biology* 40:125–147.

———. 1997b. Simple representations of biomass dynamics in structured populations. Pp. 61–79 *in* H. G. Othmer, F. R. Adler, M. A. Lewis, and J. C. Dillon, eds., *Case Studies in Mathematical Modeling: Ecology, Physiology, and Cell Biology*, Upper Saddle River, N.J.: Prentice Hall.

Nussbaumer, C., and A. Schopf. 2000. Development of the solitary larval endoparasitoid *Glyptapanteles porthetriae* (Hymenoptera: Braconidae) in its host *Lymantria dispar* (Lepidoptera: Lymantriidae). *European Journal of Entomology* 97:355–361.

Oaten, A., and W. W. Murdoch. 1975. Functional response and stability in predator-prey systems. *American Naturalist* 109:289–298.

Oksanen, L., S. D. Fretwell, J. Arruda, and P. Niemela. 1981. Exploitation ecosystems in gradients of primary productivity. *American Naturalist* 118:240–260.

Okubo, A., and S. A. Levin. 2001. *Diffusion and Ecological Problems: Modern Perspectives.* New York: Springer-Verlag.

Pacala, S. W. 1986. Neighborhood models of plant population dynamics. 2. Multispecies models of annuals. *Theoretical Population Biology* 29:262–292.

Pacala, S. W., and M. P. Hassell. 1991. The persistence of host-parasitoid associations in patchy environments. II. Evaluation of field data. *American Naturalist* 138:584–605.

Pacala, S. W., M. P. Hassell, and R. M. May. 1990. Host-parasitoid-associations in patchy environments. *Nature* 344:150–153.

Pacala, S. W., and S. A. Levin. 1997. Biologically generated spatial pattern and the coexistence of competing species. Pp. 204–232 *in* D. Tilman and P. Kareiva, eds., *Spatial Ecology: The Role of Space in Population Dynamics and Interspecific Interactions*, Princeton, N.J.: Princeton University Press.

Parma, A. M., P. Amarasekare, M. Mangel, J. Moore, W. W. Murdoch, E. Noonburg, H. P. Pascual, K. Possingham, K. Shea, C. Wilcox, and D. Yu. 1998. What can adaptive management do for our fish, forests, food and biodiversity? *Integrative Biology* 1:16–26.

Pascual, M. 1993. Diffusion-induced chaos in a spatial predator-prey system. *Proceedings of the Royal Society of London*, Series B 251:1–7.

Pascual, M., and H. Caswell. 1997. Environmental heterogeneity and biological pattern in a chaotic predator-prey system. *Journal of Theoretical Biology* 185:1–13.

Pascual, M., P. Mazzega, and S. A. Levin. 2001. Oscillatory dynamics and spatial scale: the role of noise and unresolved pattern. *Ecology* (Washington, D.C.) 82:2357–2369.

Pedata, P. A., M. S. Hunter, H. C. J. Godfray, and G. Viggiani. 1995. The population dynamics of the white peach scale and its parasitoids in a mulberry orchard in Campania, Italy. *Bulletin of Entomological Research* 85:531–539.

Persson, L., K. Leonardsson, A. M. de Roos, M. Gyllenberg, and B. Christensen. 1998. Ontogenetic scaling of foraging rates and the dynamics of a size-structured consumer-resource model. *Theoretical Population Biology* 54:270–293.

Plarre, R., K. Lieber, W. Burkholder, and J. Phillips. 1999. Host and host instar preference of *Apanteles carpatus* (Say) (Hymenoptera: Braconidae): a possible parasitoid for biological control of clothes moths (Lepidoptera: Tineidae). *Journal of Stored Products Research* 35:197–213.

Polis, G. A., and R. D. Holt. 1992. Intraguild predation: the dynamics of complex trophic interactions. *Trends in Ecology and Evolution* 7:151–154.

Polis, G. A., C. A. Myers, and R. D. Holt. 1989. The ecology and evolution of intraguild predation potential competitors that eat each other. Pp. 297–330 *in* R. F. Johnston, ed., *Annual Review of Ecology and Systematics*, vol. 20, Palo Alto, Calif.: Annual Reviews, Inc.

Raup, D. M. 1991. A kill curve for Phanerozoic marine species. *Paleobiology* 17:37–48.

Reeve, J. D. 1988. Environmental variability, migration, and persistence in host-parasitoid systems. *American Naturalist* 132:810–836.

Reeve, J. D., J. T. Cronin, and D. R. Strong. 1994a. Parasitism and generation cycles in a salt-marsh planthopper. *Journal of Animal Ecology* 63:912–920.

———. 1994b. Parasitoid aggregation and the stabilization of a salt marsh host-parasitoid system. *Ecology* 75:288–295.

Renshaw, E. 1991. *Modelling Biological Populations in Space and Time.* Cambridge University Press, Cambridge, U.K.

Richards, S. A., R. M. Roger, W. G. Wilson, and H. P. Possingham. 2000. Grazers and diggers: exploitation competition and coexistence among foragers with different feeding strategies on a single resource. *American Naturalist* 155:266–279.

Rivero, A., and J. Casas. 1999a. Incorporating physiology into parasitoid behavioral ecology: the allocation of nutritional resources. *Researches in Population Ecology* 41:39–45.

———. 1999b. Rate of nutrient allocation to egg production in a parasitic wasp. *Proceedings of the Royal Society of London*, Series B 266:1169–1174.

Rivero-Lynch, A., and H. C. J. Godfray. 1997. The dynamics of egg production, oviposition and resorption in a parasitoid wasp. *Functional Ecology* 11:184–188.

Rochat, J., and A. P. Gutierrez. 2001. Weather-mediated regulation of olive scale by two parasitoids. *Journal of Animal Ecology* 70:476–490.

Rohani, P., H. C. J. Godfray, and M. P. Hassell. 1994. Aggregation and the dynamics of host-parasitoid systems: a discrete-generation model with within-generation redistribution. *American Naturalist* 144:491–509.

Rohani, P., R. M. May, and M. P. Hassell. 1996. Metapopulations and equilibrium stability: the effects of spatial structure. *Journal of Theoretical Biology* 181:97–109.

Roland, J. 1994. After the decline: what maintains low winter moth density after successful biological control? *Journal of Animal Ecology* 63:392–398.

———. 1995. Response to Bonsall and Hassell "Identifying density-dependent processes: a comment on the regulation of winter moth." *Journal of Animal Ecology* 64:785–786.

Roland, J., and D. G. Embree. 1995. Biological control of the winter moth. *Annual Review of Entomology* 40:475–492.

Rose, M., and P. DeBach. 1992. Biological control of *Parabemisia myricae* (Kuwana) (Homoptera: Aleyrodidae) in California. *Israel Journal of Entomology* 25/26:73–95.

Rosen, D., and P. DeBach. 1979. *Species of Aphytis of the World (Hymenoptera: Aphelinidae)*. Dordrecht, Netherlands: Junk.

Rosenheim, J. A. 1996. An evolutionary argument for egg limitation. *Evolution* 50:2089–2094.

———. 1999. Characterizing the cost of oviposition in insects: a dynamic model. *Evolutionary Ecology* 13:141–165.

Rosenheim, J. A., G. E. Heimpel, and M. Mangel. 2000. Egg maturation, egg resorption and the costliness of transient egg limitation in insects. *Proceedings of the Royal Society of London*, Series B 267:1565–1573.

Rosenheim, J. A., and D. Rosen. 1991. Foraging and oviposition decisions in the parasitoid *Aphytis lingnanensis*: distinguishing the influences of egg load and experience. *Journal of Animal Ecology* 60:873–894.

Rosenzweig, M. L. 1971. Paradox of enrichment: destabilization of exploitation ecosystems in ecological time. *Science* 171:385–387.

Rosenzweig, M. L., and R. H. MacArthur. 1963. Graphical representation and stability conditions for predator-prey interactions. *American Naturalist* 97:209–223.

Ruxton, G. D., W. S. C. Gurney, and A. M. de Roos. 1992. Interference and generation cycles. *Theoretical Population Biology* 42:235–253.

Sabelis, M. W., O. Diekmann, and V. A. A. Jansen. 1991. Metapopulation persistence despite local extinction: predator-prey patch models of the Lotka-Volterra type. *Biological Journal of the Linnean Society* 42:267–284.

Sagarra, L. A., D. D. Peterkin, C. Vincent, and R. K. Stewart. 2000. Immune response of the hibiscus mealybug, *Maconellicoccus hirsutus* Green (Homoptera: Pseudococcidae), to oviposition of the parasitoid *Anagyrus kamali* Moursi (Hymenoptera: Encyrtidae). *Journal of Insect Physiology* 46:647–653.

Sait, S. M., M. Begon, and D. J. Thompson. 1994. Long-term population dynamics of the Indian meal moth *Plodia interpunctella* and its granulosis virus. *Journal of Animal Ecology* 63:861–870.

Salt, G. 1941. The effects of hosts upon insect parasites. *Biological Reviews* 16:239–264.

Sanderson, B. L., T. R. Hrabik, J. J. Magnuson, and D. M. Post. 1999. Cyclic dynamics of a yellow perch (*Perca flavescens*) population in an oligotrophic lake: evidence for intraspecific interactions. *Canadian Journal of Fisheries and Aquatic Science* 56:1534–1542.

Sarnelle, O. 1992. Nutrient enrichment and grazer effects on phytoplankton in lakes. *Ecology* 75:551–560.

Schmitt, R. J. 1996. Exploitation competition in mobile grazers: trade-offs in use of a limiting resource. *Ecology* 77:408–425.

Sequeira, R., and M. Mackauer. 1992. Covariance of adult size and developmental time in the parasitoid wasp *Aphidius ervi* in relation to the size of its host, *Acyrthosiphon pisum*. *Evolutionary Ecology* 6:34–44.

Settle, W. H., H. Ariawaan, E. T. Astuti, W. Cahyana, A. L. Hakim, D. Hindayana, A. S. Lestari, and Pakarningsih. 1996. Managing tropical rice pests through conservation of generalist natural enemies and alternative prey. *Ecology* 77:1975–1988.

Shea, K., R. M. Nisbet, W. W. Murdoch, and H. J. S. Yoo. 1996. The effect of egg limitation on stability in insect host-parasitoid population models. *Journal of Animal Ecology* 65:743–755.

Shigesada, N., and K. Kawasaki. 1997. *Biological Invasions: Theory and Practice*. Oxford: Oxford University Press.

Sirot, E., H. Ploye, and C. Bernstein. 1997. State dependent superparasitism in a solitary parasitoid: egg load and survival. *Behavioral Ecology* 8: 226–232.

Skellam, J. G. 1951. Random dispersal in theoretical populations. *Biometrika* 38:196–218.

Slatkin, M. 1974. Competition and regional coexistence. *Ecology* 55:128–134.

Snyder, W. E., and A. R. Ives. 2001. Generalist predators disrupt biological control by a specialist parasitoid. *Ecology* 82:705–716.

Srivastava, M., and R. Singh. 1995. Sex ratio adjustment by a koinobiotic parasitoid *Lysiphlebus delhiensis* (Subba Rao & Sharma) (Hymenoptera: Aphidiidae) in response to host size. *Biological Agriculture and Horticulture* 12:15–28.

Stewart-Oaten, A., and W. W. Murdoch. 1990. Temporal consequences of spatial density dependence. *Journal of Animal Ecology* 59:1027–1046.

Stiling, P. D. 1987. The frequency of density dependence in insect host-parasitoid systems. *Ecology* 68:844–856.

Stimson, J., and R. Black. 1975. Field experiments on population regulation in intertidal limpets of the genus *Acmaea*. *Oecologia* 18:111–120.

Stokkebo, S., and I. C. W. Hardy. 2000. The importance of being gravid: egg load and contest outcome in a parasitoid wasp. *Animal Behaviour* 59:1111–1118.

Takagi, M., and Y. Hirose. 1994. Building parasitoid communities: the complementary role of two introduced parasitoid species in a case of successful biological control. Pp. 437–448 *in* B. A. Hawkins and W. Sheehan, eds., *Parasitoid Community Ecology*, Oxford: Oxford University Press.

Thieme, H. In press. *Mathematical Models in Population Biology*, Princeton, N.J.: Princeton University Press.

Thomas, C. D., and I. Hanski. 1997. Butterfly metapopulations. Pp. 359–386 *in* I. Hanksi and M. E. Gilpin, eds., *Metapopulation Biology: Ecology, Genetics, and Evolution*, San Diego: Academic Press.

Thompson, J. N. 1994. *The Coevolutionary Process*. Chicago: University of Chicago Press.

————. 1999. Specific hypotheses on the geographic mosaic of coevolution. *American Naturalist* 153:S1–S14.

Tilman, D. 1980. Resources: a graphical-mechanistic approach to competition and predation. *American Naturalist* 116:362–393.

————. 1994. Competition and biodiversity in spatially structured habitats. *Ecology* 75:2–16.

Tilman, D., C. L. Lehman, and P. Kareiva. 1997. Population dynamics in spatial habitats. Pp. 3–20 *in* D. Tilman and P. Kareiva, eds., *Spatial Ecology: The Role of Space in Population Dynamics and Interspecific Interactions*, Princeton, N.J.: Princeton University Press.

Turchin, P. 1995. Population regulation: old arguments and a new synthesis. Pp. 19–40 *in* N. Cappuccino and P. W. Price, eds., *Population Dynamics: New Approaches and Synthesis*, San Diego: Academic Press.

Turchin, P., and I. Hanski. 1997. An empirically based model for latitudinal gradient in vole population dynamics. *American Naturalist* 149:842–874.

Turelli, M. 1981. Niche overlap and invasion of competitors in random environments: 1. Models without demographic stochasticity. *Theoretical Population Biology* 20:1–56.

Turnbull, A. L., and D. A. Chant. 1961. The practice and theory of biological control of insects in Canada. *Canadian Journal of Zoology* 39:697–753.

Ueno, T. 1999. Host-feeding and acceptance by a parasitic wasp (Hymenoptera: Ichneumonidae) as influenced by egg load and experience in a patch. *Evolutionary Ecology* 13:33–44.

van Baalen, M. 2000. The evolution of parasitoid egg load. Pp. 103–120 *in* M. E. Hochberg and A. R. Ives, eds., *Parasitoid Population Biology*, Princeton, N.J.: Princeton University Press.

van de Koppel, J., J. Huisman, R. van der Wal, and H. Olff. 1996. Patterns of herbivory along a productivity gradient: an empirical and theoretical investigation. *Ecology* 77:736–745.

van Lenteren, J. C., and H. J. W. van Roermund. 1999. Why is the parasitoid *Encarsia formosa* so successful in controlling whiteflies. Pp. 116–130 *in* B. A. Hawkins and H. V. Cornell, eds., *Theoretical Approaches to Biological Control*, Cambridge: Cambridge University Press.

Vet, L. E. M., A. Datema, K. Van Welzen, and H. Snellen. 1993. Clutch size in a larval-pupal endoparasitoid. *Oecologia* 95:410–415.

Volterra, V. 1926. Variations and fluctuations of the numbers of individuals in animal species living together. Reprinted 1931 *in* R. N. Chapman, ed., *Animal Ecology*, New York: McGraw Hill.

Waage, J. K. 1990. Ecological theory and the selection of biological control agents. Pp. 135–157 *in* M. Mackauer, L. E. Ehler, and J. Roland, eds., *Critical Issues in Biological Control*, Andover, U.K.: Intercept, Ltd.

Walde, S. J. 1991. Patch dynamics of a phytophagous mite population: effect of number of subpopulations. *Ecology* 72:1591–1598.

————. 1994. Immigration and the dynamics of a predator-prey interaction in biological control. *Journal of Animal Ecology* 63:337–346.

Walde, S. J., and W. W. Murdoch. 1988. Spatial density dependence in parasitoids. Pp. 441–466 *in* T. E. Mittler, ed., *Annual Review of Entomology*, Palo Alto, Calif.: Annual Reviews, Inc.

Walde, S. J., and G. N. Nachman. 1999. Dynamics of spatially structured mite populations. Pp. 163–189 *in* B. A. Hawkins and H. V. Cornell, eds., *Theoretical Approaches to Biological Control*, Cambridge: Cambridge University Press.

Warner, R. R., and P. L. Chesson. 1985. Coexistence mediated by recruitment fluctuations: a field guide to the storage effect. *American Naturalist* 125:769–787.

Watson, S., E. McCauley, and J. A. Downing. 1992. Sigmoid relationships between phosphorus algal biomass, and algal community structure. *Canadian Journal of Fisheries and Aquatic Sciences* 49:2605–2610.

Wen, B., D. K. Weaver, and J. H. Brower. 1995. Size preference and sex ratio for *Pteromalus cerealellae* (Hymenoptera: Pteromalidae) parasitizing *Sitotroga cerealella* (Lepidoptera: Gelechiidae) in stored corn. *Environmental Entomology* 24:1160–1166.

Wilson, W. G. 1998. Resolving discrepancies between deterministic population models and individual-based simulations. *American Naturalist* 151:116–134.

Wilson, W. G., C. W. Osenberg, R. J. Schmitt, and R. M. Nisbet. 1999. Complementary foraging behaviors allow coexistence of two consumers. *Ecology* 80:2358–2372.

Woodburn, T. L. 1996. Reduction of seed set in nodding thistle (*Carduus nutans*) by the seed-fly, *Urophora solstitialis*, in Australia. Pp. 165–169 *in* B. A. McPheron and G. J. Steck, eds., *Fruit Fly Pests: A World Assessment of Their Biology and Management*, Delray Beach, Fla.: St. Lucie Press.

Yoo, C. K., and M. I. Ryoo. 1989. Host preference of *Lariophagus distinguendus* Foerster (Hymenoptera: Pteromalidae) for the instars of rice weevil (*Sitophilus oryzae* (L.)) (Coleoptera: Curculionidae) and sex ratio of the parasitoid in relation to the host. *Korean Journal of Applied Entomology* 28:28–31.

Zwölfer, H. 1973. Competition and coexistence in phytophagous insects attacking the heads of *Carduus nutans* L. Pp. 74–81 *in* P. H. Dunn, ed., *II International Symposium for the Biological Control of Weeds*, Slough, U.K.: Commonwealth Institute of Biological Control.

Index

Monographs in Population Biology

Edited by Simon A. Levin and Henry S. Horn

Titles available in the series (by monograph number)

18. *The Theory of Sex Allocation*, by Eric L. Charnov
19. *Mate Choice in Plants: Tactics, Mechanisms, and Consequences*, by Mary F. Wilson and Nancy Burley
20. *The Florida Scrub Jay: Demography of a Cooperative-Breeding Bird*, by Glen E. Woolfenden and John W. Fitzpatrick
21. *Natural Selection in the Wild*, by John A. Endler
22. *Theoretical Studies on Sex Ratio Evolution*, by Samuel Karlin and Sabvin Lessard
23. *A Hierarchical Concept of Ecosystems*, by R. V. O'Neill, D. L. DeAngelis, J. B. Waide, and T.F.H. Allen
24. *Population Ecology of the Cooperatively Breeding Acorn Woodpecker*, by Walter D. Koenig and Ronald L. Mumme
25. *Population Ecology of Individuals*, by Adam Lomnicki
26. *Plant Strategies and the Dynamics and Structure of Plant Communities*, by David Tilman
27. *Population Harvesting: Demographic Models of Fish, Forest, and Animal Resources*, by Wayne M. Getz and Robert G. Haight
28. *The Ecological Detective: Confronting Models with Data*, by Ray Hilborn and Marc Mangel
29. *Evolutionary Ecology across Three Trophic Levels: Goldenrods, Gallmakers, and Natural Enemies*, by Warren G. Abrahamson and Arthur E. Weis
30. *Spatial Ecology: The Role of Space in Population Dynamics and Interspecific Interactions*, edited by David Tilman and Peter Kareiva
31. *Stability in Model Populations*, by Laurence D. Mueller and Amitabh Joshi
32. *The Unified Neutral Theory of Biodiversity and Biogeography*, by Stephen P. Hubbell
33. *The Functional Consequences of Biodiversity: Empirical Progress and Theoretical Extensions*, edited by Ann P. Kinzig, Stephen W. Pacala, and David Tilman
34. *Communities and Ecosystems: Linking the Aboveground and Belowground Components*, by David A. Wardle
35. *Complex Population Dynamics: A Theoretical/Empirical Synthesis*, by Peter Turchin
36. *Consumer-Resource Dynamics*, by William W. Murdoch, Cheryl J. Briggs, and Roger M. Nesbit